THE RADIO STATION

THE RADIO STATION

Michael C. Keith
Joseph M. Krause

c.1

Focal Press
Boston London

Focal Press is an imprint of Butterworth Publishers.

Library of Congress Cataloging-in-Publication Data

Keith, Michael C., 1945–
The radio station.

Includes bibliographies and index.
1. Radio broadcasting—United States.
I. Krause, Joseph M. II. Title.
HD8698.K45 1986 384.54′53′068 86–2250
ISBN 0–240–51747–4

Butterworth Publishers
80 Montvale Avenue
Stoneham, MA 02180

10 9 8 7 6 5 4 3 2 1

Printed in the United States of America

Like the reader and the
Poem
Radio and the listener
Create something always
Individual
And
Become unknowingly
One
In the experience

L. Bonczek

To my daughter, Marlo, and to the memory of my father.

M.C.K.

To my father, whom I dearly miss, and to my son, Daniel, who has brought so much joy into my life.

J.M.K.

CONTENTS

FOREWORD

by Rick Sklar

One of the world's preeminent broadcasters, Rick Sklar refined the contemporary hit music format into its ultimate form at radio station WABC in New York, which during his decade as programmer was the highest rated radio station in America. Mr. Sklar was a network vice president for twelve years and has served as an adjunct professor at St. John's University. He currently heads his own radio consultancy firm, Sklar Communications. His work at WABC is the subject of his book Rocking America—How the All Hit Radio Stations Took Over.

The radio industry has just undergone a period of rapid expansion. Since the government decree in the mid-sixties requiring separate FM programming in larger markets, the radio station population has doubled to ten thousand outlets. Like most other specialized businesses, stations need skilled professionals to enable them to operate successfully. But because the industry has grown so fast, it has been unable to provide the time or the training to produce enough personnel.

As a result, radio has had to improvise for its new people. Those who go directly from the classroom to the commercial broadcast operations of America find that academic experience does not fully equip them for the demands of the bottom-line-oriented, efficiency-minded operator-owners who think first of profit and later about the programming. Others starting in very small markets learn that the old "sink or swim" adage still holds true as soon as they move to larger, more demanding situations. Books like this one that draw upon the experience of successful professionals working under intense pressure to produce the ratings and a profit are one way to help equip the radio students of today to be the radio professionals of tomorrow.

Regardless of the area in radio you are considering for your career, I would advise you to read every chapter of this book thoroughly. I make this suggestion not only because you are quite likely to change career directions more than once, after you actually begin working in radio, but because of the interdependent nature of the modern radio station. Each department depends on the cooperation of the other departments for the station's goals to be achieved. For example, both the programming and the audiences that the programs have to produce for the advertiser can be affected if poor engineering results in a weak signal or shrunken coverage area. Engineering errors also can result in poor-sounding music or an awkward equipment configuration in the studio. A disc jockey who has to concentrate too much on making the equipment work cannot perform at his or her best. Advertising and promotion personnel also have a big effect on ratings. A misguided or dull advertising campaign with television promos, billboards, and ads that are ignored and contests that do not generate any excitement can weigh heavily on the program director rather than the advertising manager. It's the program director who has to answer for low shares and the ratings books.

People entering radio today face competitors who were not dreamed of ten years ago. Confronted by television and cable, video games and movies, and computer entertainment software, radio needs to stress the creativity and imagination that come into play when only one sense—hearing—is stimulated and all the information, entertainment, and sales messages must be delivered by sound alone. The radio industry faces tougher challenges than ever, but for those who can meet them, the personal and financial rewards will be great.

PREFACE

This text is the result of a mutual desire and effort to provide the student of radio with the most comprehensive account of the medium possible, from the insider's view. It is presented from the perspective of the radio professional, drawing on the insights and observations of those who make their daily living by working in the industry.

What sets this particular text apart from others written by broadcasters is that not one or two but literally dozens of radio people have contributed to this effort to disseminate factual and relevant information about the medium in a way that captures its reality. These professionals represent the top echelons of network and group-owned radio, as well as the rural daytime-only outlets spread across the country.

We have sought to create a truly practical, timely, and accessible book on commercial radio station operations. The personnel and departments that comprise a radio station are our principal focus. We begin by examining the role of station management and then move into programming, sales, news, engineering, production, and traffic, as well as other key areas that serve as the vital ingredients of any radio outlet.

Since our strategy was to draw on the experience of countless broadcast and allied professionals, our debt of gratitude is significant. It is to these individuals who contributed most directly to its making that we also dedicate this book.

In addition, we would like to express our sincere appreciation to the many other individuals and organizations that assisted us in so many important ways. These include: Arbitron Ratings Company, Auditronics Inc., Birch Radio Inc., Broadcast Electronics, Broadcasters Promotion Association, Communications Graphics, Roger Crosley and Roger Turner of Dean Junior College, Federal Communications Commission, FMR Associates, IGM Inc., Jefferson Pilot Data Systems, Mike Jones of Jones' Homes, Marketron Inc., Scott Mason of Mason Photography, McZ Limited, Radio Advertising Bureau, *Radio and Records,* Radio and Television News Directors' Association, Radio Computer Systems, Radio Corporation of America, Satellite Music Network, Society of Broadcast Engineers, Jim Steele of the National Association of Broadcasters, Annette Steiner of Dancer Fitzgerald Sample, University Press of America, and Westinghouse Broadcasting.

A special word of appreciation is owed to Lou Emond for his help in editing the draft of this manuscript and for formulating the Chapter Highlights that serve as study guides at the conclusion of each chapter.

1 State of the Fifth Estate

IN THE AIR—EVERYWHERE

Radio is ubiquitous. It is the most pervasive medium on earth. There is no patch of land, no piece of ocean surface, untouched by the electromagnetic signals beamed from the more than twenty-six thousand radio stations worldwide. Over a third of these broadcast outlets transmit in America alone. Today, ten thousand stations in this country reach 99 percent of all households, and less than 1 percent have fewer than five receivers.

Contemporary radio's unique personal approach has resulted in a shift of the audience's application of the medium: from family or group entertainer before 1950, to individual companion in the last half of the century. Although television usurped radio's position as the number one entertainment medium well over thirty years ago, radio's total reach handily exceeds that of the video medium. More people tune in to radio for its multifaceted offerings than to any other medium—print or electronic. Practically every automobile (95 percent) has a radio. "There are twice as many car radios in use (approximately 123 million) as the total circulation (62 million) of all daily newspapers," contends Kenneth Costa, vice president of marketing for the Radio Advertising Bureau (RAB). Seven out of ten adults are reached weekly by car radio.

WEEKLY: RADIO REACHES 95.8%

	Weekly Reach		Weekly Reach
Persons 12+	95.8%	Women	
Men		18+	94.1
18+	96.5%	18–24	99.5
18–24	98.6	25–34	98.1
25–34	98.4	35–49	97.9
35–49	98.0	50+	87.0
50+	92.8		
		Teens 12–17	99.7%
Adults in Households with . . .			
1–2 Persons	94.0%	5+ Persons	95.6%
3–4 Persons	96.4		

DAILY: RADIO REACHES 81.5%

	Average Daily Reach		Average Daily Reach
Persons 12+	81.5%	Women	
Men		18+	78.8%
18+	82.6%	18–24	87.0
18–24	90.3	25–34	84.0
25–34	85.7	35–49	82.5
35–49	84.5	50+	69.5
50+	74.8	Teens 12–17	88.1%
Adults in Households with . . .			
1–2 Persons	78.6%	5+ Persons	81.4%
3–4 Persons	82.3		

DAILY RADIO LISTENING: ALMOST 3½ HOURS

Daily Time Spent Listening

Persons 12+	3:21	Women	
Men		18+	3:22
18+	3:26	18–24	3:36
18–24	4:04	25–34	3:25
25–34	3:46	35–49	3:28
35–49	3:30	50+	3:10
50+	2:48		
		Teens 12–17	2:56
Adults in Households with . . .			
1–2 Persons	3:18	5+ Persons	3:35
3–4 Persons	3:25		

RADIO REACHES CONSUMERS AT HOME, AWAY, IN CARS

Percent of Listening Time by Location (Mon.–Sun., 24-Hour Avg.)

Persons 12+

59.7% At Home	18.3% In Cars	22.0% Other Places

	At Home	In Cars	Other Places
Teens 12–17	73.5%	12.8%	13.7%
Men 18+	46.5	25.1	28.4
Women 18+	68.7	13.4	17.9

FIGURE 1.1
Courtesy RAB.

The average adult spends three-and-a-half hours per day listening to radio. According to a recent Group W (Westinghouse Broadcasting) report, 84 percent of all adults listen to radio between fifteen and sixteen hours each week. A survey conducted by RADAR (Radio's All Dimensional Audience Research) concluded that 95.7 percent of all persons over twelve years old tuned in to radio. In the mid-1980s, this computes to more than 182 million Americans, and the figure continues to grow.

The number of radio receivers in use in America has risen by 40 percent since 1970, when 321 million sets provided listeners a wide range of audio services. In recent years, technological innovations in receiver design alone has contributed to the ever-increasing popularity of the medium. Bone phones, ghetto boxes, and walk-alongs, among others, have boosted receiver sales over the three billion dollar mark annually, up 26 percent since 1970. There are nine million walk-alongs in use. Radio's ability to move with its audience has never been greater.

Radio appeals to everyone and is available to all. Its mobility and variety of offerings have made it the most popular medium in history. To most of us, radio is as much a part of our day as morning coffee and the ride to work or school. It is a companion that keeps us informed about world and local events, gives us sports scores, provides us with the latest weather and school closings, and a host of other information, not to mention our favorite music, and asks for nothing in return.

It is difficult to imagine a world without such an accommodating and amusing cohort, one that not only has enriched our lives by providing us with a nonstop source of entertainment, but has also kept us abreast of happenings during times of national and global crisis. To two hundred million Americans, radio is an integral part of daily life.

A HOUSEHOLD UTILITY

Although radio seems to have been around for centuries, it is a relatively recent invention. Many people alive today lived in a world without radio—hard to imagine, yet true. The world owes a debt of gratitude to several "wireless" technologists who contributed to the development of the medium. A friendly debate continues to be waged today as to just who should rightfully be honored with the title "father of radio." There are numerous candidates who actually take us as far back as the last century. For example, there is physicist James Clerk Maxwell, who theorized the existence of electromagnetic waves, which later in the century were used to carry radio signals. Then there is German scientist Heinrich Hertz, who validated Maxwell's theory by proving that electromagnetic waves do indeed exist.

The first choice of many to be called grand patriarch of radio is Guglielmo Marconi, who is credited with devising a method of transmitting sound without the help of wires—thus "wireless telegraphy." There are a host of other inventors and innovators who can, with some justification, be considered for the title. Lee De Forest, Ambrose Fleming, Reginald Fessenden, and David Sarnoff are a few. (A further discussion of radio's preeminent technologists can be found in chapter 10.) However, of the aforementioned, perhaps the pioneer with the most substantial claim is Sarnoff. A true visionary, Sarnoff conceived of the ultimate application of Marconi's device in a now-famous memorandum. In what became known as the "radio music box" memo, Sarnoff suggested that radio receivers be mass produced for public consumption and that music, news, and information be broadcast to the households that owned the appliance. At first his proposal was all but snubbed. Sarnoff's persistence eventually paid off, and in 1919 sets were available for general purchase. Within a very few years, radio's popularity would exceed even Sarnoff's estimations.

A TOLL ON RADIO

Though not yet a household word in 1922, radio was surfacing as a medium to be reckoned with. Hundreds of thousands of Americans were purchasing the crude, battery-operated crystal sets of the day and tuning the two frequencies (750 and 833 kc) set aside by the Department of Commerce for reception of radio broadcasts. The majority of stations in the early 1920s were owned by receiver manufacturers and department stores that sold the apparatus. Newspapers and colleges owned nearly as many. Radio was not yet a commercial enterprise. Those stations not owned by parent companies often depended on public donations and grants. These

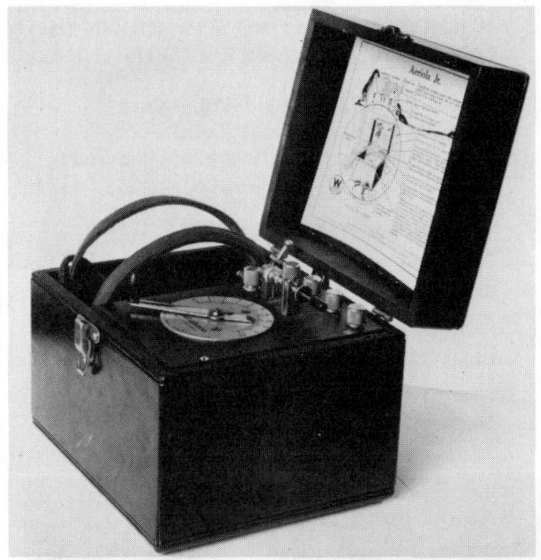

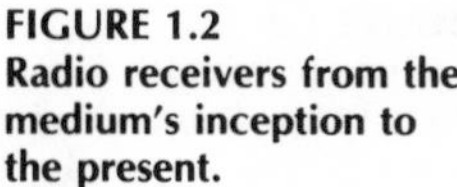

FIGURE 1.2
Radio receivers from the medium's inception to the present.

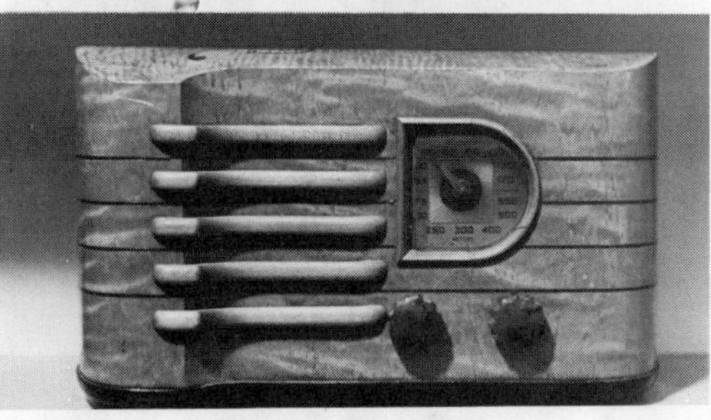

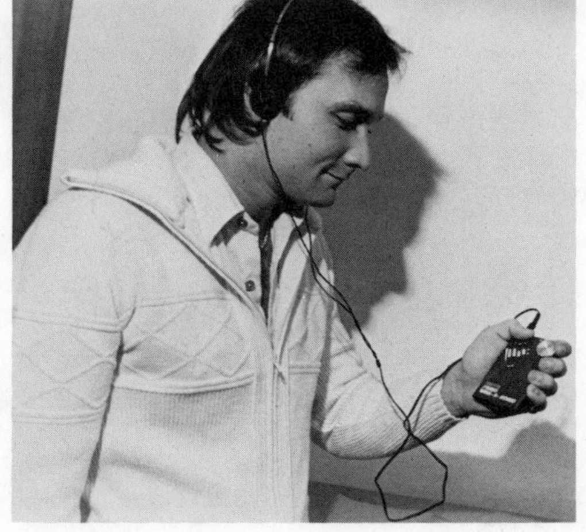

FIGURE 1.3
Age profile of walk-along users. Courtesy RAB.

8 MILLION WALK-ALONG SETS NOW IN USE; OWNERSHIP INCREASES RADIO LISTENING

6% of Persons 12+ Own a Walk-Along
75% of Walk-Alongs Include Radio
24% More Time Spent Listening to Radio Than Prior to Walk-Along Ownership

ADULTS & TEENS OWN WALK-ALONGS

Age Profile of Walk-Along Owners

18.2% Adults 35+	48.3% Adults 18–34	33.5% Teens 12–17

outlets found it no small task to continue operating. Interestingly, it was not one of these financially pinched stations that conceived of a way to generate income, but rather AT&T-owned WEAF in New York.

The first paid announcement ever broadcast lasted ten minutes and was bought by Hawthorne Court, a Queens-based real estate company. Within a matter of weeks other businesses also paid modest "tolls" to air messages over WEAF. Despite AT&T's attempts to monopolize the pay for broadcast concept, a year later in 1923 many stations were actively seeking sponsors to underwrite their expenses as well as to generate profits. Thus, the age of commercial radio was launched. It is impossible to imagine what American broadcasting would be like today had it remained a purely noncommercial medium as it has in many countries.

FIGURE 1.4
David Sarnoff, the man who put radio into the home. Courtesy RCA.

BIRTH OF THE NETWORKS

The same year that Pittsburgh station KDKA began offering a schedule of daily broadcasts, experimental network operations using telephone lines were inaugurated. As early as 1922, stations were forming chains, thereby enabling programs to be broadcast simultaneously to sev-

FIGURE 1.5
Radio fans circa 1921. Courtesy Westinghouse Electric.

eral different areas. Sports events were among the first programs to be broadcast in network fashion. Stations WJZ (later WABC) in New York and WGY in Schenectady linked for the airing of the 1922 World Series, and early in 1923 WEAF in New York and WNAC in Boston transmitted a football game emanating from Chicago. Later the same year, President Coolidge's message to Congress was aired over six stations. Chain broadcasting, a term used to describe the earliest networking efforts, was off and running.

The first major broadcast network was established in 1926 by the Radio Corporation of America (RCA) and was named the National Broadcasting Company (NBC). The network consisted of two dozen stations—several of which it had acquired from AT&T, which was encouraged by the government to divest itself of its broadcast holdings. Among the outlets RCA purchased was WEAF, which became its flagship station. Rather than form one exclusive radio combine, RCA chose to operate separate Red and Blue networks. The former comprised the bulk of NBC's stations, whereas the Blue network remained relatively small, with fewer than half a dozen outlets. Under the NBC banner, both networks would grow, the Blue network remaining the more modest of the two.

Less than two years after NBC was in operation, the Columbia Broadcasting System (initially Columbia Phonograph Broadcasting System) begain its network serviced with sixteen stations. William S. Paley, who had served as advertising manager of his family's cigar company (Congress Cigar), formed the network in 1928 and would remain its chief executive into the 1980s.

A third network emerged in 1934. The Mutual Broadcasting System went into business with affiliates in only four cities—New York, Chicago, Detroit, and Cincinnati. Unlike NBC and CBS, Mutual did not own any stations. Its primary function was that of program supplier. In 1941, Mutual led its competitors with 160 affiliates.

Although NBC initially benefited from the government's fear of a potential monopoly of communication services by AT&T, it also was forced to divest itself of a part of its holdings because of similar apprehensions. When the Federal Communications Commission (FCC) implemented more stringent chain broadcasting rules in the early 1940s, which prohibited one organization from operating two separate and distinct networks, RCA sold its Blue network, retaining the more lucrative Red network.

RADIO BROADCASTING NEWS

Vol. 3 MARCH 31, 1923 No. 13

Ethel Barrymore in "The Laughing Lady", recently broadcasted from Station WJZ. Left to Right—Alice John, Katherine Emmet, Violet Kemble Cooper, and Ethel Barrymore.

FIGURE 1.6
Radio trade paper in 1923 serving a growing industry. Courtesy Westinghouse Electric.

The FCC authorized the sale of the Blue network to Edward J. Noble in 1943. Noble, who had amassed a fortune as owner of the Lifesaver Candy Company, established the American Broadcasting Company (ABC) in 1945. In the years to come, ABC would eventually become the largest and most successful of all the radio networks.

By the end of World War II, the networks accounted for 90 percent of the radio audience and were the greatest source of individual station revenue.

CONFLICT IN THE AIR

The five years that followed radio's inception saw phenomenal growth. Millions of receivers adorned living rooms throughout the country, and over seven hundred stations were transmitting signals. A lack of sufficient regulations and an inadequate broadcast band contributed to a situation that bordered on catastrophic for the fledgling medium. Radio reception suffered greatly as the result of too many stations broadcasting, almost at will, on the same frequencies. Interference was widespread. Frustration increased among both the listening public and the broadcasters, who feared the strangulation of their industry.

Concerned about the situation, participants of the National Radio Conference in 1925 appealed to the secretary of commerce to impose limitations on station operating hours and power. The bedlam continued, since the head of the Commerce Department lacked the necessary power to implement effective changes. However, in 1926, President Coolidge urged Congress to address the issue. This resulted in the Radio Act of 1927 and the formation of the Federal Radio Commission (FRC). The five-member commission was given authority to issue station licenses, allocate frequency bands to various services, assign frequencies to individual stations, and dictate station power and hours of operation.

Within months of its inception, the FRC established the Standard Broadcast band (500–1500 kc) and pulled the plug on 150 of the existing 732 radio outlets. In less than a year, the medium that had been on the threshold of ruin was thriving. The listening public responded to the clearer reception and the increasing schedule of entertainment programming by purchasing millions of receivers. More people were tuned to their radio music boxes than ever before.

FIGURE 1.7
Today many of these pre–World War II Bakelite plastic table model receivers, which originally cost under twenty dollars, sell for over one thousand dollars as collector's items.

RADIO PROSPERS DURING THE DEPRESSION

The most popular radio show in history, "Amos 'n' Andy," made its debut on NBC in 1929, the same year that the stock market took its traumatic plunge. The show attempted to lessen the despair brought on by the ensuing Depression by addressing it with lighthearted humor. As the Depression deepened, the stars of "Amos 'n' Andy," Freeman Gosden and Charles Correll, sought to assist in the president's recovery plan by helping to restore confidence in the nation's banking system through a series of recurring references and skits. When the "Amos 'n' Andy" show aired, most of the country stopped what it was doing and tuned in. Theater owners complained that on the evening the show was broadcast ticket sales were down dramatically.

As businesses failed, radio flourished. The abundance of escapist fare that the medium offered, along with the important fact that it was provided free to the listener, enhanced radio's hold on the public.

Not one to overlook an opportunity to give his program for economic recuperation a further boost, President Franklin D. Roosevelt launched a series of broadcasts on March 12, 1933, which became known as the "fireside chats." Although the president had never received formal broadcast training, he was completely at home in front of the microphone. The audience perceived a man of sincerity, intelligence, and determination. His sensitive and astute use of the medium went a long way toward helping in the effort to restore the economy.

In the same year that Roosevelt took to the airwaves to reach the American people, he set the wheels in motion to create an independent government agency whose sole function would be to regulate all electronic forms of communication, including both broadcast and wire. To that end, the Communications Act of 1934 resulted in the establishment of the Federal Communications Commission.

As the Depression's grip on the nation weakened in the late 1930s, another crisis of awesome proportions loomed—World War II. Once again radio would prove an invaluable tool for the national good. Just as the medium completed its second decade of existence, it found itself enlisting in the battle against global tyranny. By 1939, as the great firestorm was nearing American shores, 1,465 stations were authorized to broadcast.

RADIO DURING WORLD WAR II

Before either FM or television had a chance to get off the ground, the FCC saw fit to impose a wartime freeze on the construction of new broadcast outlets. All materials and manpower were directed at defeating the enemy. Meanwhile, existing AM stations prospered and enjoyed increased stature. Americans turned to their receivers for the latest information on the war's progress. Radio took the concerned listener to the battle fronts with dramatic and timely reports from war correspondents, such as Edward R. Murrow and Eric Severied, in Europe and the South Pacific. The immediacy of the news and the gripping reality of the sounds of battle brought the war into stateside homes. This was the war that touched all Americans. Nearly everyone had a relative or knew someone involved in the effort to preserve the American way of life.

Programs that centered on concerns related to the war were plentiful. Under the auspices of the Defense Communications Board, radio set out to do its part to quash aggression and tyranny. No program of the day failed to address issues confronting the country. In fact, many programs were expressly propagandistic in their attempts to shape and influence listeners' attitudes in favor of the Allied position.

Programs with war themes were popular with sponsors who wanted to project a patriotic image, and most did. Popular commentator Walter Winchell, who was sponsored by Jergen's Lotion, closed his programs with a statement that illustrates the prevailing sentiment of the period: "With lotions of love, I remain your New York correspondent Walter Winchell, who thinks every American has at least one thing to be thankful for on Tuesday next. Thankful that we still salute a flag and not a shirt."

Although no new radio stations were constructed between 1941 and 1945, the industry saw profits double and the listening audience swell. By war's end, 95 percent of homes had at least one radio.

TELEVISION APPEARS

The freeze that prevented the full development and marketing of television was lifted within months of the war's end. Few radio broadcasters anticipated the dilemma that awaited them. In 1946 it was business as usual for the medium, which enjoyed new-found prestige as the consequence of its valuable service during the war. Two years later, however, television was the new celebrity on the block, and radio was about to experience a significant decline in its popularity.

While still an infant in 1950, television succeeded in gaining the distinction of being the number one entertainment medium. Not only did radio's audience begin to migrate to the TV screen, but many of the medium's entertainers and sponsors jumped ship as well. Profits began to decline, and the radio networks lost their prominence.

In 1952, as television's popularity continued to eclipse radio's, three thousand stations of the faltering medium were authorized to operate. Several media observers of the day predicted that television's effect would be too devastating for the older medium to overcome. Many radio station owners around the country sold their facilities. Some reinvested their money in television, while others left the field of broadcasting entirely.

A NEW DIRECTION

A technological breakthrough by Bell Laboratory scientists in 1948 resulted in the creation of the transistor. This innovation provided radio manufacturers the chance to produce miniature portable receivers. The new transistors, as they were popularly called, enhanced radio's mobility. Yet the medium continued to flounder throughout the early 1950s as it attempted to formulate a strategy that would offset the effects of television. Many radio programmers felt that the only way to hold onto their dwindling audience was to offer the same material, almost program for program, aired by television. Ironically, television had appropriated its programming approach from radio, which no longer found the system viable.

By 1955, radio revenues had dropped to 90 million dollars, and it was apparent to all that the medium had to devise another way to attract a following. Prerecorded music became a mainstay for many stations that had dropped their network affiliations in the face of decreased program schedules. Gradually, music became the primary product of radio stations and the disc jockey (deejay, jock), their new star.

RADIO ROCKS AND ROARS

The mid-fifties saw the birth of the unique cultural phenomenon known as rock 'n' roll, a term invented by deejay Alan Freed to describe a new form of music derived from rhythm and blues. The new sound took hold of the nation's youth and helped return radio to a position of prominence.

In 1955, Bill Haley's recording of "Rock Around the Clock" struck paydirt and sold over a million copies, thus ushering in a new era in contemporary music. The following year Elvis Presley tunes dominated the hit charts. Dozens of stations around the country began to focus their playlists on the newest music innovation. The Top 40 radio format, which was conceived about the time rock made its debut, began to top the ratings charts. In its original form Top 40 appealed to a much larger cross section of the listening public because of the diversity of its offerings. At first artists such as Perry Como, Les Paul and Mary Ford, and Doris Day were more common than the rockers. Then the growing penchant of young listeners for the doo-wop sound figured greatly in the narrowing of the Top 40 playlist to mostly rock 'n' roll records. Before long the Top 40 station was synonymous with rock and teens.

A few years passed by before stations employing the format generated the kind of profits their ratings seemed to warrant. Many advertisers initially resisted spending money on stations that primarily attracted kids. By 1960, however, rock stations could no longer be denied since they led their competitors in most cities. Rock and radio formed the perfect union.

FM'S ASCENT

Rock eventually triggered the wider acceptance of FM, whose creator, Edwin H. Armstrong, set out to produce a static-free alternative to the AM band. In 1938 he accomplished his objective, and two years later the FCC authorized FM broadcasting. However, World War II thwarted the implementation of Armstrong's innovation. It was not until 1946 that FM stations were under construction. Yet FM's launch was less than dazzling. Television was on the minds of most Americans, and the prevailing attitude was that a new radio band was hardly necessary.

Over 600 FM outlets were on the air in 1950, but by the end of the decade the number had shrunk by one hundred. Throughout the 1950s and early 1960s, FM stations directed their programming to special-interest groups. Classical and soft music were offered by many. This conservative, if not somewhat highbrow, programming helped foster an elitist image. FM became associated with the intellectual or, as it was sometimes referred to, the "egghead" community. Some FM stations purposely expanded on their snob appeal image in an attempt to set themselves apart from popular, mass appeal radio. This, however, did little to fill their coffers.

FM remained the poor second cousin to AM throughout the 1960s, a decade that did, however, prove transitional for FM. Many FM licenses were held by AM station operators who sensed that someday the new medium might take off. An equal number of FM licensees used the unprofitable medium for tax write-off purposes. Although many AM broadcasters possessed FM frequencies, they often did little when it came to programming them. Most chose to simulcast their AM broadcasts. It was more cost effective during a period when FM drew less then 10 percent of the listening audience.

In 1961 the FCC authorized stereo broadcasting on FM. This would prove to be a benchmark in the evolution of the medium. Gradually more and more recording companies were pressing stereo disks. The classical music buff was initially considered the best prospect for the new product. Since fidelity was of prime concern to the classical music devotee, FM stations that could afford to go stereo did so. The "easy listening" stations soon followed suit.

Another benchmark in the development of FM occurred in 1965 when the FCC passed legislation requiring that FM broadcasters in cities whose populations exceeded one hundred thousand break simulcast with their AM counterparts for at least 50 percent of their broadcast day. The commission felt that simply duplicating an AM signal did not constitute efficient use of an FM frequency. The FCC also thought that the move would help foster growth in the medium, which eventually proved to be the case.

The first format to attract sizable audiences to FM was Beautiful Music, a creation of program innovator Gordon McLendon. The execution of the format made it particularly adaptable to automation systems, which many AM/FM combo operations resorted to when the word came down from Washington that simulcast days were over. Automation kept staff size and production expenses to a minimum. Many stations assigned FM operation to their engineers, who kept the system fed with reels of music tapes and cartridges containing commercial material. Initially, the idea was to keep the FM as a form of garnishment for the more lucrative AM operation. In other words, at combo stations the FM was thrown in as a perk to attract advertisers—two stations for the price of one. To the surprise of more than a few station managers, the FM side began to attract impressive numbers. The more-music, less-talk (meaning fewer commercials) stereo operations made money. By the late 1960s, FM claimed a quarter of the radio listening audience, a 120 percent increase in less than five years.

Contributing to this unprecedented rise in popularity was the experimental Progressive format, which sought to provide listeners with an alternative to the frenetic, highly commercial AM sound. Rather than focus on the best-selling songs of the moment, as was the tendency on AM, these stations were more interested in giving airtime to album cuts that normally never touched the felt of studio turntables. The Progressive or album rock format slowly chipped away at Top 40's ratings numbers and eventually earned itself part of the radio audience.

The first major market station to choose a daring path away from the tried and true chart hit format was WOR-FM in New York. On July 30, 1966, the station broke from its AM side and embarked upon a new age in contemporary music programming. Other stations around the country were not long in following their lead.

The FM transformation was to break into full stride in the early and mid-1970s. Stereo component systems were a hot consumer item and the preferred way to listen to music, including rock. However, the notion of Top 40 on FM was still alien to most. FM listeners had long regarded their medium as the alternative to the pulp and punch presentation typical of the Standard Broadcast band. The idea of contemporary hit stations on FM offended the sensibilities of a portion of the listening public. Nonetheless, Top 40 began to make its debut on FM and for many license holders it marked the first time they enjoyed sizable profits. By the end of the decade, FM's profits would triple, as would its share of the audience. After three decades of living in the shadow of AM, FM achieved parity in 1979 when it equaled AM's listenership. The following year it moved ahead. In the mid-1980s, studies demonstrated that FM now attracts as much as 70 percent of the radio audience.

With more than a thousand more commercial AM stations (4,726) than FM stations (3,490) and only a third of the audience, the older medium is now faced with a unique challenge that could determine its very survival. In an attempt to retain a share of the audience, many AM stations have dropped music in favor of news and talk. WABC-AM's shift from Musicradio to Talkradio in 1982 clearly illustrated the metamorphosis that AM was undergoing. WABC had long been the nation's foremost leader in th pop-rock music format.

In a further effort to avert the FM sweep, dozens of AM stations have gone stereo and many more plan to do the same. AM broadcasters are hoping this will give them the competitive edge they urgently need. Music will likely return to the AM side, bringing along with it some unique format approaches. A number of radio consultants believe that the real programming innovations in the next few years will occur at AM stations. "Necessity is the mother of invention," says Dick Ellis, radio format specialist for Peters Productions in San Diego, California. "Expect some very exciting and interesting things to happen on AM in the not too distant future." Ellis is not alone in his predictions. Other media experts expect the one-time giant to make an impressive recovery.

PROLIFERATION AND FRAG-OUT

Specialization, or narrowcasting as it has come to be called, salvaged the medium in the early 1950s. Before that time radio bore little resemblance to its sound during the age of television. It was the video medium that copied radio's approach to programming during its golden age. "Sightradio," as television was sometimes ironically called, drew from the older electronic medium its programming schematic and left radio hovering on the edge of the abyss. Gradually radio station managers realized they could not combat the dire effects of television by programming in a like manner. To survive they had to change. To attract listeners they had to offer a different type of service. The majority of stations went to spinning records and presenting short newscasts. Sports and weather forecasts became an industry staple.

Initially, most outlets aired broad appeal music. Specialized forms, such as jazz, rhythm and blues, and country, were left off most playlists, except in certain regions of the country. Eventually these all-things-to-all-people stations were challenged by what is considered to be the first popular attempt at format specialization. As legend now has it, radio programmer Todd Storz and his assistant Bill Steward of KOWH-AM in Omaha, Nebraska, decided to limit their station's playlist to only those records that currently enjoyed high sales. The idea for the scheme struck them at a local tavern as they pending money to play mostly gs on the jukebox. Their pro- t became known as "Top 40." f executing their new format, ne ratings. Word on their suc- iring other stations around the ie pop record approach. They ss.

960s other formats had evolved, tiful Music, which was intro- n Francisco station KABL, and ch first aired over XETRA located kico. Both formats were the progeny of Gordon McLendon and were successfully copied across the country.

The diversity of musical styles that evolved in the mid-1960s, with the help of such disparate performers as the Beatles and Glen Campbell, gave rise to a myriad of format variations. While some stations focused on 1950s rock 'n' roll ("blasts from the past," "oldies but goodies"), others stuck to current hits and still others chose to play more obscure rock album cuts. The 1960s saw the advent of the radio formats of Soft Rock and Acid and Psychedelic hard rock. Meanwhile, Country, whose popularity had been confined mostly to areas of the South and Midwest, experienced a sudden growth in its acceptance through the crossover appeal of artists such as Johnny Hartford, Bobbie Gentry, Bobby Goldsboro, Johnny Cash, and, in particular, Glen Campbell, whose sophisticated country-flavor songs topped both the Top 40 and Country charts.

As types of music continued to become more diffused in the 1970s, a host of new formats came into use. The listening audience became more and more fragmented. "Frag-out," a term coined by radio consultant Kent Burkhart, posed an ever-increasing challenge to program directors whose job it was to attract a large enough piece of the radio audience to keep their stations profitable.

The late 1970s and early 1980s saw the rise and decline of the Disco format, which eventually evolved into Urban Contemporary, and a wave of interest in synthesizer-based electropop. Formats such as Soft Rock faded from the scene only to be replaced by a narrower form of Top 40 called Contemporary Hit. New formats continue to surface with almost predictable regularity. Among the most recent batch are All-Comedy, Children's Radio, All-Sports, Eclectic-Oriented Rock, All-Weather, and All-Beatles.

Although specialization saved the industry from an untimely end over three decades ago, the proliferation in the number of radio stations (which nearly quadrupled since 1950) competing for the same audience has brought about the age of hyperspecialization. Today there are more than one hundred format variations in the radio marketplace, as compared to a handful when radio stations first acknowledged the necessity of programming to a preselected segment of the audience as the only means to remain in business. (For a more detailed discussion on radio formats, see chapter 3.)

FIGURE 1.8
Some popular radio program formats from 1960 to the present. New formats are constantly evolving.

Acid Rock
Adult Contemporary
Album-Oriented Rock
Arena Rock
Beautiful Music
Big Band
Black
Bluegrass
Bubble Gum
Classical
Contemporary Country
Contemporary Hits
Country and Western
Chicken Rock
Dance
Disco
Easy Listening
English Rock
Ethnic
Folk Rock
Free Form
Jazz
Mellow Rock
Middle-of-the-Road
Motown
News
News/Talk
New Wave
Nostalgia
Oldies
Pop
Progressive
Punk Rock
Religious
Rhythm and Blues
Soft Rock
Southern Rock
Standards
Talk
Top 40
Urban Contemporary
Urban Country

PROFITS IN THE AIR

Although radio has been unable to regain the share of the national advertising dollar it attracted before the arrival of television, it does earn far more today than it did during its so-called heyday. About 7 percent of all money spent on advertising goes to radio. This computes to billions of dollars.

Despite the enormous gains since WEAF introduced the concept of broadcast advertising, radio cannot be regarded as a get-rich-quick scheme. Many stations walk a thin line between profit and loss. While some major market radio stations demand and receive over a thousand dollars for a one-minute commercial, an equal number sell time for the proverbial "dollar a holler."

While the medium's earnings have maintained a progressive growth pattern, it also has experienced periods of recession. These financial slumps or dry periods have almost all occurred since 1950. Initially television's effect on radio's revenues was devastating. The medium began to recoup its losses when it shifted from its reliance on the networks and national advertisers to local businesses. Today 70 percent of radio's revenues come from local spot sales as compared to half that figure in 1948.

By targeting specific audience demographics, the industry remained solvent. In the first half of the 1980s, a typical radio station earned fifty thousand dollars annually in profits. As the medium regained its footing after the staggering blow administered it by television, it experienced both peaks and valleys financially. In 1961, for example, the FCC reported that more radio stations recorded losses than in any previous period since it began keeping records of such things. Two years later, however, the industry happily recorded its greatest profits ever. In 1963 the medium's revenues exceeded 636 million dollars. In the next few years earnings would be up 60 percent, surpassing the 1.5 billion mark, and would leap another 150 percent between 1970 and 1980. FM profits have tripled since 1970 and have significantly contributed to the overall industry figures.

The segment of the industry that has found it most difficult to stay in the black is the AM daytimer. These radio stations are required by the FCC to cease broadcasting around sunset so as not to interfere with other AM stations. Of the twenty-three hundred daytimers in operation, nearly a third have reported losses at one time or another over the past decade. The unique problem facing daytime-only broadcasters has been further aggravated by FM's dramatic surge in popularity. The nature of their license gives daytimers subordinate status to full-time AM operations, which have found competing no easy trick, especially in the light of FM's success. Because of the lowly status of the daytimer in a marketplace that has become increasingly thick with rivals, it is extremely difficult for these stations to prosper, although some do very well. Many daytimers have opted for specialized forms of programming to attract advertisers. For example, religious and ethnic formats have proven successful. As a consequence of the formidable obstacles facing the AM daytime operation, many have been put up for sale, and asking prices have been as low as fifty thousand dollars.

Meanwhile, the price for FM stations has skyrocketed since 1970. Many full-time metro market AMs have sold for multimillions, for the simple reason that they continue to appear in the top of their respective ratings surveys.

In general, individual station profits have not kept pace with profits industry-wide due to the rapid growth in the number of outlets over the past two decades. To say the least, competition is keen and in many markets downright fierce. It is common for thirty or more radio stations to vie for the same advertising dollars in large cities, and the introduction of other media, such as cable, intensifies the skirmish over sponsors.

RADIO AND GOVERNMENT REGULATIONS

Almost from the start it was recognized that radio could be a unique instrument for the public good. This point was never made more apparent than in 1912, when a young wireless operator named David Sarnoff picked up the distress signal from the sinking *Titanic* and relayed the message to ships in the vicinity, which then came to the rescue of those still alive. The survivors were the beneficiaries of the first attempt at regulating the new medium. The Wireless Ship Act of 1910 required that ships carrying fifty or more passengers have wireless equipment on board. Sarnoff's effective use of the medium from an experimental station on Nantucket Island resulted in the saving of seven hundred lives.

Radio's first practical application was as a means of communicating from ship to ship and from ship to shore. During the first decade of this century, Marconi's wireless invention was seen primarily as a way of linking the ships at sea with the rest of the world. Until that time, when ships left port they were beyond any conventional mode of communications. The wireless was a boon to the maritime services, including the Navy, which equipped each of its warships with the new device.

Coming on the heels of the *Titanic* disaster, the Radio Act of 1912 sought to expand the general control of radio on the domestic level. The secretary of commerce and labor was appointed to head the implementation and monitoring of the new legislation. The primary function of the act was to license wireless stations and operators. The new regulations empowered the Department of Commerce and Labor to impose fines and revoke the licenses of those who operated outside the parameters set down by the communications law.

Growth of radio on the national level was curtailed by World War I, when the government saw fit to take over the medium for military purposes. However, as the war raged on, the same young wireless operator, David Sarnoff, who had been instrumental in saving the lives of passengers on the ill-fated *Titanic,* was hard at work on a scheme to drastically modify the scope of the medium, thus converting it from an experimental and maritime communications apparatus to an appliance designed for use by the general public. Less than five years after the war's end, receivers were being bought by the millions, and radio as we know it today was born.

As explained earlier, the lack of regulations dealing with interference nearly resulted in the premature end of radio. By 1926 hundreds of stations clogged the airways, bringing pandemonium to the dial. The Radio Act of 1912 simply did not anticipate radio's new application. It was the Radio Act of 1927 that first approached radio as a mass medium. The Federal Regulatory Commision's five commissioners quickly implemented a series of actions that restored the fledgling medium's health.

The Communications Act of 1934 charged a seven-member commission with the responsibility of ensuring the efficient use of the airways—which the government views as a limited resource that belongs to the public and is leased to broadcasters. Over the years the FCC has concentrated its efforts on maximizing the usefulness of radio for the public's benefit. Consequently, broadcasters have been required to devote a portion of their airtime to programs that address important community and national issues. In addition, broadcasters have had to promise to serve as a constant and reliable source of information, while retaining certain limits on the amount of commercial material scheduled.

The FCC has steadfastly sought to keep the medium free of political bias and special-interest groups. In 1949 the commission implemented regulations making it necessary for stations that present a viewpoint to provide an equal amount of airtime to contrasting or opposing viewpoints. The Fairness Doctrine obliged broadcasters "to afford reasonable opportunity for the discussion of conflicting views of public importance." Later it also stipulated that stations notify persons when attacks were made on them over the air.

Although broadcasters generally acknowledge the unique nature of their business, many have felt that the government's involvement has exceeded reasonable limits in a society based upon a free enterprise system. Since it is their money, time, and energy they are investing, broadcasters feel they should be afforded greater opportunity to determine their own programming.

In the late 1970s, a strong movement headed by Congressman Lionel Van Deerlin sought to reduce the FCC's role in broadcasting, in order to allow the marketplace to dictate how the industry conducted itself. Van Deerlin actually proposed that the Broadcast Branch of the commission be abolished and a new organization with much less authority created. His bill was defeated, but out of his and others' efforts came a new attitude concerning the government's hold on the electronic media.

President Reagan's antibureaucracy, free enterprise philosophy gave impetus to the deregulation move already under way when he assumed office. The FCC, headed by Chairman Mark Fowler, expanded on the deregulation proposal that had been initiated by his predecessor, Charles Ferris. The deregulation bill eliminated the requirement that radio stations devote a portion of their airtime (8 percent for AM and 6 percent for FM) to nonentertainment programming of a public affairs nature. In addition, stations no longer had to undergo the lengthy process of ascertainment of community needs as a condition of license renewal, and guidelines pertaining to the amount of time devoted to commercial announcements was eliminated. The rule requiring stations to maintain detailed program logs was also abolished. A simplified postcard license renewal form was adopted, and license terms were extended from three to seven years. In a further step the commission raised the ceiling on the number of broadcast outlets a company or individual could own—from seven AM, seven FM, and seven television stations to twelve each.

The extensive updating of the Communications Act of 1934 was based on the belief that the marketplace should serve as the primary regulator. Opponents of the reform feared that

Air Concert "Picked Up" By Radio Here

Victrola music, played into the air over a wireless telephone, was "picked up" by listeners on the wireless receiving station which was recently installed here for patrons interested in wireless experiments. The concert was heard Thursday night about 10 o'clock, and continued 20 minutes. Two orchestra numbers, a soprano solo—which rang particularly high and clear through the air—and a juvenile "talking piece" constituted the program.

The music was from a Victrola pulled up close to the transmitter of a wireless telephone in the home of Frank Conrad, Penn and Peebles avenues, Wilkinsburg. Mr. Conrad is a wireless enthusiast and "puts on" the wireless concerts periodically for the entertainment of the many people in this district who have wireless sets.

Amateur Wireless Sets, made by the maker of the Set which is in operation in our store, are on sale here $10.00 up.

West Basement

FIGURE 1.9 History-making ad in 1920. Courtesy Westinghouse Electric.

with their new-found freedom radio stations would quickly turn their backs on community concerns and concentrate their full efforts on fattening their pocketbooks. Proponents of deregulation applauded the FCC's actions, contending that the listening audience would indeed play a vital role in determining the programming of radio stations, since the medium had to meet the needs of the public in order to prosper.

While the government continues to closely scrutinize the actions of the radio industry to ensure that it operates in an efficient and effective manner, it is no longer perceived as the fearsome, omnipresent Big Brother it once was. Today broadcasters more fully enjoy the fruits of a laissez-faire system of economy.

JOBS IN RADIO

The radio industry employs over one hundred thousand people and the number is certain to grow as many more stations go on the air. Since 1972 forty thousand individuals have found full-time employment in radio. Today, opportunities for women and minorities are greater than ever. Radio has until fairly recently been a male-dominated profession. In 1975 men in the industry outnumbered women nearly four to one. But that has changed. Now women are being hired more than ever before, and not just for office positions. Women have made significant inroads into programming, sales, and management positions, and there is no reason to think that this trend will not continue. It will take a while, however, before an appropriate proportion of women and minorities are working in the medium. The FCC's insistence on equal opportunity employment within the broadcast industry and the continued proliferation of radio facilities make prospects good for all who are interested in broadcasting careers.

A common misconception is that a radio station consists primarily of deejays with few other job options available. Wrong! Nothing could be further from the truth. Granted, disc jockeys comprise part of a station's staff, but many other employees are necessary to keep the station on the air.

An average-size station in a medium market employs between eighteen and twenty-six people, and on-air personnel comprise about a third of that figure. Stations are usually broken down into three major areas: sales, programming, and engineering. Each area, in particular the first two, requires a variety of people for positions that demand a wide range of skills. Subsequent chapters in this book will bear this out.

Proper training and education are necessary to secure a job at most stations, although many will train people to fill the less demanding po-

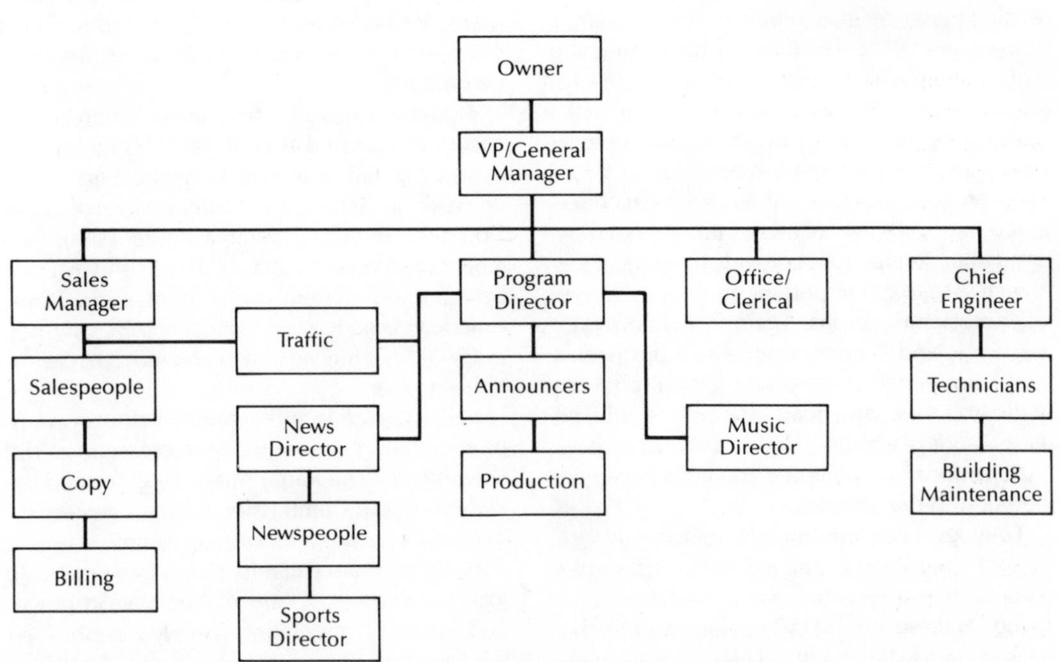

FIGURE 1.10 Organizational flow chart for a medium market radio station.

sitions. Over a thousand schools and colleges offer courses in radio broadcasting, and most award certificates or degrees. As in most other fields today, the more credentials a job candidate possesses, the better he or she looks to a prospective employer.

Perhaps no other profession weighs practical, hands-on experience as heavily as radio does. This is especially true in the on-air area. On the programming side, it is the individual's sound that wins the job, not the degree. However, it is the formal training and education that usually contribute most directly to the quality of the sound that the program director is looking for when hiring. In reality, only a small percentage of radio announcers have college degrees, but statistics have shown that those who do stand a better chance of moving into managerial positions.

Many station managers look for the college-educated person, particularly for the areas of news and sales. Before 1965 the percentage of radio personnel with college training was relatively low. But the figure has gradually increased as more and more colleges add broadcasting curricula. Thousands of communications degrees are conferred annually, thus providing the radio industry a pool of highly educated job candidates. Today, college training is a plus when one is searching for employment in radio.

The job application or résumé that lists practical experience in addition to formal training is most appealing. The majority of colleges with radio curricula have college stations. These small outlets provide the aspiring broadcaster with a golden opportunity to gain some much-needed on-air experience. Some of the nation's foremost broadcasters began their careers at college radio stations. Many of these same schools have internship programs that provide the student with the chance to get important on-the-job training at professional stations. Again, experience is experience, and it does count to the prospective employer. Small commercial stations often are willing to hire broadcast students to fill part-time and vacation slots. This constitutes "professional" experience and is an invaluable addition to the résumé.

Entry-level positions in radio seldom pay well. In fact, many small market stations pay near minimum wage. However, the experience gained at these low-budget operations more than makes up for the small salaries. The first year or two in radio constitutes the dues-paying period, a time in which a person learns the ropes. The small radio station provides inexperienced people the chance to become involved in all facets of the business. Rarely does a new employee perform only one function. For example, a person hired to deejay will often prepare and deliver newscasts, write and produce commercials, and may even sell airtime.

In order to succeed in a business as unique as radio, a person must possess many qualities, not the least of which are determination, skill, and the ability to accept and benefit from constructive criticism. A career in radio is like no other, and the rewards, both personal and financial, can be exceptional. "It's a great business," says Lynn Christian, vice president of Century Broadcasting Corporation. "No two days are alike. After thirty years in radio, I still recommend it over other career opportunities."

CHAPTER HIGHLIGHTS

1. The average adult spends three-and-a-half hours per day listening to radio. Radio is the most available source of entertainment, companionship, and information.

2. Guglielmo Marconi is generally considered the father of radio, although David Sarnoff is a likely contender.

3. As early as 1922, the Department of Commerce set aside two frequencies for radio broadcasts. WEAF in New York aired the first commercial.

4. Station networks, first called "chain broadcasting," operated as early as 1922. Radio Corporation of America (RCA) formed the first major network in 1926, the National Broadcasting Company (NBC). Columbia Broadcasting System (CBS) was formed in 1928, and Mutual Broadcasting System (MBS) followed in 1934. American Broadcasting Company (ABC), formed in 1945, became the largest and most successful radio network.

5. Early station proliferation led to overlapping signals. Signal quality decreased, as did listenership. The Radio Act of 1927 formed the Federal Radio Commission (FRC), a five-member board authorized to issue station licenses, allocate frequency bands, assign frequencies to individual stations, and dictate station power and hours of operation. The FRC established the Standard Broadcast band (500–1500 kc).

6. Radio prospered during the Depression by

providing cost-free entertainment and escape from the harsh financial realities. "Amos 'n' Andy," which made its debut in 1929, was the most popular radio show in history. President Franklin D. Roosevelt's fireside chats began on March 12, 1933. The Communications Act of 1934 established the seven-member Federal Communications Commission (FCC).

7. World War II led the FCC to freeze construction of new broadcast outlets; therefore, existing AM outlets prospered.

8. Two years after the war's end, television usurped the entertainment leadership.

9. The Bell Lab scientists' invention of the transistor in 1948 helped save radio by providing portability. Music became radio's mainstay.

10. The Top 40 radio format, conceived about the same time as the emergence of rock music, became the most popular format with younger audiences.

11. Edwin H. Armstrong developed the static-free FM band in 1938. The FCC authorized stereo FM broadcasting in 1961, and in 1965 it separated FM from AM simulcasts in cities with populations in excess of one hundred thousand. Beautiful Music was the first popular format on FM, which relied heavily on automation. The Progressive format focused on album cuts rather than Top 40. By the mid-1980s FM possessed 70 percent of the listening audience.

12. "Narrowcasting" is specialized programming. "Frag-out" refers to the fragmentation of audience because of numerous formats.

13. Seventy percent of radio's revenues come from local spot sales.

14. In 1949 the FCC formulated the Fairness Doctrine, which obligates broadcasters to present opposing points of view.

15. The radio industry employs over one hundred thousand people, with women and minorities making significant gains in recent years. A combination of practical experience and formal training is the best preparation for a career in broadcasting.

SUGGESTED FURTHER READING

Aitkin, Hugh G.J. *Syntony and Spark.*, New York: John Wiley and Sons, 1976.

Archer, G.L. *History of Radio to 1926*. New York: Arno Press, 1971.

Aronoff, Craig E., ed. *Business and the Media*. Santa Monica, Calif.: Goodyear Publishing Company, 1979.

Baker, W.J. *A History of the Marconi Company*. New York: St. Martin's Press, 1971.

Barnouw, Erik. *A Tower of Babel: A History of Broadcasting in the United States to 1933*, vol. 1. New York: Oxford University Press, 1966.

———. *The Golden Web: A History of Broadcasting in the United States 1933 to 1953*, vol. 2. New York: Oxford University Press, 1968.

———. *The Image Empire: A History of Broadcasting in the United States from 1953*, vol. 3. New York: Oxford University Press, 1970.

———. *The Sponsor: Notes on a Modern Potentate*. New York: Oxford University Press, 1978.

Bergreen, Laurence. *Look Now, Pay Later: The Rise of Network Broadcasting*. Garden City, N.Y.: Doubleday, 1980.

Bittner, John R. *Broadcasting and Telecommunications*, 2nd ed. Englewood Cliffs, N.J.: Prentice-Hall, 1985.

———. *Professional Broadcasting: A Brief Introduction*. Englewood Cliffs, N.J.: Prentice-Hall, 1981.

Blake, Reed H., and Haroldsen, E.O. *A Taxonomy of Concepts in Communications*. New York: Hastings House, 1975.

Broadcasting Yearbook. Washington, D.C.: Broadcasting Publishing, 1935 to date, annual.

Browne, Bartz, and Coddington (consultants). *Radio Today—and Tomorrow*. Washington, D.C.: National Association of Broadcasters, 1982.

Campbell, Robert. *The Golden Years of Broadcasting*. New York: Charles Scribner's Sons, 1976.

Cantril, Hadley. *The Invasion from Mars*. New York: Harper & Row, 1966.

Chapple, Steve, and Garofalo, R. *Rock 'n' Roll Is Here to Pay*. Chicago: Nelson-Hall, 1977.

Cox Looks at FM Radio. Atlanta: Cox Broadcasting Corporation, 1976.

Delong, Thomas A. *The Mighty Music Box*. Los Angeles: Amber Crest Books, 1980.

Dreher, Carl. *Sarnoff: An American Success*. New York: Quadrangle, 1977.

Dunning, John. *Tune In Yesterday*. Englewood Cliffs, N.J.: Prentice-Hall, 1976.

Edmonds, I.G. *Broadcasting for Beginners*. New York: Holt, Rinehart, and Winston, 1980.

Erickson, Don. *Armstrong's Fight for FM Broad-*

casting. Birmingham: University of Alabama, 1974.

Fang, Irving E. *Those Radio Commentators*. Ames: Iowa State University Press, 1977.

Fornatale, Peter, and Mills, J.E. *Radio in the Television Age*. New York: Overlook Press, 1980.

Foster, Eugene S. *Understanding Broadcasting*. Reading, Mass.: Addison-Wesley, 1978.

Hall, Claud, and Hall, Barbara. *This Business of Radio Programming*. New York: Hastings House, 1978.

Hasling, John. *Fundamentals of Radio Broadcasting*. New York: McGraw-Hill, 1980.

Head, Sydney W., and Sterling, C.H. *Broadcasting in America: A Survey of Television, Radio, and the New Techologies*, 4th ed. Boston: Houghton Mifflin, 1982.

Hilliard, Robert, ed. *Radio Broadcasting: An Introduction to the Sound Medium*, 3rd ed. New York: Longman, 1985.

Lazarsfeld, Paul F., and Kendall, P.L. *Radio Listening in America*. Englewood Cliffs, N.J.: Prentice-Hall, 1948.

Leinwall, Stanley. *From Spark to Satellite*. New York: Charles Scribner's Sons, 1979.

Levinson, Richard. *Stay Tuned*. New York: St. Martin's Press, 1982.

Lewis, Peter, ed. *Radio Drama*. New York: Longman, 1981.

Lichty, Lawrence W., and Topping, M.C. *American Broadcasting: A Source Book on the History of Radio and Television*. New York: Hastings House, 1976.

MacDonald, J. Fred. *Don't Touch That Dial: Radio Programming in American Life, 1920–1960*. Chicago: Nelson-Hall, 1979.

McLuhan, Marshall. *Understanding Media: The Extensions of Man*. New York: McGraw-Hill, 1964.

Paley, William S. *As It Happened: A Memoir*. Garden City, N.Y.: Doubleday, 1979.

Pierce, John R. *Signals*. San Francisco: W.H. Freeman, 1981.

Pusateri, C. Joseph. *Enterprise in Radio*. Washington, D.C.: University Press of America, 1980.

Radio Facts. New York: Radio Advertising Bureau, 1985.

Routt, Ed. *The Business of Radio Broadcasting*. Blue Ridge Summit, Pa.: Tab Books, 1972.

Seidle, Ronald J. *Air Time*. Boston: Holbrook Press, 1977.

Settle, Irving. *A Pictorial History of Radio*. New York: Grosset and Dunlap, 1960.

Sipemann, Charles A. *Radio's Second Chance*. Boston: Little, Brown, 1946.

Sklar, Rick. *Rocking America: How the All-Hit Radio Stations Took Over*. New York: St. Martin's Press, 1984.

Smith, F. Leslie. *Perspectives on Radio and Television: An Introduction to Broadcasting in the United States*. New York: Harper & Row, 1979.

Sterling, Christopher H. *Electronic Media*. New York: Praeger, 1984.

Wertheim, Arthur F. *Radio Comedy*. New York: Oxford University Press, 1979.

Whetmore, Edward J. *The Magic Medium: An Introduction to Radio in America*. Belmont, Calif.: Wadsworth Publishing, 1981.

2 Station Management

NATURE OF THE BUSINESS

It has been said that managing a radio station is like running a business that is a combination of theater company and car dealership. The medium's unique character requires that the manager deal with a broad mix of people, from on-air personalities to secretaries and from sales personnel to technicians. Few other businesses can claim such an amalgam of employees. Even the station manager of the smallest outlet with as few as six or eight employees must direct individuals with very diverse backgrounds and goals. For example, radio station WXXX in a small Maine community may employ three to four full-time air people, who likely were brought in from other areas of the country. The deejays have come to WXXX to begin their broadcasting careers with plans to gain some necessary experience and move on to larger markets. Within a matter of a few months the station will probably be looking for replacements.

Frequent turnover of on-air personnel at small stations is a fact of life. As a consequence, members of the air staff often are regarded as transients or passers-through by not only the community but also the other members of the station's staff. Less likely to come and go are a station's administrative and technical staff. Usually they are not looking toward the bright lights of the bigger markets, since the town in which the station is located is often home to them. A small station's sales department may experience some turnover but usually not to the degree that the programming department does. Salespeople also are likely to have been recruited from the ranks of the local citizenry, whereas air personalities more typically come from outside the community.

Running a station in a small market presents its own unique challenges (and it should be noted that half of the nation's radio outlets are located in communities with fewer than twenty-five thousand residents), but stations in larger markets are typically faced with stiffer competition and fates that often are directly tied to ratings. In contrast to station WXXX in the small Maine community, where the closest other station is fifty miles away and therefore no competitive threat, an outlet located in a metropolitan area may share the airwaves with thirty or more other broadcasters. Competition is intense, and radio stations in large urban areas usually suc-

	AM	FM		AM	FM
WABC	770		WLIR		92.7
WADB		95.9	WLIX	540	
WADO	1280		WLTW		106.7
WALK	1370	97.5	WMCA	570	
WAPP		103.5	WMCX		88.1
WAWZ	1380	99.1	WMGQ		98.3
WBAB		102.3	WMJY		107.1
WBAI		99.5	WMTR	1250	
WBAU		90.3	WNBC	660	
WBGO		88.3	WNCN		104.3
WBJB		90.5	WNEW	1130	102.7
WBLI		106.1	WNJR	1430	
WBLS		107.5	WNNJ	1360	
WBRW	1170		WNYC	830	93.9
WCBS	880	101.1	WNYE		91.5
WCTC	1450		WNYG	1440	
WCTO		94.3	WNYM	1330	
WCWP		88.1	WNYU		89.1
WDHA		105.5	WOBM	1170	92.7
WERA	1590		WOR	710	
WEVD		97.9	WPAT	930	93.1
WFAS	1230	103.9	WPIX		101.9
WFDU		89.1	WPLJ		95.5
WFME		94.7	WPOW	1330	
WFMU		91.1	WPRB		103.3
WFUV		90.7	WPST		97:5
WGBB	1240		WPUT	1510	
WGLI	1290		WQXR	1560	96.3
WGRC	1300		WRAN	1510	
WGSM	740		WRCN	1570	103.9
WHBI		105.9	WRFM		105.1
WHLI	1100		WRHU		88.7
WHN	1050		WRKL	910	
WHPC		90.3	WRKS		98.7
WHTG	1410	106.3	WRTN		93.5
WHTZ		100.3	WRVH		105.5
WHUD		100.7	WSBH		95.3
WHWH	1350		WSIA		88.9
WICC	600		WSKQ	620.0	
WINS	1010		WSOU		89.5
WIXL		103.7	WSUS		102.3
WJDM	1530		WTHE	1520	
WJLK	1310	94.3	WUSB		90.1
WKCR		89.9	WVIP	1310	106.3
WKDM	1380		WVOX	1460	
WKJY		98.3	WVRM		89.3
WKMB	1070		WWDJ	970	
WKRB		90.9	WWRL	1600	
WKTU		92.3	WXMC	1310	
WKWZ		88.5	WYNY		97.1
WKXW		101.5	WYRS		96.7
WLIB	1190		WZFM		107.1
WLIM	1580				

FIGURE 2.1
Newspaper listing of radio stations in a major market. Large cities such as New York, Chicago, Los Angeles, and Philadelphia often have over sixty stations.

ceed or fail based upon their showing in the latest listener surveys. The metro market station manager must pay close attention to what surrounding broadcasters are doing, while striving to maintain the best product possible in order to retain a competitive edge and prosper.

Meanwhile, the government's perception of the radio station's responsibility to its consumers also sets it apart from other industries. Since its inception, radio's business has been Washington's business, too. Station managers, unlike the heads of most other enterprises, have had to conform to the dictates of a federal agency especially conceived for the purpose of overseeing their activities. With the failure to satisfy the FCC's expectations possibly resulting in stiff penalties, such as fines and even the loss of an operating license, radio station managers have been obliged to keep up with a fairly prodigious volume of rules and regulations.

In the 1980s deregulation actions designed to unburden the broadcaster of what has been regarded by many as unreasonable government intervention have made the life of the station manager somewhat less complicated. Nevertheless, the government continues to play an important role in American radio, and managers who value their license wisely invest energy and effort in fulfilling federal conditions. After all, a radio station without a frequency is just a building with a lot of expensive equipment.

The listener's perception of the radio business, even in this day and age when nearly every community with a small business district has a radio station, is often unrealistic. The portrayal by film and television of the radio station as a hotbed of zany characters and bizarre antics has helped foster a misconception. This is not to suggest that radio stations are the most conventional places to work. Because it is the station's function to provide entertainment to its listeners, it must employ creative people, and where these people are gathered, be it a small town or a large city, the atmosphere is sure to be charged. "The volatility of the air staff's emotions and the oscillating nature of radio itself actually distinguishes our business from others," observes J. Salter, general manager, WFKY, Frankfort, Kentucky.

Faced with an audience whose needs and tastes sometimes change overnight, today's radio station has become adept at shifting gears as conditions warrant. What is currently popular in music, fashion, and leisure-time activities will be nudged aside tomorrow by something new. This, says Randy Lane, general manager of WABB, Mobile, Alabama, forces radio stations to stay one step ahead of all trends and fads. "Being on the leading edge of American culture makes it necessary to undergo more changes and updates than is usually the case in other businesses. Not adjusting to what is currently in vogue can put a station at a distinct disadvantage. You have to stay in touch with what is happening in your own community as well as the trends and cultural movements occurring in other parts of the country."

The complex internal and external factors that derive from the unusual nature of the radio business make managing today's station a formidable challenge. Perhaps no other business demands as much from its managers. Conversely, few other businesses provide an individual with as much to feel good about. It takes a special kind of person to run a radio station.

THE MANAGER AS CHIEF COLLABORATOR

There are many schools of thought concerning the approach to managing a radio station. For example, there are the standard X, Y, and Z models or theories of management. The first theory embraces the idea that the general manager is the captain of the vessel, the primary authority, with solemn, if not absolute, control of the decision-making process. The second theory casts the manager in the role of collaborator or senior advisor. The third theory forms a hybrid of the preceding two; the manager is both coach and team player, or chief collaborator. Of the three models, broadcast managers tend to favor the third approach.

Lynn Christian, who has served as general manager of several radio stations, preferred working for a manager who used the hybrid model rather than the purely authoritarian model. "Before I entered upper management, I found that I performed best when my boss sought my opinion and delegated responsibility to me. I believe in department head meetings and the full disclosure of projects within the top organization of the station. If you give someone the title, you should be prepared to give that person some authority, too. I respect the integrity of my people, and if I lose it, I replace them quickly. In other words, 'You respect me, and I'll respect you,' is the way I have always managed."

Randy Bongarten, president of NBC Radio, concurs with Christian and adds, "Management styles have to be adaptive to individual situations so as to provide what is needed at the time. In general, the collaborator or team leader approach gets the job done. Of course, I don't think there is any one school of management that is right one hundred percent of the time."

Jim Arcara, executive vice president of Capitol Cities Communication, parent company of ABC, also is an advocate of the hybrid management style. "It's a reflection of what is more natural to me as well as my company. Employees are capable of making key decisions, and they should be given the opportunity to do so. An effective manager also delegates responsibility."

As a station manager, Randy Lane of WABB finds that a collaborative atmosphere is more productive. "My basic management philosophy involves conferencing with department heads before making any decisions. I believe in involving members of the staff in solving problems that affect the workplace. The team strategy has always worked best for me."

The manager/collaborator approach has gained in popularity in the past two decades. Radio functions well in a teamlike atmosphere. Since practically every job in the radio station is designed to support and enhance the air product, establishing a connectedness among what is usually a small band of employees tends to yield the best results, contends Jane Duncklee, general manager, WBOS-FM, Boston. "I strongly believe that employees must feel that they are a valid part of what is happening and that their input has a direct bearing on those decisions which affect them and the operation as a whole. I try to hire the best people possible and then let them do their jobs with a minimum of interference and a maximum of support."

Marlin R. Taylor, founder and creative director of Bonneville Broadcasting System and former manager of several major market radio outlets, including WRFM, New York, and WBCN, Boston, believes that the manager using the collaborative system of management gets the most out of employees. "When a staff member feels that his or her efforts and contributions make a difference and are appreciated, that person will remain motivated. This kind of employee works harder and delivers more. Most people, if they enjoy the job they have and like the organization they work for, are desirous of improving their level of performance and contributing to the health and well-being of the station. I really think that many station managers should devote even more time and energy to people development."

Paul Aaron, general manager of KFBK/KAER, Sacramento, California, believes that managers must first assert their authority, that is, make it clear to all that they are in charge, before the transition to collaborator can take place. "It's a sort of process or evolution. Actually, when you come right down to it, any effective management approach includes a bit of both the authoritarian and collaborative concepts. The situation at the station will have a direct impact on the management style I personally deem most appropriate. As the saying goes, 'different situations call for different measures.' When assuming the reins at a new station, sometimes it's necessary to take a more dictatorial approach until the organization is where you feel it should be. Often a lot of cleanup and adjustments are necessary before there can be a greater degree of equanimity. Ultimately, however, there should be equanimity."

Surveys have shown that most broadcast ex-

FIGURE 2.2 WMJX/WMEX General Manager Bill Campbell meets on a regular basis with members of his staff. Like most successful managers, he maintains good communications with station personnel.

ecutives view the chief collaborator or hybrid management approach as compatible with their needs. "It has pretty much become the standard modus operandi in this industry. A radio manager must direct as well as invite input. To me it makes sense, in a business in which people are the product, to create an atmosphere that encourages self-expression, as well as personal and professional growth. After all, we are in the communications business. Everyone's voice should at least be heard," contends Lynn Christian.

WHAT MAKES A MANAGER

As in any other profession, the road to the top is seldom a short and easy one. It takes years to get there, and dues must be paid along the way. To begin with, without a genuine affection for the business and a strong desire to succeed, it is unlikely that the position can ever be attained. Furthermore, without the proper training and experience the top job will remain elusive. So then what goes into becoming a radio station manager? First, a good foundation is necessary. Formal education is a good place to start. Hundreds of institutions of higher learning across the country offer programs in broadcast operations. The college degree has achieved greater importance in radio over the last decade or two and, as in most other industries today, it has become a standard credential for those vying for management slots. Anyone entering broadcasting in the 1980s with aspirations to operate a radio station should acquire as much formal training as possible. Station managers with master's degrees are not uncommon. However, a bachelor's degree in communications gives the prospective station manager a good foundation from which to launch a career.

In a business that stresses the value of practical experience, seldom, if ever, does an individual land a management job directly out of college. In fact, most station managers have been in the business at least fifteen years. "Once you get the theory nailed down you have to apply it. Experience is the best teacher. I've spent thirty years working in a variety of areas in the medium. In radio, in particular, hands-on experience is what matters," says Richard Bremkamp, Jr., general manager, WRCH/WRCQ, Hartford, Connecticut.

To Roger Ingram, general manager of WZPL-FM, Indianapolis, experience is what most readily opens the door to management. "While a degree is kind of like a union card in this day and age, a good track record is what wins the management job. You really must possess both."

Jane Duncklee of WBOS-FM began her ascent to station management by logging commercials for airplay and eventually moved into other areas. "For the past seventeen years I have been employed by Champion Broadcasting Systems. During that time I have worked in every department of the radio station from traffic, where I started, to sales, programming, engineering, and finally management on both the local and corporate levels."

Many radio station managers are recruited from the sales area rather than programming. Since the general manager's foremost objective is to generate a profit, station owners usually feel more confident in hiring someone with a solid sales or business background. Consequently, three out of four radio managers have made their living at some point selling airtime. It is a widely held belief that this experience best prepares an individual for the realities encountered in the manager's position. "I spent over a decade and a half in media sales before becoming a station manager. In fact, my experience on the radio level was exclusively confined to sales and then for only eight months. After that I moved into station management. Most of my radio-related sales experience took place on the national level with station rep companies," recalls Norm Feuer, president, Viacom Communications.

KAAY's Carl Evans holds that a sales background is especially useful, if not necessary, to general managers. "I spent a dozen years as a

FIGURE 2.3
Typical classified ad for a station manager in an industry trade magazine.

HELP WANTED—GENERAL MANAGER

Broadcast Group looking for superb talent, not promises. Successful candidate must be college educated, have outstanding references, be self-motivated, and possess leadership qualities. Must also be sales, programming, and bottom line oriented. Send résumé and letter stating your goals, starting salary, management philosophy, and how you can achieve a position of sales and ratings dominance in our growing company. Send response to Box 22.

station account executive, and prior to entering radio I represented various product lines to retailers. The key to financial success in radio exists in an understanding of retailing."

It is not uncommon for station managers to have backgrounds out of radio, but almost invariably their experience comes out of the areas of sales, marketing, and finance. Paul Aaron of KFBK/KAER worked as a fund raiser for the United Way of America before entering radio, and contends that many managers come from other fields in which they have served in positions allied to sales, if not in sales itself. "Of those managers who have worked in fields other than radio, most have come to radio via the business sector. There are not many former biologists or glass blowers serving as station managers," says Aaron.

While statistics show that the station salesperson has the best chance of being promoted to the station's head position (more general managers have held the sales manager's position than any other), a relatively small percentage of radio's managers come from the programming ranks. "I'm more the exception than the rule. I have spent all of my career in the programming side, first as a deejay at stations in Phoenix, Denver, and Pittsburgh, and then as program director for outlets in Kansas City and Chicago. I'll have to admit, however, that while it certainly is not impossible to become a GM by approaching it from the programming side, resistance exists," admits WABB's Randy Lane.

The programmer's role is considered by many in the industry to be more an artistic function than one requiring a high degree of business savvy. However accurate or inaccurate this assessment is, the result is that fewer managers are hired with backgrounds exclusively confined to programming duties. Programmers have reason to be encouraged, however, since a trend in favor of hiring program directors has surfaced in the 1980s, and predictions suggest that it will continue.

The most attractive candidate for a station management position is the one whose experience has involved both programming and sales responsibilities. No general manager can fully function without an understanding and appreciation of what goes into preparing and presenting the air product, nor can he or she hope for success without a keen sense of business and finance.

Today's highly competitive and complex radio market requires that the person aspiring to management have both formal training, preferably a college degree in broadcasting, and experience in all aspects of radio station operations, in particular sales and programming. Ultimately, the effort and energy an individual invests will bear directly on the dividends he or she earns, and there is not a single successful station manager who has not put in fifteen-hour days. The station manager is expected to know more and do more than anyone else, and rightfully so, since he or she is the person who stands to gain the most.

THE MANAGER'S DUTIES AND RESPONSIBILITIES

A primary objective of the station manager is to operate in a manner that generates the most profit, while maintaining a positive and productive attitude among station employees. This is more of a challenge than at first it may seem, claims KSKU's Cliff Shank. "In order to meet the responsibility that you are faced with daily, you really have to be an expert in so many areas—sales, marketing, finance, legal matters, technical, governmental, and programming. It helps if you're an expert in human nature, too." Jane Duncklee of WBOS-FM puts it this way: "Managing a radio station requires that you divide yourself equally into at least a dozen parts and be a hundred percent whole in each situation."

Bill Campbell, general manager of WMJX/WMEX, Boston, says that the theme that runs throughout Tom Peters's book *In Search of Excellence* is one that is relative to the station manager's task. "The idea in Peters's book is that you must make the customer happy, get your people involved, and get rid of departmental waste and unnecessary expenditures. A station should be a lean and healthy organism."

Station managers generally must themselves answer to a higher authority. The majority of radio stations, roughly 90 percent, are owned by companies and corporations which both hire the manager and establish financial goals or projections for the station. It is the station manager's job to see that corporate expectations are met and, ideally, exceeded. Managers who fail to operate a facility in a way that satisfies the corporate hierarchy may soon find themselves looking for another job.

Fewer than 10 percent of the nation's stations are owned by individuals or partnerships. At these radio outlets the manager still must meet the expectations of the station owner(s). In some cases, the manager may be given more latitude or responsibility in determining the station's fate, while in others the owner may play a greater role in the operation of the station.

It is a basic function of the manager's position to formulate station policy and see that it is implemented. To ensure against confusion, misunderstanding, and possible unfair labor practices which typically impede operations, a station policy book is often distributed to employees. The stations' position on a host of issues, such as hiring, termination, salaries, raises, promotions, sick leave, vacation, benefits, and so forth usually are contained therein. As standard practice a station may require that each new employee read and become familiar with the contents of the policy manual before actually starting work. Job descriptions, as well as organization flow charts, commonly are outlined to make it abundantly clear to staff members who is responsible for what. A well-conceived policy book may contain a statement of the station's programming philosophy with an explanation of the format it employs. The more comprehensive a policy book, the less likely there will be confusion and disruption.

Hiring and retaining good people is another key managerial function. "You have some pretty delicate egos to cope with in this business. Radio attracts some very bright and highly talented people, sometimes with erratic temperaments. Keeping harmony and keeping people are among the foremost challenges facing a station manager," claims Viacom's Norm Feuer.

Carl Evans of KAAY-AM/FM agrees with Feuer, adding "Once you hire the best person for the job, and that is no easy task, you must be capable of inspiring that person to do his or her best. Actually, if you are unable to motivate your people, the station will fail to reach its potential. Hire the best people you can and nurture them."

As mentioned earlier, managers of small market radio stations are confronted with a unique set of problems when it comes to hiring and holding onto qualified people, especially on-air personnel. "In our case, finding and keeping a professional-sounding staff with our somewhat limited budget is an ongoing problem. This is true at most small market stations, however," observes WFKY's J. G. Salter.

The rural station is where the majority of newcomers gain their experience. Because salaries are necessarily low and the fledgling air person's ambitions are usually high, the rate of turnover is significant. Managers of small outlets spend a great deal of time training people. "It is a fact of the business that radio people, particularly deejays, usually learn their trade at the 'out-of-the-way,' low-power outlet. To be a manager at a small station, you have to be a teacher, too. But it can be very rewarding despite the obvious problem of having to rehire to fill positions so often. We deal with many beginners. I find it exciting and gratifying, and no small challenge, to train newcomers in the various aspects of radio broadcasting," says Salter.

Randy Lane enjoys the instructor's role also, but notes that the high turnover rate affects product continuity. "With air people coming and going all the time, it can give the listening public the impression of instability. The last thing a station wants to do is sound schizophrenic. Establishing an image of dependability is crucial to any radio station. Changing air people every other month doesn't help. As a station manager, it is up to you to do the best you can with the resources at hand. In general, I think small market managers do an incredible job with what they have to work with."

While managers of small market stations must wrestle with the problems stemming from diminutive budgets and high employee turnover, those at large stations must grapple with the difficulties inherent in managing larger budgets, bigger staffs, and, of course, stiffer competition. "It's all relative, really. While the small town station gives the manager turnover headaches, the metropolitan station manager usually is caught up in the ratings battle, which consumes vast amounts of time and energy. Of course, even larger stations are not immune to turnover," observes WRCH's Bremkamp.

It is up to the manager to control the station's finances. A knowledge of bookkeeping and accounting procedures is necessary. "You handle the station's purse-strings. An understanding of budgeting is an absolute must. Station economics is the responsibility of the GM. The idea is to control income and expenses in a way that yields a sufficient profit," says Roger Ingram of WZPL.

The manager allocates and approves spending in each department. Heads of departments must work within the budgets they have helped establish. Budgets generally cover the expenses involved in the operation of a particular area within the station for a specified period of time, such as a six- or twelve-month period. No manager wants to spend more than is absolutely required. A thorough familiarity with what is involved in running the various departments within a station prevents waste and overspending. "A manager has to know what is going on in programming, engineering, sales, actually every little corner of the station, in order to run a tight ship and make the most revenue possible. Of course, you should never cut corners simply for the sake of cutting corners. An operation must spend in order to make. You have to have effective cost control in all departments. That doesn't mean damaging the product through undernourishment either," says KAAY's Evans.

To ensure that the product the station offers is the best it can be, the station manager must keep in close touch with every department. Since the station's sound is what wins listeners, the manager must work closely with the program director and engineer. Both significantly contribute to the quality of the air product. The program director is responsible for what goes on the air, and the engineer is responsible for the way it sounds. Meanwhile, the marketing of the product is vital. This falls within the province of the sales department. Traditionally, the general manager works more closely with the station's sales manager than with anyone else.

An excellent air product attracts listeners, and listeners attract sponsors. It is as basic as that. The formula works when all departments in a station work in unison and up to their potential, contends Bonneville's Marlin Taylor. "In radio our product is twofold—the programming we send over our frequency and the listening audience we deliver to advertisers. A station's success is linked to customer/listener satisfaction, just like a retail store's. If you don't have what the consumer desires, or the quality doesn't meet his standards, he'll go elsewhere and generally won't return."

In a fast-moving, dynamic industry like radio, where both cultural and technological innovations have an impact on the way a station operates, the manager must stay informed and keep one eye toward the future. Financial projections must be based on data that include the financial implications of prospective and predicted events. An effective manager anticipates change and develops appropriate plans to deal with it. Industry trade journals *(Broadcasting, Radio and Records, Advertising Age, Radio Only)* and conferences conducted by organizations such as the National Association of Broadcasters, National Radio Broadcasters Association, and the Radio Advertising Bureau help keep the station manager informed of what tomorrow may bring.

Station consultants and "rep companies," which sell local station airtime to national advertising agencies, also support the manager in his efforts to keep on top of things. "A station manager must utilize all that is available to stay in touch with what's out there. Foresight is an essential ingredient for any radio manager. Hindsight is not enough in an industry that operates with one foot in the future," says Lynn Christian, who summarizes the duties and responsibilities of a station manager: "To me the challenges of running today's radio station include building and maintaining audience ratings, attracting and keeping outstanding employees, increasing gross revenues annually, and creating a positive community image for the station—not necessarily in that order." WRCH's Bremkamp is more laconic. "It boils down to one sentence: Protect the license and turn a profit."

ORGANIZATIONAL STRUCTURE

Radio stations come in all sizes and generally are classified as being either small, medium, or large (metro) market outlets. The size of the community that a station serves usually reflects the size of its staff. That is to say, the station in a town of five thousand residents may have as few as six full-time employees. It is a question of economics. However, some small market radio outlets have staffs that rival those of larger market stations because their income warrants it. Few small stations earn enough to have elaborate staffs, however. Out of financial necessity, an employee may serve in several capacities.

n's manager also may assume the sales manager, and announcers e news responsibilities. Meanwhile, ncluding the station's secretary, may ercial copy. The key word at the is flexibility, since each member s expected to perform numerous

arket stations are located in more densely populated areas. An outlet in a city with a population of between one hundred thousand and a half million may be considered medium sized. Albany, New York; Omaha, Nebraska; and Albuquerque, New Mexico are typical medium markets. Greater competition exists in these markets, more than in the small market where only one or two stations may be vying for the listening audience. Each of the medium markets cited have over a dozen stations.

The medium market radio station averages twelve to twenty employees. While an overlapping of duties does occur even in the larger station, positions usually are more limited to specific areas of responsibility. Seldom do announcers substitute as newspeople. Nor do sales people fill airshifts as is frequently the case at small outlets. (Reexamine fig. 1.10, a medium market flow chart.)

Large (also referred to as "major" or "metro") market radio stations broadcast in the nation's most populated urban centers. New York, Los Angeles, and Chicago rank first, second, and third, respectively. The top twenty radio markets also include cities such as Houston, Philadelphia, Boston, Detroit, and San Francisco. Competition is greatest in the large markets, where as many as seventy stations may be dividing the audience pie.

Major market stations employ as many as fifty to sixty people and as few as twenty, depending on the nature of their format. Stations relying on automation, regardless of the size of their market, usually employ fewer people.

What follows are brief descriptions of department head duties and responsibilities. They are expanded upon in later chapters. These individuals are regarded as station middle management and generally report directly to the station's manager.

Operations Manager

Second only to the general manager in level of authority, the operations manager is sometimes considered the station's assistant manager. The operations manager's duties include:

- Supervising administrative (office) staff
- Helping to develop station policies and see that they are implemented
- Handling departmental budgeting

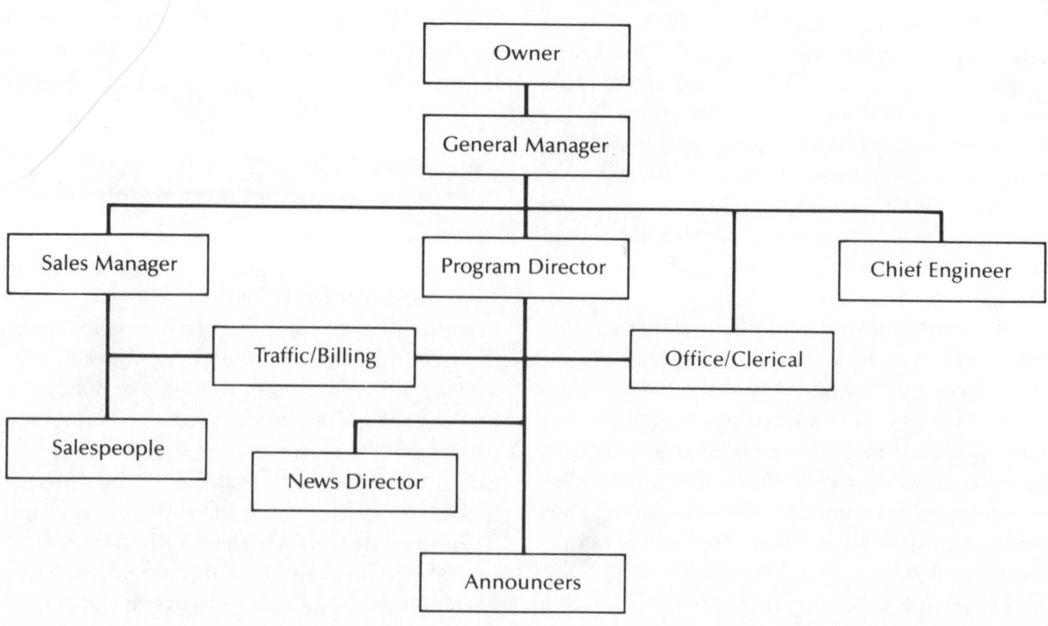

FIGURE 2.4
Organizational structure of a small market radio station.

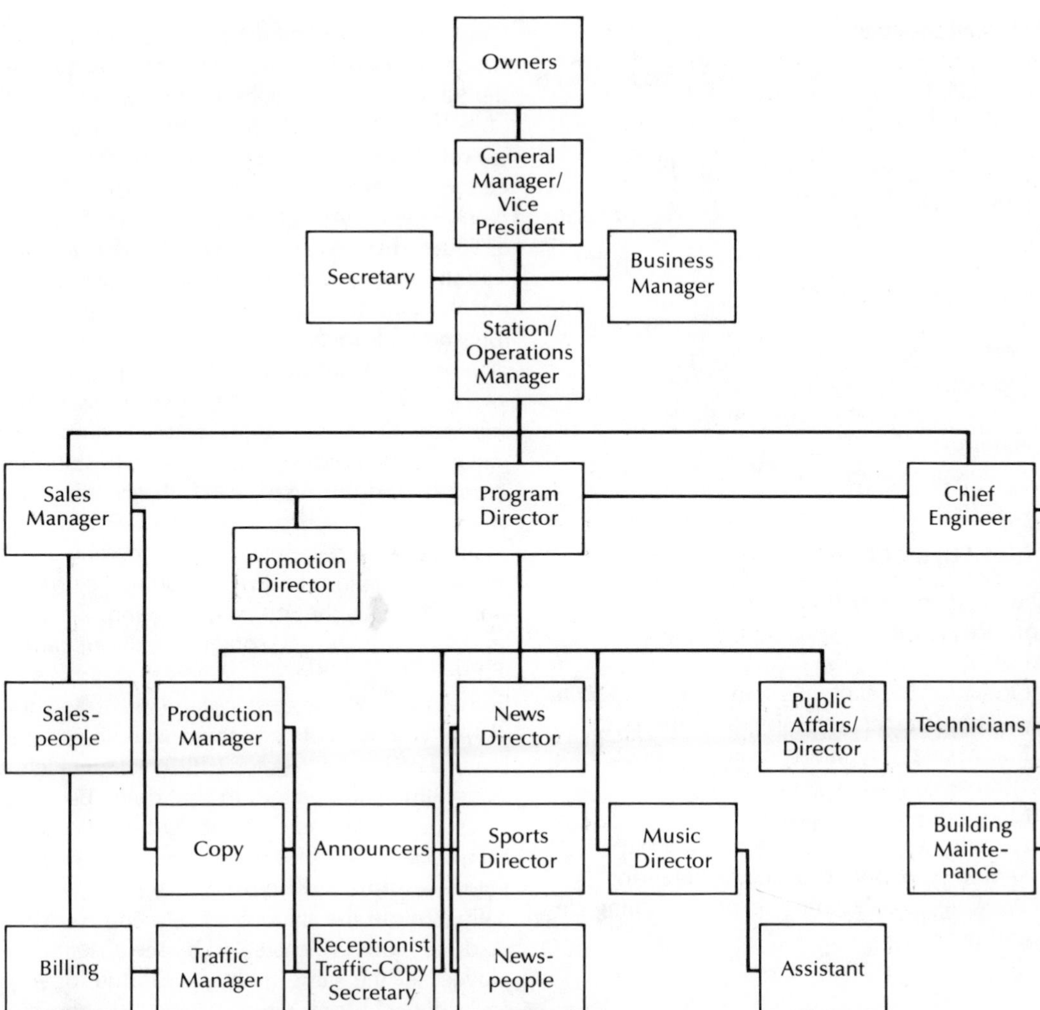

FIGURE 2.5 Organizational structure of a larger market radio station.

- Keeping abreast of government rules and regulations pertaining to entire operation
- Working as the liaison with the community to ensure that the station provides appropriate service and maintains its "good guy" image

Duties tend to be more skewed toward programming and may even include the job of station programmer. In cases such as this, the operations manager's duties are designed with a primary emphasis on on-air operations. Assistant program directors are commonly appointed to work with the operations manager to accomplish programming goals. Not all stations have established this position, preferring the department head to station manager approach.

Program Director

One of the three key department head positions at a radio station, the program director is responsible for:

- Developing and executing format
- Hiring and managing air staff
- Establishing the schedule of airshifts
- Monitoring the station to ensure consistency and quality of product
- Keeping abreast of competition and trends that may affect programming
- Maintaining the music library
- Complying with FCC rules and regulations
- Directing the efforts of the news and public affairs areas

Sales Manager

The sales manager's position is a pivotal one at any radio station, involving:

- Generating station income by directing the sale of commercial airtime
- Supervising sales staff
- Working with the station's rep company to attract national advertisers
- Assigning lists of retail accounts and local advertising agencies to salespeople
- Establishing sales quotas
- Coordinating on-air and in-store sales promotions
- Developing sales materials and rate cards

Chief Engineer

The chief engineer's job is a vital station function. Responsibilities include:

- Operating the station within prescribed technical parameters established by the FCC
- Purchasing, repairing, and maintaining equipment
- Monitoring signal fidelity
- Adapting studios for programming needs
- Setting up remote broadcast operations
- Working closely with the programming department

WHOM MANAGERS HIRE

Managers want to staff their stations with the most qualified individuals. Their criteria go beyond education and work experience to include various personality characteristics. "A strong résumé is important, but the type of person is what really decides it for me. The goal is to hire someone who will fit in nicely with the rest of the station's members. A station is a bit like a family in that it is never too large and people are in fairly close contact with one another. Integrity, imagination, intelligence, and a desire to succeed are the basic qualities that I look for when hiring," says Christopher Spruce, general manager of WZON-AM, Bangor, Maine, a station that is owned by novelist Stephen King. (The "ZON" in the call letters was adopted from the 1960s television program "Twilight Zone.")

Ambition and a positive attitude are attributes that rank high among most managers. "People with a sincere desire to be the best, to win, are the kind who really bolster an operation. Our objective at Viacom is to be number one at what we do, so any amount of negativity or complacency is viewed as counterproductive," says Norm Feuer. Lynn Christian, who once served as Feuer's boss while manager of a Miami radio station, says that competitiveness and determination are among the qualities he, too, looks for, and adds loyalty and dedication to the list. "Commitment to the station's philosophy and goals must exist in every employee or there are soft spots in the operation. In today's marketplace you have to operate from a position of strength, and this takes a staff that is with you all the way."

Stability and reliability are high on any manager's list. Radio is known as a nomadic profession, especially the programming end. Deejays tend to come and go, sometimes disappearing in the night. "A manager strives to staff his station with dependable people. I want an employee who is stable and there when he or she is supposed to be," states WBOS's Duncklee.

A station can ill afford to shut down because an air person fails to show up for his scheduled shift. When a no-show occurs, the station's continuity is disrupted both internally and externally. Within the station, adjustments must be made to fill the void created by the absent employee. At the same time, a substitute deejay often detracts from efforts to instill in the audience the feeling that it can rely on the station.

Managers are wary of individuals with fragile or oversized egos. "No prima donnas, please! I want a person who is able to accept constructive criticism and use it to his or her advantage without feeling that he is being attacked. The ability to look at oneself objectively and make the adjustments that need to be made is very important," observes KSKU's Cliff Shank. Carl Evans echoes his sentiments. "Open-mindedness is essential. An employee who feels that he can't learn anything from anyone or is never wrong is usually a fly in the ointment."

Honesty and candor are universally desired qualities, says Bremkamp. "Every station manager appreciates an employee who is forthright and direct, one who does not harbor ill feelings or unexpressed opinions and beliefs. An air of openness keeps things from seething and possibly erupting. I prefer an employee to come to me and say what is on his or her mind, rather than keep things sealed up inside. The lid eventually pops and then you have a real problem on your hands."

Self-respect and esteem for the organization are uniquely linked, claims WFKY's Salter. "If an individual does not feel good about himself, he cannot feel good about the world around him. This is not a question of ego, but rather one of appropriate self-perception. A healthy self-image and attitude in an employee makes that person easier to manage and work with. The secret is identifying potential problems or weaknesses in prospective employees before you sign them on. It's not easy without a degree in Freud."

Other personal qualities that managers look for in employees are patience, enthusiasm, discipline, creativeness, logic, and compassion. "You look for as much as you can during the interview and, of course, fill in the gaps with phone calls to past employers. After that, you hope that your decision bears fruit," says WZON's Spruce.

THE MANAGER AND THE PROFIT MOTIVE

Earlier we discussed the unique nature of the radio medium and the particular challenges that face station managers as a consequence. Radio, indeed, is a form of "show business," and both words of the term are particularly applicable since the medium is at once theater and retailer. Radio provides entertainment to the public and, in turn, sells the audience it attracts to advertisers.

The general manager is answerable to many: the station's owners, listeners, and sponsors. However, the party the manager must first please in order to keep his or her own job is the station's owner and, more often than not, this person's number one concern is profit. As in any business, the more money the manager generates the happier the owner.

Critics have chided the medium for what they argue is an obsessive preoccupation with making money that has resulted in a serious shortage of high-quality, innovative programming. They lament too much sameness. Meanwhile, station managers often are content to air material that draws the kind of audience that advertisers want to reach.

Marlin Taylor observes that too many managers overemphasize profit at the expense of the operation. "Not nearly enough of the radio operators in this country are truly committed to running the best possible stations they can, either because that might cost them more money or they simply don't understand or care what it means to be the best. In my opinion, probably no more than 20 percent of the nation's stations are striving to become the IBM of radio, that is, striving for true excellence. Many simply are being milked for what the owners and managers can take out of them."

The pursuit of profits forces the station manager to employ the programming format that will yield the best payday. In several markets certain formats, such as classical and jazz, which tend to attract limited audiences, have been dropped in favor of those which draw greater numbers. In some instances, the actions of stations have caused outcries by unhappy and disenfranchised listeners who feel that their programming needs are being disregarded. Several disgruntled listener groups have gone to court in an attempt to force stations to reinstate abandoned formats. Since the government currently avoids involvement in programming decisions, leaving it up to stations to do as they see fit, little has come of their protests.

The dilemma facing today's radio station manager stems from the complexity of having to please numerous factions while still earning enough money to justify his or her continued existence at the station. Marlin Taylor has suggested that stations reinvest more of their profits as a method of upgrading the overall quality of the medium. "Overcutting can have deleterious effects. A station can be too lean, even anemic. In other words, you have to put something in to get something out. Too much draining leaves the operation arid and subject to criticism by the listening public. It behooves the station manager to keep this thought in mind and, if necessary, impress it upon ownership. The really successful operations know full well that money has to be spent to nurture and develop the kind of product that delivers both impressive financial returns and listener praise."

While it is the manager who must deal with bottom-line expectations, it is also the manager who is expected to maintain product integrity.

WDMO BROADCASTING CO., INC. INCOME STATEMENT WDMO-AM FEB PERIOD, 1985												AS OF 1/25/85
****** CURRENT MONTH *****							****** YEAR TO DATE ******					
BUDGET	%	THIS YEAR	%	LAST YEAR	%		BUDGET	%	THIS YEAR	%	LAST YEAR	%
						REVENUE						
34,200	86.4	31,206	92.4	33,789		NATIONAL SALES	70,200	91.2	60,684	93.0	65,254	
83,600	80.0	71,505	111.2	64,321		LOCAL SALES	171,600	85.5	137,226	104.1	131,763	
2,375	111.4	2,715	111.4	2,438		NETWORK SALES	4,875	114.3	5,430	116.0	4,683	
808	99.5	825	113.8	725		PRODUCTION SALES	1,658	102.2	1,650	120.0	1,375	
475	116.7		.0	475		TRADE SALES	975	.0	1,138	136.2	835	
121,458	82.7	106,251	104.4	101,748		GROSS REVENUE	249,308	87.5	206,127	101.1	203,910	
17,250	78.9	13,352	.0			AGENCY COMMISSIONS	33,830	77.4	26,705	.2		
104,208	83.3	92,898	91.3	101,748		TOTAL NET REVENUE	215,478	89.1	179,423	88.0	203,910	
						OPERATING EXPENSES						
10,560	45.8	3,557	61.8	5,751		ENGINEERING	20,097	33.7	9,200	80.8	11,384	
14,320	102.2	11,362	.0			PROGRAMMING	28,640	79.3	29,279	.2		
8,690	28.7	2,379	.0			NEWS	17,380	27.4	4,993	.0		
9,355	99.4	5,942	.0			SALES	18,710	63.5	18,601	.0		
15,480	56.0	7,183	.0			ADMINISTRATIVE	30,960	46.4	17,352	.0		
58,405	68.6	30,421	529.0	5,751		TOTAL OPERATING EXPENSES	115,787	51.1	79,425	697.7	11,384	
45,803	100.3	62,477	65.1	95,997		OPERATING INCOME	99,691	136.4	99,998	51.9	192,526	
	.0	84	.0			OTHER INCOME		.0	251-	.0		
9,200	86.8	6,443	.0			OTHER EXPENSES	18,400	70.0	15,964	.0		
36,603	103.1	56,119	58.5	95,997		INCOME BEFORE TAXES	81,291	153.3	83,783	43.5	192,526	
	.1		.1			PROVISION FOR TAXES		.1		.1		
36,603	103.1	56,119	58.5	95,997		NET INCOME/LOSS	81,291	153.3	83,783	43.5	192,526	

Station's income statement containing budget. The station manager must be adept at directing station finances.

The effective manager takes pride in the unique role radio plays in society and does not hand it over to advertisers, notes WRCH's Bremkamp. "You have to keep close tabs on your sales department. They are out to sell the station, sometimes one way or the other. Overly zealous salespeople can, on occasion, become insensitive to the station's format in their quest for ad dollars. Violating the format is like mixing fuel oil with water. You may fill your tank for less money, but you're not likely to get very far. The onus is placed on the manager to protect the integrity of the product while making a dollar. Actually, doing the former usually takes care of the latter."

Conscientious station managers are aware of the obligations confronting them, and sensitive to the criticism that crass commercialism can produce a desert or "wasteland" of bland and uninspired programming. They are also aware that while gaps and voids may exist in radio programming and that certain segments of the population may not be getting exactly what they want, it is up to them to produce enough income to pay the bills and meet the ownership's expectations.

THE MANAGER AND THE COMMUNITY

In the early 1980s, the FCC reduced the extent to which radio stations must become involved in community affairs. Ascertainment procedures requiring that stations determine and address community issues have all but been eliminated. If a station chooses to do so, it may spin the hits twenty-four hours a day and virtually divorce itself from the concerns of the community. However, a station that opts to function independently of the community to which it is licensed may find itself on the outside looking in. This is especially true of small market stations which, for practical business reasons, traditionally have cultivated a strong connection with the community.

A station manager is aware that it is important to the welfare of his or her organization to behave as a good citizen and neighbor. While the sheer number of stations in vast metropolitan areas makes it less crucial that a station exhibit civic-mindedness, the small market radio outlet often finds that the level of business it generates is relative to its community involvement. Therefore, maintaining a relationship with the town leaders, civic groups, and religious leaders, among others, enhances a station's visibility and status and ultimately affects business. No small market station can hope to operate autonomously and attract the majority of local advertisers. Stations that remain aloof in the community in which they broadcast seldom realize their full revenue potential.

Cognizant of the importance of fostering an image of goodwill and civic-mindedness, the station manager seeks to become a member in good standing in the community. Radio managers often actively participate in groups or associations, such as the local Chamber of Commerce, Jaycees, Kiwanis, Rotary Club, Optimists, and others, and encourage members of their staff to become similarly involved. The station also strives to heighten its status in the community by devoting airtime to issues and events of local importance and by making its microphones available to citizens for discussions of matters pertinent to the area. In so doing, the station becomes regarded as an integral part of the community, and its value grows proportionately.

Surveys have shown that over a third of the managers of small market radio stations are native to the area their signal serves. This gives them a vested interest in the quality of life in their community and motivates them to use the power of their medium to further improve living conditions.

Medium and large market station managers realize, as well, the benefits derived from participating in community activities. "If you don't localize and take part in the affairs of the city or town from which you draw your income, you're operating at a disadvantage. You have to tune in your audience if you expect them to do likewise," says WRCH's Bremkamp.

The manager has to work to bring the station and the community together. Neglecting this responsibility lessens the station's chance for prosperity, or even survival. Community involvement is a key to success.

THE MANAGER AND THE GOVERNMENT

Earlier in this chapter, station manager Richard Bremkamp cited protecting the license as one of the primary functions of the general manager. By "protecting" the license he meant conforming to the rules and regulations established by the FCC for the operation of broadcast facilities. Since failure to fulfill the obligations of a license may result in punitive actions, such as reprimands, fines, and even the revocation of the privilege to broadcast, managers have to be aware of the laws affecting station operations and see to it that they are observed by all concerned.

The manager delegates responsibilities to department heads who are directly involved in the areas affected by the commission's regulations. For example, the program director will attend to the legal station identification, station logs, program content, and a myriad of other concerns of interest to the government. Meanwhile, the chief engineer is responsible for meeting technical standards, while the sales manager is held accountable for the observance of certain business and financial practices. Other members of the station also are assigned various responsibilities applicable to the license. Of course, in the end it is the manager who must guarantee that the station's license to broadcast is, indeed, protected.

Contained in Title 47, Part 73 of the *Code of Federal Regulations (CFR)* are all the rules and regulations pertaining to radio broadcast operations. The station manager keeps the annual update of this publication accessible to all those employees involved in maintaining the license. A copy of the *CFR* may be obtained through the Superintendent of Documents, Government Printing Office, Washington, D.C. 20402. A modest fee is charged. Specific inquiries concerning the publication can be addressed to the Director, Office of the Federal Register, National Archives and Records Service, General Services Administration, Washington, D.C. 20408.

To give an idea of the scope of this document, as well as the government's dicta concerning broadcasters, the *CFR*'s index to radio broadcast services is included as an appendix to this chapter. Immediately apparent from a perusal of the index is the preponderance of items that come under the auspices of the engineering department. Obviously, the FCC is concerned with many areas of radio operation, but its focus on the technical aspect is prodigious. Many other items involve programming and sales, and, of course, responsibilities, also overlap into other areas of the station.

To reiterate, although the station manager shares the duties involved in complying with the FCC's regulations with other staff members, he holds primary responsibility for keeping the station out of trouble and on the air.

In chapter 1 it was noted that many of the rules and regulations pertaining to the daily operation of a radio station have undergone revision or have been rescinded. Since the *CFR* is published annually, certain parts may become obsolete during that period. John E. Byrne, director of the Office of the Federal Register, suggests that the *Federal Register*, from which the *CFR* derives its information, be consulted monthly. "These two publications must be used together to determine the latest version of any

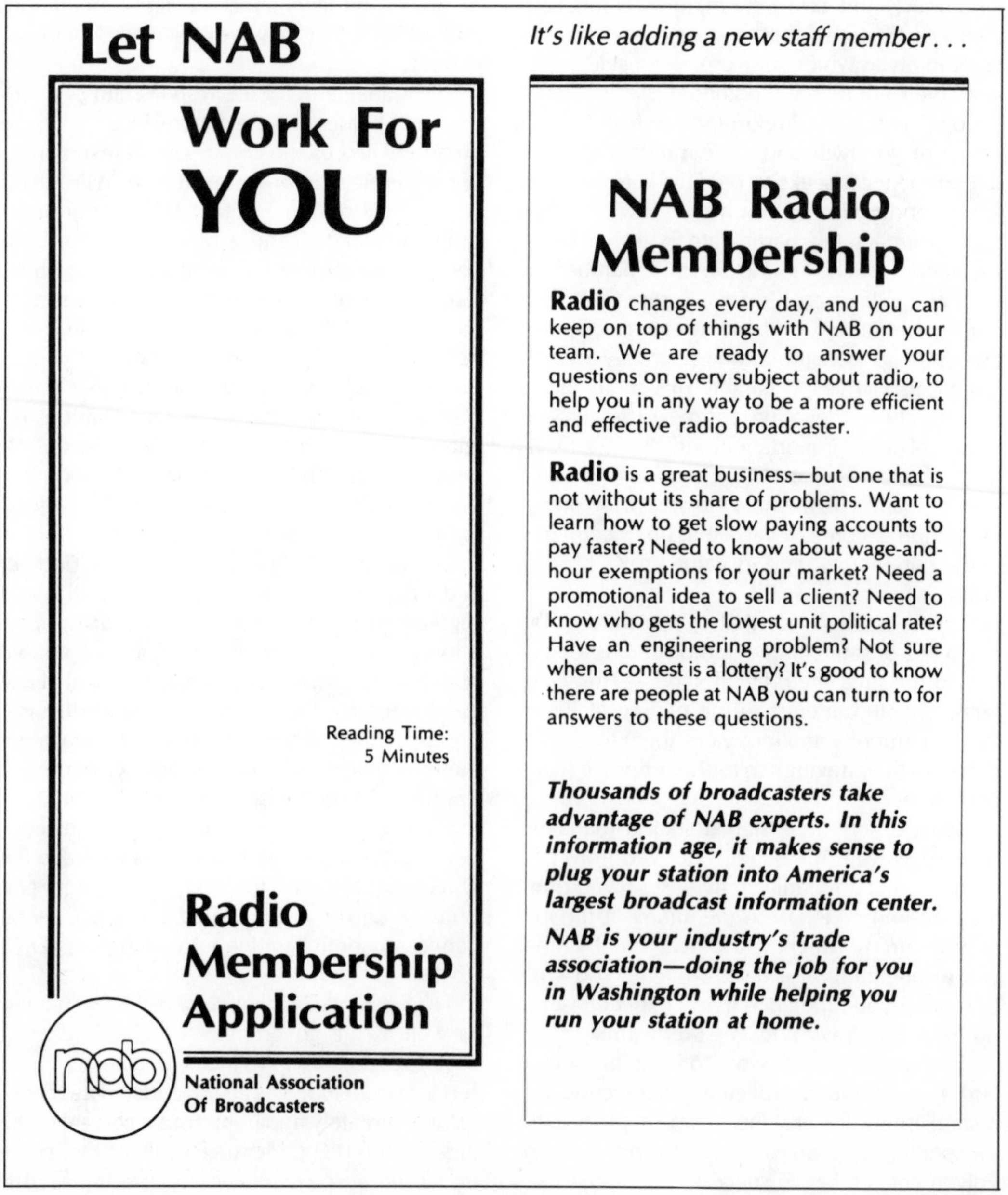

FIGURE 2.7
NAB membership application. More commercial radio stations belong to the NAB than any other industry association. Courtesy NAB.

Let NAB
Work For
YOU

Reading Time:
5 Minutes

Radio
Membership
Application

nab
National Association
Of Broadcasters

It's like adding a new staff member . . .

NAB Radio
Membership

Radio changes every day, and you can keep on top of things with NAB on your team. We are ready to answer your questions on every subject about radio, to help you in any way to be a more efficient and effective radio broadcaster.

Radio is a great business—but one that is not without its share of problems. Want to learn how to get slow paying accounts to pay faster? Need to know about wage-and-hour exemptions for your market? Need a promotional idea to sell a client? Need to know who gets the lowest unit political rate? Have an engineering problem? Not sure when a contest is a lottery? It's good to know there are people at NAB you can turn to for answers to these questions.

Thousands of broadcasters take advantage of NAB experts. In this information age, it makes sense to plug your station into America's largest broadcast information center.

NAB is your industry's trade association—doing the job for you in Washington while helping you run your station at home.

given rule," says Byrne. A station may subscribe to the *Federal Register* or visit the local library.

Since the FCC may, at any time during normal business hours, inspect a radio station to see that it is in accord with the rules and regulations, a manager makes certain that everything is in order. An FCC inspection checklist is contained in the *CFR*, and industry organizations, such as the National Association of Broadcasters (NAB) and the National Radio Broadcasters Association (NRBA), provide member stations with similar checklists. Occasionally, managers run mock inspections in preparation for the real thing. A state of preparedness prevents embarrassment and problems.

THE MANAGER AND UNIONS

The unions most active in radio are the American Federation of Television and Radio Artists (AFTRA), the National Association of Broadcast

Keeping you informed . . .

NAB Publications

RadioActive, NAB's monthly radio magazine, features radio promotions, sales and management strategies, computer information, and other radio topics.

Radio members also get NAB's weekly newsletter *Highlights*. It keeps you up-to-date on rule changes and NAB's lobbying efforts. Each month, the NAB *Info/Pak* contains detailed information on every topic from computers and engineering to legal and research. *Com/Tech Reports* cover new technologies and their marketplace impact. Through these reports, you will learn how you can make money with SCAs, cellular radio, channel leasing, and other opportunities.

Radio publications* from NAB are specifically targetted to inform you of programming trends, sales techniques, computer software, and hundreds of other topics.

All of the above items are included in your NAB dues.

Radio's annual financial report, employees' compensation and fringe benefit report, computer software directory and newsletter can help you in the financial planning of your station.

* There is a surcharge on some NAB publications.

NAB Lobbying

Radio broadcasters have benefitted from NAB's intensive lobbying efforts in the following areas. Only NAB, with its staff and resources, can address *every* issue.

Radio Deregulation. NAB proposed de-reg in 1978. The FCC adopted it in 1981. You no longer need to worry about documenting ascertainment, limiting your commercial minutes, checking program time percentages, or logging requirements. NAB's staff worked three years for these gains, then NAB's lawyers defended dereg against court appeals which could have wiped out those gains.

Seven-Year License Terms. The longer license term your station now enjoys is the result of years of effort by NAB on Capitol Hill.

Format Freedom. Only NAB could have mustered the hundreds of thousands of dollars in legal fees it took to take the format freedom case all the way to the Supreme Court. We did it, and we won.

Radio Marti. Hundreds of broadcasters could be experiencing even more Cuban interference by now if NAB had not engineered a major compromise in Congress.

Class IV Power Increase/Daytimer Extended Hours. Only an organization with NAB's stature could have successfully worked with the State Department, the Canadian and Mexican governments, and broadcasters in those countries to get the new North American spectrum agreements. This has paved the way for Class IV power increases, and has already resulted in extended hours for many daytimers. These agreements are a result of talks begun by NAB in 1979.

FIGURE 2.7 *continued*

Employees and Technicians (NABET), and the International Brotherhood of Electrical Workers (IBEW). Major market radio stations are the ones most likely to be unionized. The overwhelming majority of American stations are nonunion, and, in fact, in the 1980s union membership declined.

Dissatisfaction with wages and benefits, coupled with a desire for greater security, often are what motivate station employees to vote for a union. Managers seldom encourage the presence of a union since many believe that unions impede and constrict their ability to control the destiny of their operations. However, a small percentage of managers believe that the existence of a union may acutally stabilize the working environment and cut down on personnel turnover.

It is the function of the union to act as a bargaining agent working in "good faith" with station employees and management to upgrade and improve working conditions. The focus of

FIGURE 2.7
continued

Phone Rates. NAB has asked the FCC to protect broadcasters from excessive phone rate increases resulting from the AT&T breakup. We are providing data to show the Commission that radio needs reasonable phone rates in order to continue providing the kind of service the public expects.

Performers' Royalty Fees. NAB has saved you at least one percent of your yearly gross five times in the last six years by turning back music industry efforts to establish a royalty for people who perform music in addition to the existing royalties for those who write it.

Docket 80-90. NAB did extensive engineering tests to show the FCC the damaging effects of spacing reductions in the original 80-90 FM proposal. Many FM stations stood to lose as much as one-third of their coverage. The final form of 80-90 will protect existing service, while allowing many existing stations to upgrade their facilities.

10 kHz Spacing. NAB conducted extensive tests and organized a nationwide lobbying campaign to defeat this ill-conceived effort to load the AM band with more stations and deal severe damage to existing service.

NAB has also protected the value of your license in the last five years by:

Blocking efforts to break up AM-FM combos;

Killing efforts to ban advertising of saccharine products;

Killing a proposal which would have required 45 days' public notice of a proposed sale;

Successfully lobbying against efforts in Congress to impose spectrum fees on radio;

Stopping efforts to require public lists of all employees, ranked by salary; and

Stopping efforts to require PSAs in high-listener time periods.

Helping you succeed . . .

NAB Services

Radio Conventions twice yearly. The all-radio *only* Convention in the fall brings broadcasters from all over the USA together. The Radio Convention and Programming Conference is all-radio "nuts and bolts." NAB's Annual Spring Convention and International Exposition is the largest in the world.

Radio sales and management seminars in convenient locations throughout the USA can help you. NAB's Station Services Department annually sponsors Sales & Management Seminars at Notre Dame University. Other management seminars on a variety of specific topics are held from time to time in various locations or by video teleconference.

NAB Insurance

Radio broadcasters save hundreds and even thousands of dollars with NAB's insurance programs.

- Libel/Slander Coverage
- Property and Casualty
- Group Life
- Long-Term Disability

NAB insurance packages are designed by broadcasters *for* broadcasters, and are specifically designed for the needs of your radio station.

Phone NAB Services *right now* to learn how much you could be saving.
The toll-free number is (800) 368-5644.

union efforts usually is in the areas of salary, sick leave, vacation, promotion, hiring, termination, working hours, and retirement benefits, among others.

A unionized station appoints or elects a shop steward who works as a liaison between the union, which represents the employees, and the station's management. Employees may lodge complaints or grievances with the shop steward, who will then review the union's contract with the station and proceed accordingly. Station managers are obliged to work within the agreement that they, along with the union, helped formulate.

As stated, unions are a fact of life in many major markets. They are far less prevalent elsewhere, although unions do exist in some medium and even small markets. Most small operations would find it impractical, if not untenable, to function under a union contract. Union demands would quite likely cripple most marginal or small profit operations.

FIGURE 2.7
continued

Radio Membership Application

We hereby apply for membership in the National Association of Broadcasters and agree to abide by the Bylaws of the Association. A copy of these Bylaws is available on request.

Call Letters ______ AM ______ FM
CP# ______ Air Date ______
Frequency ______ AM ______ FM
Network ______ AM ______ FM
Executive In Charge ______
Nickname ______
Title ______
Licensee ______

Location:
Address ______
City of License ______
State ______ Zip ______
Phone ______

Mailing:
P.O. Box ______
City, State, Zip ______
Station Type ______
Class ______ AM ______ FM
Group Name ______
Group Executive ______
Address ______
______

Monthly Dues $ ______
(See reverse side)
NAB Member Plaque $ ______
(Plaques cost $25.00 each)
Total Enclosed $ ______
Please attach check for first month's dues.

☐ Yes, NAB may contact this station on behalf of TARPAC.
Signature ______
Title ______
Date ______

Annual membership dues include a subscription to *RadioActive* at $26 per year and *NAB Highlights* at $52 per year. Subscriptions run for the duration of membership.

Let us work for you . . .

NAB Radio Dues

1. Complete the Radio Membership application on the other side.
2. Return signed application in the postage-paid envelope.
3. Enclose minimum of one month's dues.

Class	Annual Revenues	Monthly Dues	Check One
A	$ 0 - $ 100,000	$ 34.00	____
B	100,001 - 125,000	38.00	____
C	125,001 - 150,000	48.00	____
D	150,001 - 175,000	57.00	____
E	175,001 - 200,000	61.00	____
F	200,001 - 250,000	70.00	____
G	250,001 - 300,000	79.00	____
H	300,001 - 350,000	88.00	____
I	350,001 - 400,000	99.00	____
J	400,001 - 450,000	106.00	____
K	450,001 - 500,000	123.00	____
L	500,001 - 600,000	145.00	____
M	600,001 - 700,000	166.00	____
N	700,001 - 800,000	192.00	____
O	800,001 - 900,000	212.00	____
P	900,001 - 1,000,000	237.00	____
Q	1,000,001 - 1,250,000	275.00	____
R	1,250,001 - 1,500,000	336.00	____
S	1,500,001 - 1,750,000	389.00	____
T	1,750,001 - 2,000,000	422.00	____
U	2,000,001 - 2,500,000	456.00	____
V	2,500,001 - 3,000,000	489.00	____
W	3,000,001 - 3,500,000	504.00	____
X	3,500,001 - and over	525.00	____

Radio Dues are voluntarily reclassified annually.

NAB membership dues are tax deductible.

Broadcast revenue is defined as gross revenue (including network income) less agency commission.

For AM/FM combination operation in the same community, dues are based on the combined revenues of the two stations. Group discounts apply.

Approved by the NAB Board of Directors: ______

Managers who extend employees every possible courtesy and operate in a fair and reasonable manner are rarely affected by unions, whose prime objective is to protect and ensure the rights of station workers.

THE MANAGER AND INDUSTRY ASSOCIATIONS

Every year dozens of broadcast trade organizations conduct conferences and seminars intended to generate greater industry awareness and unity. At these gatherings, held at various locations throughout the country, radio managers and station personnel exchange ideas and share experiences, which they bring back to their stations.

The largest broadcast industry trade organization is the National Association of Broadcasters, which was originally conceived out of a need to improve operating conditions in the 1920s. Initially a lobbying organization, the NAB has expanded considerably in scope. The primary objective of the organization is to support and promote the stability and development of the industry. The industry's second largest trade association, the National Radio Broadcasters Association, shares a similar aim. Both these industry support groups enjoy widespread membership. Many stations belong to both organizations.

Thousands of radio stations also are members of the Radio Advertising Bureau (RAB), which was founded in 1951, a time when radio's fate was in serious jeopardy due to the rise in television's popularity. "The RAB is designed to serve as the sales and marketing arm of America's commercial radio industry. Members include radio stations, broadcast groups, networks, station representatives, and associated industry organizations in every market, in all fifty states," explains RAB's Kenneth J. Costa.

Dozens of other broadcast trade organizations focus their attention on specific areas within the radio station, and regional and local broadcast organizations are numerous. Listed below is a partial rundown of national organizations that support the efforts of radio broadcasters. A more comprehensive list may be found in *Broadcasting Yearbook,* the definitive industry directory, or by consulting local area directories.

National Association of Broadcasters—1771 N Street, N.W., Washington, D.C. 20036.

National Radio Broadcasters Association—1750 DeSales Street, N.W., Suite 500, Washington, D.C. 20036.

Radio Advertising Bureau—485 Lexington Avenue, New York, NY 10017.

National Association of Farm Broadcasters—Box 119, Topeka, KS 66601.

American Women in Radio and Television—1321 Connecticut Avenue, N.W., Washington, D.C. 20036.

Broadcast Education Association—1771 N Street, N.W., Washington, D.C. 20036.

Broadcast Pioneers—320 W. 57th Street, New York, NY 10019.

Clear Channel Broadcasting Service—1776 K Street, N.W., Washington, D.C. 20006.

Daytime Broadcasters Association—Box 730, Worthington, MN 56187–0730.

National Association of Black Owned Broadcasters—1730 M Street, N.W., Suite 708, Washington, D.C. 20036.

National Religious Broadcasters—CN 1926, Morristown, NJ 07960.

Radio Network Association—575 Lexington Avenue, Room 505, New York, NY 10022.

Radio-Television News Directors Association—1735 De Sales Street, N.W., Washington, D.C. 20036.

Society of Broadcast Engineers—Box 50844, Indianapolis, IN 46250.

Association of Black Broadcasters—3 E. 4th Street, Cincinnati, OH 45202.

The NAB, NRBA, RAB, and others charge a station membership fee and commonly base the amount on the size of the radio outlet. Some organizations require individual membership fees, which often are absorbed by the radio station as well.

CHAPTER HIGHLIGHTS

1. Radio's unique character requires that station managers deal with a wide variety of talents and personalities.

2. The authoritarian approach to management implies that the general manager makes all of the policy decisions. The collaborative approach allows the general manager to involve other station staff in the formation of policy. The hybrid or chief collaborator approach combines elements of both the authoritarian and collaborative management models. The chief collaborator management approach is most prevalent in radio today.

3. To attain management status an individual needs a solid formal education and practical experience in many areas of station operation—especially sales.

4. Key managerial functions include: operating in a manner that produces the greatest profit, meeting corporate expectations, formulating station policy and seeing to its implementation, hiring and retaining good people, inspiring staff to do their best, training new employees, maintaining communication with all departments to assure an excellent air product, and keeping an eye toward the future.

5. The operations manager, second only to the general manager, supervises administrative staff, helps develop and implement station policy, handles departmental budgeting, functions as regulatory watchdog, and works as liaison with the community. The program director is responsible for format, hires and manages air staff, schedules airshifts, monitors air product quality, keeps abreast of competition, maintains the music library, complies with FCC rules, and directs the efforts of news and public affairs. The sales manager heads the sales staff, works with the station's rep company, assigns account lists to salespeople, establishes sales quotas, coordinates sales promotions, and develops sales materials and rate cards. The chief engineer operates within the FCC technical parameters; purchases, repairs, and maintains equipment; monitors signal fidelity; adapts studios for programming needs; sets up remote broadcasts; and works closely with programming.

6. Managers hire individuals who possess: a solid formal education, strong professional experience, ambition, a positive attitude, reliability, humility, honesty, self-respect, patience, enthusiasm, discipline, creativity, logic, and compassion.

7. Radio provides entertainment to the public and, in turn, sells the audience it attracts to advertisers. It is the station manager who must ensure a profit, but he or she must also maintain product integrity.

8. To foster a positive community image, the station manager becomes actively involved in the community and devotes airtime to community concerns—even though the FCC recently reduced a station's obligation to do so.

9. Although the station manager delegates responsibility for compliance with FCC regulations to appropriate department heads, the manager is ultimately responsible for "protecting" the license. Title 47, Part 73 of the *Code of Federal Regulations* contains the rules pertaining to radio broadcast operations. Updates of regulations are listed monthly in the *Federal Register*.

10. The American Federation of Television and Radio Artists (AFTRA), the National Association of Broadcast Employees and Technicians (NABET), and the International Brotherhood of Electrical Workers (IBEW) are the unions most active in radio.

11. The National Association of Broadcasters (NAB), the National Radio Broadcasters Association (NRBA), and the Radio Advertising Bureau (RAB) are the largest broadcast trade industry organizations.

APPENDIX: *CODE OF FEDERAL REGULATIONS* INDEX

The index to the *Code of Federal Regulations,* Title 47, Part 73, Subparts A, B, G, and H, shown here, deals specifically with commercial radio operations. The department heads most affected by the sections listed in the index are noted in parenthesis after each entry. GM = General Manager, PD = Program Director, E = Engineer, SM = Sales Manager.

SUBPART A—AM BROADCAST STATIONS

73.1 Scope. (All dept. heads)
73.14 AM broadcast definitions. (All)
73.21 Classes of AM broadcast channels and stations. (All)
73.22 Assignment of Class II-A stations. (GM, E)
73.24 Broadcast facilities; showing required. (GM, E)
73.25 Clear channels; Classes I and II stations. (GM, E)
73.26 Regional channels; Classes III-A and III-B stations. (GM, E)
73.27 Local channels; Class IV stations. (GM, E)
73.28 Assignment of stations to channels. (GM, E)
73.29 Class IV stations on regional channels. (GM, E)
73.33 Antenna systems; showing required. (E)
73.35 Multiple ownership. (GM)
73.37 Applications for broadcast facilities; showing required. (GM, E)
73.40 AM transmission system performance requirements. (E)
73.44 AM transmission system emission limitations. (E)
73.45 AM antenna systems. (E)
73.49 AM transmission system installation and safety requirements. (E)
73.51 Determining operating power. (E)
73.53 Requirements for type acceptance of antenna monitors. (E)
73.54 Antenna resistance and reactance measurements. (E)

73.57 Remote reading antenna and common point ammeters. (E)
73.58 Indicating instruments. (E)
73.61 AM directional antenna field measurements. (E)
73.62 Directional antenna system tolerances. (E)
73.66 Remote control authorizations. (E)
73.67 Remote control operation. (E)
73.68 Sampling systems for antenna monitors. (E)
73.69 Antenna monitors. (E)
73.72 Operating during the experimental period. (E)
73.88 Blanketing interference. (E)
73.93 AM operator requirements. (E, PD)
73.99 Presunrise service authorization (PSRA) and Postsunset service authorization (PSSA). (E, GM, PD)

Other Operating Requirements

73.127 Use of multiplex transmission. (E, GM, PD)
73.128 AM stereophonic broadcasting. (E, PD)
73.132 Territorial exclusivity. (PD)
73.140 Use of automatic transmission systems (ATS). (E)
73.142 Automatic transmission system facilities. (E)
73.144 Fail-safe transmitter control for automatic transmission systems. (E)
73.146 Automatic transmission system monitoring and alarm points. (E)
73.150 Directional antenna systems. (E)
73.151 Field strength measurements to establish performance of directional antennas. (E)
73.152 Modification of directional antenna data. (E)
73.153 Field strength measurements in support of applications or evidence at hearings. (E)
73.154 Directional antenna partial and skeleton proof of performance field strength measurements. (E)
73.157 Special antenna test authorizations. (E)
73.158 Directional antenna monitoring points. (E)
73.160 Vertical plane radiation characteristics, f(θ). (E)
73.181 Introduction to AM technical standards. (E)
73.182 Engineering standards of allocation. (E)
73.183 Groundwave signals. (E)
73.184 Groundwave field strength charts. (E)
73.185 Computation of interfering signal. (E)
73.186 Establishment of effective field at one mile. (E)
73.187 Limitation on daytime radiation. (E)
73.188 Location of transmitters. (E)
73.189 Minimum antenna heights or field strength requirements. (E)
73.190 Engineering charts. (E)

SUBPART B—FM BROADCAST STATIONS

73.201 Numerical designation of FM broadcast channels. (E)
73.202 Table of Allotments. (E, GM)
73.203 Availability of channels. (E, GM)
73.204 International agreements and other restrictions of use of channels. (GM, E)
73.205 Zones. (E)
73.206 Classes of stations and permissible channels. (GM, E)
73.207 Minimum distance separation between stations. (GM, E)
73.208 Reference points and distance computations. (E)
73.209 Protection from interference. (E)
73.211 Power and antenna height requirements. (E)
73.212 Administrative changes in authorizations. (E)
73.213 Stations at spacings below the minimum separations. (E)
73.220 Restrictions on use of channels. (GM, E)
73.232 Territorial exclusivity. (GM, PD)
73.239 Use of common antenna site. (E)
73.240 Multiple ownership. (GM)
73.242 Duplication of AM and FM programming. (PD)
73.258 Indicating instruments. (E)
73.265 FM operator requirements. (E, PD)
73.267 Determining operating power. (E)
73.274 Remote control authorizations. (E)
73.275 Remote control operation. (E)
73.277 Permissible transmissions. (E)
73.293 Use of FM multiplex subcarriers. (E)
73.295 FM subsidiary communications services. (GM, PD, E)
73.297 FM stereophonic sound broadcasting. (E, PD)
73.310 FM technical definitions. (E)
73.311 Field strength contours. (E)
73.312 Topographic data. (E)
73.313 Prediction of coverage. (E)
73.314 Field strength measurements. (E)
73.315 Transmitter location. (E)
73.316 FM antenna systems. (E)
73.317 Transmission system requirements. (E)
73.319 FM multiplex subcarrier technical standards. (E)
73.322 FM stereophonic sound transmission standards. (E)
73.330 Use of modulation monitors at auxiliary transmitters. (E)
73.333 FM engineering charts. (E)
73.340 Use of automatic transmission systems (ATS). (E)
73.342 Automatic transmission system facilities. (E)
73.344 Fail-safe transmitter control for automatic transmission systems. (E)
73.346 Automatic transmission system monitoring and alarm points. (E)

SUBPART G—EMERGENCY BROADCAST SYSTEM

Scope and Objectives

73.901 Scope of subpart. (GM, PD, E)
73.902 Objectives of subpart. (GM, PD, E)

Definitions

73.903 Emergency Broadcast System (EBS). (PD, E, GM)
73.904 Licensee. (GM, E)
73.905 Emergency Action Notification (EAN). (GM, PD, E)
73.906 Attention signal. (E, PD)
73.907 Emergency Action Termination. (E, PD)
73.908 EBS Checklist. (PD, E)
73.909 Standard Operating Procedures (SOP's). (E, PD)
73.910 Authenticator word lists. (PD, E)
73.911 Basic Emergency Broadcast System Plan. (GM, PD, E)
73.912 NIAC order. (GM, PD, E)
73.913 Emergency Broadcast System Authorization. (GM, PD, E)
73.914 Primary Station (Primary). (GM, PD, E)
73.915 Primary Relay Station (Pri Relay). (GM, PD, E)
73.916 Common Program Control Station (CPCS). (GM, PD, E)
73.917 Originating Primary Relay Station (Orig Pri Relay). (GM, PD, E)
73.918 Non-participating Station (Non-EBS). (GM, PD, E)
73.919 State Relay Network. (GM, PD, E)
73.920 Operational (Local) Area. (GM, PD, E)
73.921 State Emergency Broadcast System Operational Plan. (GM, PD, E)
73.922 Emergency Broadcast System programing priorities. (GM, PD, E)

Participation

73.926 Participation in the Emergency Broadcast System. (GM, PD, E)
73.927 Participation by communications common carriers. (GM, PD, E)

Emergency Actions

73.931 Dissemination of Emergency Action Notification. (GM, PD, E)
73.932 Radio Monitoring and Attention Signal transmission requirement. (PD, E)
73.933 Emergency Broadcast System operation during a National level emergency. (GM, PD, E)

Day-to-Day Emergency Operation

73.935 Day-to-Day emergencies posing a threat to the safety of life and property; State Level and Operational (Local) Area Level Emergency Action Notification. (GM, PD, E)
73.936 Emergency Broadcast System operation during a State level emergency. (GM, PD, E)
73.937 Emergency Broadcast System operation during an Operational (Local) Area Level emergency. (GM, PD, E)

EBS Attention Signal Equipment

73.940 Encoder devices. (E)
73.941 Decoder devices. (E)
73.942 Acceptability of EBS Attention Signal equipment. (E)
73.943 Individual construction of encoders and decoders. (E)

Tests

73.961 Tests of the Emergency Broadcast System procedures. (E, PD)
73.962 Closed Circuit Tests of approved National level interconnecting systems and facilities of the Emergency Broadcast System. (E)

SUBPART H—RULES APPLICABLE TO ALL BROADCAST STATIONS

73.1001 Scope. (All dept. heads)
73.1010 Cross reference to rules in other parts.
73.1020 Station license period. (GM, PD, E)
73.1030 Notifications concerning interference to radio astronomy, research and receiving installations. (E)
73.1120 Station location. (GM, E)
73.1125 Station main studio location. (E)
73.1130 Station program origination. (E)
73.1150 Transferring a station. (GM)
73.1201 Station identification. (PD)
73.1202 Retention of letters received from the public. (GM, PD)
73.1205 Fraudulent billing practices. (GM, SM)
73.1206 Broadcast of telephone conversations. (PD, E)
73.1207 Rebroadcasts. (PD)
73.1208 Broadcast of taped, filmed or recorded material. (PD)
73.1209 References to time. (PD)
73.1210 TV/FM dual-language broadcasting in Puerto Rico. (GM, PD)
73.1211 Broadcast of lottery information. (PD, GM)
73.1212 Sponsorship identification; list retention; related requirements. (GM, PD)
73.1213 Antenna structure, marking and lighting. (E)
73.1215 Specifications for indicating instruments. (E)
73.1216 Licensee-conducted contests. (GM, PD)
73.1225 Station inspection by FCC. (GM, PD, E)
73.1226 Availability to FCC of station logs and records. (GM, PD, E)
73.1230 Posting of station and operator licenses. (PD, E)
73.1250 Broadcasting emergency information. (PD, E)
73.1510 Experimental authorizations. (E)
73.1515 Special field test authorizations. (E)
73.1520 Operation for tests and maintenance. (E)
73.1530 Portable test stations. [Definition] (E)
73.1540 Carrier frequency measurements. (E)
73.1545 Carrier frequency departure tolerances. (E)
73.1550 Extension meters. (E)
73.1560 Operating power tolerance. (E)
73.1570 Modulation levels: AM, FM, and TV aural. (E)
73.1580 Transmission system inspections. (E)
73.1590 Equipment performance measurements. (E)
73.1610 Equipment tests. (E)
73.1615 Operation during modification of facilities. (GM, E)
73.1620 Program tests. (E)
73.1635 Special temporary authorizations (STA). (GM, E)
73.1650 International broadcasting agreements. (GM, E)
73.1660 Type acceptance of broadcast transmitters. (E)
73.1665 Main transmitters. (E)
73.1670 Auxiliary transmitters. (E)
73.1675 Auxiliary antennas. (E)
73.1680 Emergency antennas. (E)
73.1690 Modification of transmission systems. (E)
73.1700 Broadcast day. (GM, PD, E)
73.1705 Time of operation. (PD, E, GM)
73.1710 Unlimited time. (GM, PD, E)
73.1715 Share time. (GM, PD, E)
73.1720 Daytime. (GM, PD, E)
73.1725 Limited time. (GM, PD, E)
73.1730 Specified hours. (GM, PD, E)
73.1735 AM station operation presunrise and postsunset. (GM, PD, E)
73.1740 Minimum operating schedule. (GM, PD, E)
73.1745 Unauthorized operation. (GM, PD, E)
73.1750 Discontinuance of operation. (GM, PD, E)
73.1800 General requirements relating to the station log. (GM, PD)
73.1820 Station log. (PD, E)
73.1835 Special technical records. (E)
73.1840 Retention of logs. (PD, E)
73.1850 Public inspection of program logs. (GM, PD)
73.1860 Transmitter duty operators. (E)
73.1870 Chief operators. (E)
73.1910 Fairness Doctrine. (GM, PD)
73.1920 Personal attacks. (GM, PD)
73.1930 Political editorials. (GM, PD)
73.1940 Broadcasts by candidates for public office. (GM, PD)
73.2080 Equal employment opportunities. (GM)
73.3500 Application and report forms. (GM, E)
73.3511 Applications required. (GM, E)
73.3512 Where to file; number of copies. (GM, E)
73.3513 Signing of applications. (GM)
73.3514 Content of applications. (GM)
73.3516 Specification of facilities. (GM)
73.3517 Contingent applications. (GM)
73.3518 Inconsistent or conflicting applications. (GM)
73.3519 Repetitious applications. (GM)
73.3520 Multiple applications. (GM)
73.3522 Amendment of applications. (GM)
73.3525 Agreements for removing application conflicts. (GM)
73.3526 Local public inspection file of commercial stations. (GM, PD, E)
73.3533 Application for construction permit or modification of construction permit. (GM, E)
73.3534 Application for extension of construction permit or for construction permit to replace expired construction permit. (GM, E)
73.3536 Application for license to cover construction permit. (GM, E)
73.3537 Application for license to use former main antenna as an auxiliary. (E)
73.3538 Application to make changes in an existing station. (GM, E)
73.3539 Application for renewal of license. (GM, E)
73.3540 Application for voluntary assignment or transfer of control. (GM)
73.3541 Application for involuntary assignment of license or transfer of control. (GM)
73.3542 Application for temporary or emergency authorization. (GM, E)
73.3543 Application for renewal or modification of special service authorization. (GM, E)
73.3544 Application to obtain a modified station license. (GM, E)
73.3545 Application for permit to deliver programs to foreign stations. (GM, PD)
73.3548 Applications to operate by remote control. (E)
73.3549 Requests for extension of authority to operate without required monitors, indicating instruments and EBS attention signal devices. (GM, E)
73.3550 Requests for new or modified call sign assignments. (GM, PD)
73.3561 Staff consideration of applications requiring Commission action. (GM)
73.3562 Staff consideration of applications not requiring action by the Commission. (GM)

73.3564 Acceptance of applications. (GM)
73.3566 Defective applications. (GM)
73.3568 Dismissal of applications. (GM)
73.3570 AM broadcast station applications involving other North American countries. (GM)
73.3571 Processing of AM broadcast station applications. (GM)
73.3573 Processing FM broadcast and FM translator station applications. (GM)
73.3578 Amendments to applications for renewal, assignment or transfer of control. (GM)
73.3580 Local public notice of filing of broadcast applications. (GM)
73.3584 Petitions to deny. (GM)
73.3587 Procedure for filing informal objections. (GM)
73.3591 Grants without hearing. (GM)
73.3592 Conditional grant. (GM)
73.3593 Designation for hearing. (GM)
73.3594 Local public notice of designation for hearing. (GM)
73.3597 Procedures on transfer and assignment applications. (GM)
73.3598 Period of construction. (GM, E)
73.3599 Forfeiture of construction permit. (GM, E)
73.3601 Simultaneous modification and renewal of license. (GM, E)
73.3603 Special waiver procedure relative to applications. (GM)
73.3605 Retention of applications in hearing status after designation for hearing. (GM)
73.3612 Annual employment report. (GM)
73.3613 Filing of contracts. (GM, PD)
73.3615 Ownership reports. (GM)
73.4000 Listing of FCC policies. (GM, E)
73.4005 Advertising—refusal to sell. (GM, SM)
73.4010 Advertising time—amount of. (GM, SM, PD)
73.4015 Alcoholic beverage advertising. (GM, SM, PD)
73.4030 Astrology material, broadcasts of. (PD)
73.4040 Audience ratings: Licensee distortion. (GM, PD)
73.4045 Barter agreements. (GM, SM)
73.4060 Citizens agreements. (GM, PD)
73.4065 Combination advertising rates; joint sales practices. (GM, SM)
73.4070 Commercials, false, misleading and deceptive. (GM, SM, PD)
73.4075 Commercials, loud. (PD, SM, E)
73.4080 Commercials, program length. (SM, PD)
73.4082 Comparative broadcast hearings—specialized programming formats. (PD)
73.4085 Conflict of interest. (GM, PD)
73.4091 Direct broadcast satellites. (PD, E)
73.4094 Dolby encoder. (E)
73.4095 Drug lyrics. (PD)
73.4097 EBS attention signal tests on automated programming systems. (E, PD)
73.4100 Financial qualifications; new AM and FM stations. (GM)
73.4102 FAA communications, broadcast of. (E)
73.4104 FM assignment policies and procedures. (GM, PD)
73.4105 Foreign language programs. (PD, GM)
73.4110 Format changes of stations. (PD, GM)
73.4115 Fraudulent billing practices. (GM, SM)
73.4120 Harassing and threatening phone calls (resulting from station broadcasts). (GM, PD)
73.4125 Horse racing information broadcasts. (PD, GM)
73.4126 Horse racing information transmitted via FM broadcast subcarriers. (PD, GM)
73.4130 Horse racing: Off-track and parimutuel betting advertising. (GM, PD)
73.4135 Interference to TV reception by FM stations. (E)
73.4140 Minority ownership; tax certificates and distress sales. (GM)
73.4145 Musical format service companies' agreements with broadcasters. (GM, PD)
73.4150 Musical recordings; repetitious broadcasts. (GM, PD)
73.4154 Network/AM, FM station affiliation agreements. (GM, PD)
73.4155 Network clipping. (GM, PD)
73.4157 Network signals which adversely affect affiliate broadcast service. (GM, PD, E)
73.4160 Night time service areas, class II and III AM stations; computation. (E, GM)
73.4165 Obscene language. (GM, PD)
73.4170 Obscene lyrics. (GM, PD)
73.4180 Payment disclosure: Payola, plugola, kickbacks. (GM, PD)
73.4185 Political broadcasting, the law of. (GM, PD)
73.4190 Political candidate authorization notice and sponsorship identification. (GM, PD)
73.4200 Polls, call-in, on radio and TV stations. (GM, PD)
73.4205 Private interest broadcasts by licensees to annoy and harass others. (GM, PD, E)
73.4210 Procedure Manual: "The Public and Broadcasting." (GM)
73.4215 Program matter: Supplier identification. (GM, PD)
73.4220 Promise versus performance: Commercial announcements. (GM, PD, SM)
73.4230 Sales contracts, failure to perform. (GM, SM)
73.4240 Sirens and like emergency sound effects in announcements. (PD)
73.4242 Sponsorship identification rules, applicability of. (SM, PD)
73.4246 Stereophonic pilot subcarrier use during monophonic programming. (E)
73.4250 Subliminal perception. (GM, PD)
73.4255 Tax certificates: Issuance of. (GM)
73.4260 Teaser announcements. (GM, PD)
73.4265 Telephone conservation broadcasts (network and like sources). (GM, PD)
73.4267 Time brokerage. (GM)
73.4275 Tone clusters; audio attention-getting devices. (GM, PD, E)
73.4280 Violation of laws of U.S.A. by station applicants; Commission policy. (GM, PD)

SUGGESTED FURTHER READING

Agor, Weston H. *Intuitive Management.* Englewood Cliffs, N.J.: Prentice-Hall, 1984.

Aronoff, Craig E., ed. *Business and the Media.* Santa Monica, Calif.: Goodyear Publishing Company, 1979.

Albrecht, Karl G. *Successful Management Objectives.* Englewood Cliffs, N.J.: Prentice-Hall, 1979.

Appleby, Robert C. *The Essential Guide to Management.* Englewood Cliffs, N.J.: Prentice-Hall, 1981.

Boyatzis, Richard E. *The Competent Manager.* New York: John Wiley and Sons, 1983.

Brown, Arnold. *Supermanaging.* New York: McGraw-Hill, 1984.

Cole, Barry, and Oettinger, M. *Reluctant Regulators: The FCC and the Broadcast Audience.* Reading, Mass.: Addison-Wesley, 1978.

Coleman, Howard W. *Case Studies in Broadcast Management.* New York: Hastings House, 1978.

Davis, Dennis K. *Mass Communication and Everyday Life.* Belmont, Calif.: Wadsworth Publishing, 1981.

Ellmore, R. Terry. *Broadcasting Law and Reg-*

ulation. Blue Ridge Summit, Pa.: Tab Books, 1982.

Goodworth, Clive T. *How to Be a Super-Effective Manager: A Guide to People Management*. London: Business Books, 1984.

Johnson, Joseph S., and Jones, K. *Modern Radio Station Practices,* 2nd ed. Belmont, Calif.: Wadsworth Publishing, 1978.

Kahn, Frank J., ed. *Documents of American Broadcasting,* 4th ed. Englewood Cliffs, N.J.: Prentice-Hall, 1984.

Kobert, Norman. *The Aggressive Management Style*. Englewood Cliffs, N.J.: Prentice-Hall, 1981.

Krasnow, Erwin G., and Longley, L.D. *The Politics of Broadcast Regulations*. New York: St. Martin's Press, 1978.

Lichty, Lawrence W., and Topping, Malachi C. *American Broadcasting: A Source Book on the History of Radio and Television*. New York: Hastings House, 1976.

McCavitt, William E., and Pringle, Peter. *Electronic Media Management*. Boston: Focal Press, 1986.

McCormack, Mark H. *What They Don't Teach You at Harvard Business School*. New York: Bantam, 1984.

Miner, John B. *The Management Process: Theory, Research, and Practice*. New York: Macmillan, 1978.

Pember, Don R. *Mass Media in America*. Chicago: Science Research Association, 1981.

Quaal, Ward L., and Brown, James A. *Broadcast Management,* 2nd ed. New York: Hastings House, 1976.

Robinson, Sol. *Broadcast Station Operating Guide*. Blue Ridge Summit, Pa.: Tab Books, 1969.

Routt, Ed. *The Business of Radio Broadcasting*. Blue Ridge Summit, Pa.: Tab Books, 1972.

Schwartz, Tony. *Media, The Second God*. New York: Praeger, 1984.

Townsend, Robert. *Further Up the Organization*. New York: Knopf, 1984.

3 Programming

PROGRAM FORMATS

Programming a radio station has become an increasingly complex task. There are twice the number of stations today competing for the audience's attention than existed in the 1960s, and more enter the fray almost daily. Other media have proliferated as well, resulting in a further distraction of radio's customary audience. The government's laissez-faire, "let the marketplace dictate," philosophy concerning commercial radio programming gives the station great freedom in deciding the nature of its air product, but determining what to offer the listener, who is often presented with dozens of alternatives, involves intricate planning.

The basic idea, of course, is to air the type of format that will attract a sizable enough piece of the audience demographic to satisfy the advertiser. Once a station decides upon the format it will program, it then must know how to effectively execute it.

What follows are brief descriptions of the most frequently employed formats in radio today. There are a host of other formats, or subformats, over one hundred, in fact. Many are variations of those listed.

FIGURE 3.1
Format chart shows the popularity of formats around the country. The chart changes with each ratings sweep. Reprinted with permission from *Radio and Records*.

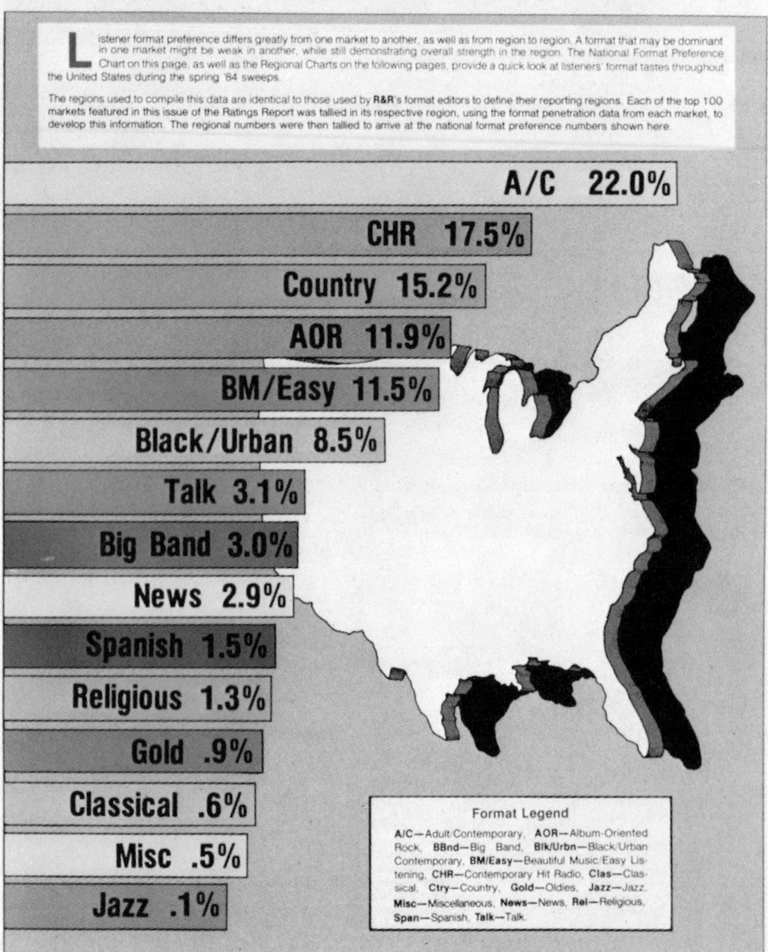
FORMAT PREFERENCE CHARTS
Listener format preference differs greatly from one market to another, as well as from region to region. A format that may be dominant in one market might be weak in another, while still demonstrating overall strength in the region. The National Format Preference Chart on this page, as well as the Regional Charts on the following pages, provide a quick look at listeners' format tastes throughout the United States during the spring '84 sweeps.
The regions used to compile this data are identical to those used by R&R's format editors to define their reporting regions. Each of the top 100 markets featured in this issue of the Ratings Report was tallied in its respective region, using the format penetration data from each market, to develop this information. The regional numbers were then tallied to arrive at the national format preference numbers shown here.
A/C 22.0%
CHR 17.5%
Country 15.2%
AOR 11.9%
BM/Easy 11.5%
Black/Urban 8.5%
Talk 3.1%
Big Band 3.0%
News 2.9%
Spanish 1.5%
Religious 1.3%
Gold .9%
Classical .6%
Misc .5%
Jazz .1%
Format Legend
A/C—Adult/Contemporary, AOR—Album-Oriented Rock, BBnd—Big Band, Blk/Urbn—Black/Urban Contemporary, BM/Easy—Beautiful Music/Easy Listening, CHR—Contemporary Hit Radio, Clas—Classical, Ctry—Country, Gold—Oldies, Jazz—Jazz, Misc—Miscellaneous, News—News, Rel—Religious, Span—Spanish, Talk—Talk.

Adult Contemporary (A/C)

Adult Contemporary is the most widely employed format in the 1980s. It is very strong in the twenty-five to forty-nine age group, which makes it particularly appealing to advertisers, since this demographic group has the largest disposable income. Also, some advertisers spend money on A/C stations simply because they like the format themselves. The Adult Contemporary format is also one of the most effective in attracting women listeners.

A/C outlets emphasize current (since 1970s) pop standards, sans raucous or pronounced beats—in other words, no hard rock. Some A/C stations could be described as soft rockers. However, the majority mix in enough ballads and easy listening sounds to justify their title. The main thrust of this format's programming is the music. More music is aired by deemphasizing chatter. Music is commonly presented in uninterrupted sweeps or blocks, perhaps ten to twelve minutes in duration, followed by a brief recap of artists and song titles. High-profile morning talent or teams have become popular at A/C stations in the 1980s. Commercials generally are clustered at predetermined times, and midday and evening deejay talk often is limited to brief informational announcements. News and sports are secondary to the music.

Contemporary Hit Radio (CHR)

Also known as Top 40, CHR stations play only those records which currently are the fastest selling. CHR's narrow playlists are designed to draw teens and young adults. The heart of this format's demographic is the twelve- to eighteen-year-old, although in the mid-1980s it enjoyed a broadening of its core audience. The format is characterized by its swift, often unrelenting pace. Silence, known as "dead air," is the enemy. The idea is to keep the sound hot and tight to keep the kids from station hopping, which is no small task since many markets have at least two hit-oriented stations. CHR deejays have undergone several shifts in status since the inception of the chart music format in the 1950s. Initially, pop deejay personalities played an integral role in the air sound. However, in the mid-1960s the format underwent a major change when deejay presence was significantly reduced. Programming innovator Bill Drake decided that the Top 40 sound needed to be refurbished and tightened. Thus deejay talk and even the number of commercials scheduled each hour were cut back in order to improve flow. Despite criticism that the new sound was too mechanical, Drake's technique succeeded at strengthening the format's hold on the listening audience.

In the mid and late 1970s the deejay's role on hit stations began to regain its former prominence, but in the 1980s the format underwent a further renovation that resulted in a narrowing of its playlist and a decrease in deejay presence. "Super" or "Hot" hit stations, as they also are called, are among the most popular in the country and can be found either near or at the top of the rating charts in their markets.

News is of secondary importance on CHR stations. In fact, news programming is considered by many program directors to be a tune-out factor. "Kids don't like news," they claim. However, despite deregulation, which has freed stations of nonentertainment program requirements, most retain at least a modicum of news out of a sense of obligation. CHR stations are very promotion minded and contest oriented.

Country

Since the 1960s the Country format has been adopted by more stations than any other. Although seldom a leader in the ratings race, its appeal is exceptionally broad. An indication of country music's rising popularity is the fact that there are over ten times as many full-time Country stations today than there were twenty years ago. Although the format is far more prevalent in the South and Midwest, most medium and large markets in the North have Country stations. Due to the diversity of approaches within the format (there are more subformats in Country than in any other)—for example, traditional, middle-of-the-road, contemporary hit, and so on—the Country format attracts a broad age group, appealing to young as well as older adults.

Country radio has always been particularly popular among blue-collar workers. According to the Country Music Association and the Organization of Country Radio Broadcasters, the Country music format is drawing a more upscale audience in the 1980s than it has in the past. Most Country stations are AM, but the format is becoming prevalent on FM, too. Depending on the approach they employ, Country outlets may emphasize or deemphasize air personalities, include news and public affairs fea-

Adult / Contempora

N&A Begins on Page 66

1	1	CHICAGO/You're The Inspiration (WB)
7	2	WHAM!/Careless Whisper (Columbia)
3	3	FOREIGNER/I Want To Know What Love Is (Atlantic)
6	4	STEVE PERRY/Foolish Heart (Columbia)
5	5	RAY PARKER JR./Jamie (Arista)
2	6	JACK WAGNER/All I Need (Qwest/WB)
4	7	JERMAINE JACKSON/Do What You Do (Arista)
9	8	STEVIE WONDER/Love Light In Flight (Motown)
11	9	DIANA ROSS/Missing You (RCA)
12	10	ELTON JOHN/In Neon (Geffen)
14	11	B. STREISAND w/K. CARNES/Make No Mistake... (Columbia)
8	12	BOB SEGER & SILVER BULLET BAND/Understanding (Capitol)
18	13	GEORGE BENSON/20/20 (WB)
10	14	JULIAN LENNON/Valotte (Atlantic)
13	15	HONEYDRIPPERS/Sea Of Love (Es Paranza/Atlantic)
19	16	MANHATTAN TRANSFER/Baby Come Back To Me... (Atlantic)
BREAKER	17	BILLY JOEL/Keeping The Faith (Columbia)
BREAKER	18	KENNY ROGERS/Crazy (RCA)
17	19	LINDA RONSTADT/Skylark (Asylum)
BREAKER	20	CULTURE CLUB/Mistake No. 3 (Virgin/Epic)
15	21	PAUL McCARTNEY/No More Lonely Nights (Columbia)
16	22	RICK SPRINGFIELD w/RANDY CRAWFORD/Taxi Dancing (RCA)
BREAKER	23	REO SPEEDWAGON/Can't Fight This Feeling (Epic)
25	24	MADONNA/Like A Virgin (Sire/WB)
DEBUT	25	PHILIP BAILEY w/PHIL COLLINS/Easy Lover (Columbia)

FIGURE 3.2
National Adult Contemporary playlist indicates songs most often aired by stations employing this format. Reprinted with permission from *Radio and Records*, Winter 1984.

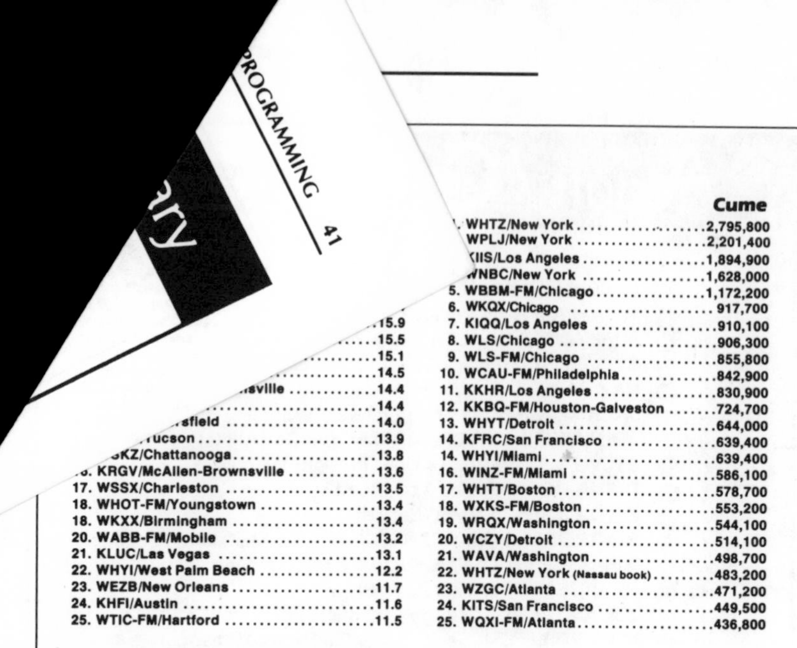

[illegible]	15.9
[illegible]	15.5
[illegible]	15.1
[illegible]	14.5
…nsville	14.4
[illegible]	14.4
…rsfield	14.0
…Tucson	13.9
…SKZ/Chattanooga	13.8
…. KRGV/McAllen-Brownsville	13.6
17. WSSX/Charleston	13.5
18. WHOT-FM/Youngstown	13.4
18. WKXX/Birmingham	13.4
20. WABB-FM/Mobile	13.2
21. KLUC/Las Vegas	13.1
22. WHYI/West Palm Beach	12.2
23. WEZB/New Orleans	11.7
24. KHFI/Austin	11.6
25. WTIC-FM/Hartford	11.5

	Cume
…. WHTZ/New York	2,795,800
WPLJ/New York	2,201,400
…KIIS/Los Angeles	1,894,900
…WNBC/New York	1,628,000
5. WBBM-FM/Chicago	1,172,200
6. WKQX/Chicago	917,700
7. KIQQ/Los Angeles	910,100
8. WLS/Chicago	906,300
9. WLS-FM/Chicago	855,800
10. WCAU-FM/Philadelphia	842,900
11. KKHR/Los Angeles	830,900
12. KKBQ-FM/Houston-Galveston	724,700
13. WHYT/Detroit	644,000
14. KFRC/San Francisco	639,400
14. WHYI/Miami	639,400
16. WINZ-FM/Miami	586,100
17. WHTT/Boston	578,700
18. WXKS-FM/Boston	553,200
19. WRQX/Washington	544,100
20. WCZY/Detroit	514,100
21. WAVA/Washington	498,700
22. WHTZ/New York (Nassau book)	483,200
23. WZGC/Atlanta	471,200
24. KITS/San Francisco	449,500
25. WQXI-FM/Atlanta	436,800

FIGURE 3.3
One of the main attractions of the CHR format is its ability to build large cumes. While the format has a high repetition factor and lower quarter-hour shares, its ability to garner a substantial cume attracts many broadcasters to the format. Reprinted with permission from *Radio and Records*.

tures, or confine their programming almost exclusively to music.

Easy Listening

The Beautiful Music station of the 1960s and 1970s has become the Easy Listening station of the 1980s. Playlists in this format have been carefully updated in an attempt to attract a somewhat younger audience. The term "Beautiful Music" was exchanged for "Easy Listening" in an effort to dispel the geriatric image the former term seemed to convey. Easy Listening is the ultimate "wall-to-wall" music format. Talk of any type is kept minimal, although many stations in this format concentrate on news and information during morning drivetime.

Instrumentals of established songs are a mainstay at Easy Listening stations, which also share a penchant for lush orchestrations featuring plenty of strings. These stations boast a devoted audience. Station hopping is uncommon. Efforts to draw younger listeners into the Easy Listening fold have been moderately successful, but most of the format's primary adherents are over fifty. Music syndicators provide prepackaged (canned) programming to approximately half of the nation's Easy Listening stations, and over three quarters of the outlets within this format utilize automation systems to varying degrees. Easy Listening has held strong in several markets; WGAY FM/AM in Washington, D.C., and WLKW-FM in Providence, Rhode Island, have topped the ratings for years, although the format has lost ground in the 1980s to Adult Contemporary and other adult-appeal formats.

Album-Oriented Rock (AOR)

The birth of the AOR format in the mid-1960s was the result of a basic disdain for the highly formulaic Top 40 sound that prevailed at the time. In the summer of 1966, WOR-FM, New York, introduced Progressive radio, the forerunner of AOR. As an alternative to the superhyped, ultracommercial sound of the hit song station, WOR-FM programmed an unorthodox combination of nonchart rock, blues, folk, and jazz. In the 1970s the format concentrated its attention almost exclusively on album rock, while becoming less free-form and more formulaic and systematic in its programming approach.

While AOR has done well in garnering the eighteen- to thirty-four-year-old male, this format has done poorly in winning female listeners. This has proven to be a sore spot with certain advertisers. In the 1980s, the format lost its prominence due, in part, to the meteoric rebirth of hit radio.

Generally, AOR stations broadcast their music in sweeps, or at least segue two or three songs. A large airplay library is typical, in which three to seven hundred cuts may be active. Depending on the outlet, the deejay may or may not have "personality" status. In fact, the more-music/less-talk approach particularly common at Easy Listening stations is emulated by many album rockers. Consequently, news plays a very minor part in the station's programming efforts.

AOR stations are very lifestyle oriented and invest great time and energy developing promotions relevant to the interests and attitudes of their listeners.

News and/or Talk

There are News, News/Talk, News Plus, and Talk formats, and each is distinct and unique unto itself. News stations differ from the others in that they devote their entire air schedule to the presentation of news and news-related stories and features. The All-News format was introduced by Gordon McLendon at XETRA (known as "XTRA") in Tijuana, Mexico, in the early 1960s. Its success soon inspired the spread of the format in the United States. Due to the enormous expense involved in the presentation of a purely News format, requiring three to four times the staff and budget of most music operations, the format has been confined to larger markets able to support the endeavor.

The News/Talk format is a hybrid. It combines extensive news coverage with blocks of two-way telephone coversations. These stations commonly daypart or segmentalize their programming by presenting lengthened newscasts during morning and afternoon drivetime hours and conversation in the midday and evening periods. The News/Talk combo was conceived by KGO in San Francisco in the 1960s and has gradually gained in popularity so that it now leads both the strictly News and the Talk formats.

Talk radio began at KABC-AM, Los Angeles, in 1960. However, talk shows were familiar to listeners in the 1950s, since a number of adult music stations devoted a few hours during evenings or overnight to call-in programs. The motivation behind most early Talk programming stemmed from a desire to strengthen weak time slots while satisfying public affairs programming requirements. Like its nonmusic siblings, Talk became a viable format in the 1960s and does well in the 1980s, although it too has suffered due to greater competition. In contrast to All-News, which attracts a slightly younger and more upscale audience, All-Talk amasses a large following among blue-collar workers and retirees.

The newest of the news and information-oriented formats calls itself News Plus. While its emphasis is on news, it fills periods with music, often Adult Contemporary in flavor. News Plus stations also may carry a heavy schedule of sporting events. This combination has done well in several medium and large markets.

News and/or Talk formats are primarily located on the AM band, where they have become increasingly prevalent since FM has captured all but a few of radio's music listeners. Nonmusic formats are beginning to appear on FM, however, and AM's adaptation to stereo is providing some impetus to the movement.

Oldies/Nostalgia

While the Oldies/Nostalgia formats are not identical, they both derive the music they play from years gone by. Whereas the Nostalgia station, sometimes referred to as "Big Band," constructs its playlist around tunes popular as far back as the 1940s and 1950s, the Oldies outlet focuses its attention on the pop tunes of the late 1950s and early 1960s. A typical oldies quarter-hour might consist of songs by Elvis Presley, the Everly Brothers, Brian Hyland, and the Ronettes. In contrast, a Nostalgia quarter-hour might consist of tunes from the pre-rock era performed by artists like Frankie Lane, Les Baxter, the Mills Brothers, and Tommy Dorsey, to name only a few.

Nostalgia radio caught on in the late 1970s, the concept of programmer Al Ham. Nostalgia is a highly syndicated format, and most stations go out-of-house for programming material. Because much of the music predates stereo processing, AM outlets are most apt to carry the Nostalgia sound. Music is invariably presented in sweeps and, for the most part, deejays maintain a low profile. Similar to Easy Listening, Nostalgia pushes its music to the forefront and keeps other program elements at an unobtrusive distance. In the 1980s, Easy Listening/Beautiful Music stations have lost some listeners to this format, which claims a 4.5 percent share of the radio audience.

The Oldies format was first introduced in the 1960s by programmers Bill Drake and Chuck Blore. While Nostalgia's audience tends to be around the age of fifty, most Oldies listeners are in their thirties and forties. Unlike Nostalgia, most Oldies outlets originate their own programming, and very few are automated. In contrast with its vintage music cousin, Oldies allows greater deejay presence. At many Oldies stations, air personalities play a key role. Music is rarely broadcast in sweeps, and commercials, rather than being clustered, are inserted in a random fashion between songs.

The Oldies audience has dwindled slightly in recent years due to a competition from Adult Contemporary, which occasionally mixes vintage with current tunes. Nostalgia has gradually picked up listeners since its inception. Although neither format has achieved the top of the ratings charts, these are modest formats that can claim loyal audiences.

Urban Contemporary (UC)

The "melting pot" format, Urban Contemporary attracts large numbers of Hispanic and Black listeners, as well as White. As the term suggests, stations employing this format are usually located in metropolitan areas with large, heterogeneous populations. UC was born in the early

1980s, the offspring of the short-lived Disco format, which burst on the scene in 1978. What characterizes UC the most is its upbeat, danceable sound and deejays who are hip, friendly, and energetic. Although UC outlets stress danceable tunes, their playlists generally are anything but narrow. However, a particular sound may be given preference over another depending on the demographic composition of the population in the area that the station serves. For example, UC outlets may play greater amounts of music with a Latin or rhythm-and-blues flavor, while others may air larger proportions of light jazz, reggae, or new rock. Several AM stations around the country have adopted the UC format; however, it is more likely to be found on the FM side, where it has taken numerous stations to the forefront of their market's ratings.

UC has had an impact on Black stations, which have experienced erosion in their youth numbers. Many Black stations have countered by broadening their playlists to include artists not traditionally programmed. Because of their high-intensity, fast-paced sound, UC outlets can give a Top 40 impression, but in contrast they commonly segue songs or present music in sweeps and give airplay to lengthy cuts, sometimes six to eight minutes long. While Top 40 or CHR stations seldom program cuts lasting more than four minutes, UC outlets find long cuts compatible with their programming approach. Remember, UCs are very dance oriented. Newscasts play a minor role in this format, which caters to a target audience aged eighteen to thirty-four. Contests and promotions are important program elements.

Classical

Although there are fewer than three dozen full-time commercial Classical radio stations in the country, no other format can claim a more loyal following. Despite small numbers and soft ratings, most Classical stations do manage to generate a modest income. Over the years, profits have remained relatively minute in comparison to other formats. However, member stations of the Concert Music Broadcasters Association reported ad revenue increases of up to 40 percent in the first half of the 1980s. Due to its upscale audience, blue-chip accounts find the format an effective buy. This is first and foremost an FM format, and it has broadcast over the megahertz band for almost as long as it has existed.

In many markets commercial Classical stations have been affected by public radio outlets programming classical music. Since commercial Classical stations must break to air the sponsor messages that keep them operating, they must adjust their playlists accordingly. This may mean shorter cuts of music during particular dayparts—in other words, less music. The noncommercial Classical outlet is relatively free of such constraints and thus benefits as a result. A case in point is WCRB-FM in Boston, the city's only full-time Classical station. While it attracts most of the area's Classical listeners throughout the afternoon and evening hours, it loses many to public radio WGBH's "Morning Pro Musica," which is aired seven days a week between seven and noon. At least in part, public radio's success consists in programming having fewer interruptions.

Classical stations target the twenty-five- to forty-nine-year old, higher-income, college-educated listener. News is presented at sixty- to ninety-minute intervals and generally runs from five to ten minutes. The format is characterized by a conservative, straightforward air sound. Sensationalism and hype are avoided, and on-air contests and promotions are as rare as announcer chatter.

Religious

Live broadcasts of religious programs began while the medium was still in its experimental stage. In 1919 the U.S. Army Signal Corps aired a service from a chapel in Washington, D.C. Not long after that, KFSG in Los Angeles and WMBI in Chicago began to devote themselves to religious programming. Soon dozens of other radio outlets were broadcasting the message of God. In the 1980s, over six hundred stations broadcast religious formats on a full-time basis, and another fifteen hundred air at least six hours of religious features on a weekly basis.

Religious broadcasters typically follow one of two programming approaches. One includes music as part of its presentation, while the other does not. The Religious station that features music often programs contemporary tunes containing a Christian or life-affirming perspective. This approach also includes the scheduling of blocks of religious features and programs. Nonmusic Religious outlets concentrate on inspirational features and complementary talk and informational shows.

Religious broadcasters claim that their spiritual messages reach nearly half of the nation's radio audience, and the American Research Corporation in Irvine, California, contends that over 25 percent of those tuned to Religious stations attend church more frequently. Two-thirds of the country's Religious radio stations broadcast over AM frequencies.

Ethnic

Afro-Americans constitute the largest minority in the nation, thus making Black the most prominent ethnic format. Over three hundred radio stations gear themselves to the needs and desires of Black listeners. WDIA-AM in Memphis claims to be the first station to program exclusively to a nonwhite audience. It introduced the format in 1947. Initially growth was gradual in this format, but in the 1960s, as the Motown sound took hold of the hit charts and the Black Pride movement got under way, more Black stations entered the airways.

At its inception, the Disco craze in the 1970s brought new listeners to the Black stations, which shortly saw their fortunes change when all-Disco stations began to surface. Many Black outlets witnessed an exodus of their younger listeners to the Disco stations. This prompted a number of Black stations to abandon their more traditional playlists, which consisted of rhythm and blues, gospel, and soul tunes, for exclusively Disco music. When Disco perished in the early 1980s, the Urban Contemporary format took its place. Today, Progressive Black stations, such as WBLS-FM, New York, combine dance music with soulful rock and contemporary jazz, and many have transcended the color barrier by including certain white artists on their playlists. In fact, many Black stations employ white air personnel in efforts to broaden their demographic base. WILD-AM in Boston, long considered the city's "Black" station, is an example of this trend. "We have become more of a general appeal station than a purely ethnic one. We've had to in order to prosper. We strive for a distinct, yet neuter or deethnicized, sound on the air. The Black format has changed considerably over the year," notes WILD program director Elroy Smith. The old-line R & B and gospel stations still exist and can be found mostly in the South.

Hispanic or Spanish-language stations constitute the second largest ethnic format. KCOR-AM, San Antonio, became the first all-Spanish station in 1947, just a matter of months after WDIA-AM in Memphis put the Black format on the air. Cities with large Latin populations are able to support the format, and in some metropolitan areas with vast numbers of Spanish-speaking residents, such as New York, Los Angeles, and Miami, several radio outlets are exclusively devoted to Hispanic programming.

Programming approaches within the format are not unlike those prevalent at Anglo stations. That is to say, Spanish-language radio stations also modify their sound to draw a specific demographic. For example, many offer contemporary music for younger listeners and more traditional music for older listeners.

Spanish media experts predict that there will be a significant increase in the number of Hispanic stations through the 1980s and 1990s. They anticipate that most of this growth will take place on the AM band.

Hundreds of other radio stations countrywide apportion a significant piece of their air schedules (over twenty hours weekly), if not all, to foreign-language programs in Portuguese, German, Polish, Greek, Japanese, and so on. Over twenty stations broadcast exclusively to American Indians and Eskimos.

Middle-of-the-Road (MOR)

The Middle-of-the-Road format (also called Variety, General Appeal, Diversified) is ambiguously referred to because it defies more specific labeling. Succinctly, this is the "not too anything" format, meaning that the music it airs is not too current, not too old, not too loud, and not too soft.

MOR has been called the "bridge" format because of its "all things to all people" programming. However, its audience has decreased over the years, particularly since 1980, due to the rise in popularity of more specialized formats. According to radio program specialist Dick Ellis (Peters Productions), MOR now has a predominantly over-forty age demographic, several years older than just a decade ago. In some major markets the format continues to do well in the ratings mainly because of strong on-air personalities. Washington's WMAL has been the only AM outlet in the city consistently to appear among the top five, frequently as number one. This is true also of other MORs around the country. But this is not the format that it once was. Since its inception in the 1950s, up through the 1970s, stations working the MOR sound have often dominated their markets. Yet the Soft Rock and Oldies formats in the 1970s, the updating of Easy Listening, and particularly the ascendency of Adult Contemporary have conspired to significantly erode MOR's numbers.

MOR is the home of the on-air personality. Perhaps no other format gives its air people as much latitude and freedom. This is not to suggest that MOR announcers may do as they please. They, like any other, must abide by format and programming policy, but MOR personalities often serve as the cornerstone of their station's air product. Some of the best-known deejays in the country have come from the MOR milieu. It would then follow that the music is rarely, if ever, presented in sweeps or even segued. Deejay patter occurs between each cut of music, and announcements are inserted in the same way. News and sports play another vital function at these stations. During drive periods, MOR often presents lengthened blocks of news, replete with frequent traffic reports, weather updates, and the latest sports information. Many MOR outlets are affiliated with major-league teams. With few exceptions, MOR is an AM format. Although it has noted slippage in recent years, it will likely continue to bridge whatever gaps may exist in a highly specialized radio marketplace.

The preceding is an incomplete list of the myriad of radio formats that serve the listening public. The program formats mentioned constitute the majority of the basic format categories prevalent today. Tomorrow? Who knows. Radio is hardly a static industry, but one subject to the whims of popular taste. When something new captures the imagination of the American public, radio responds, and often a new format is conceived.

THE PROGRAMMER

Program directors (PDs) are radiophiles. They live the medium. Most admit to being smitten by radio at an early age. "It's something that is in your blood and grows to consuming proportions," admits Peter Falconi, program director, WGAN AM/FM, Portland, Maine. Brian Mitchell, program director of WQIK AM/FM, Jacksonville, Florida, recalls an interest in the medium as a small child and for good reason. "I was born into a broadcasting family. My father is a station owner and builder. During my childhood radio was the primary topic at the dinner table. It fed the flame that I believe was already ignited anyway. Radio fascinated me from the start."

The customary route to the programmer's job involves deejaying and participation in other on-air-related areas, such as copywriting, production, music, and news. It would be difficult to state exactly how long it takes to become a program director. It largely depends on the individual and where he or she happens to be. There are instances where newcomers have gone into programming within their first year in the business. When this happens it is likeliest to occur in a small market where turnover may be high. On the other hand, it is far more common to spend years working toward this goal, even in the best of situations. "Although my father owned the station, I spent a long time in a series of jobs before my appointment to programmer. Along the way, I worked as station janitor, and then got into announcing, production, and eventually programming," recounts Mitchell.

One of the nation's foremost air personalities, Dick Fatherly, whom *Billboard Magazine* has described as a "long-time legend," now heads programming at KTOP-AM, Topeka, Kansas. Fatherly spent years as a deejay before making the transition. "In the twenty-five years that I've been in this business, I have worked as a jock, newsman, production director, and even sales rep. Eventually I ended up in program management. During my career I have worked at WABC, WICC, WFUN, WHB, to mention a few. Plenty of experience, you might say," comments Fatherly.

Experience contributes most toward the making of the station's programmer. However, individuals entering the field with hopes of becoming a PD do well to acquire as much formal training as possible. The programmer's job has become an increasingly demanding one as the result of expanding competition. "A good knowledge of research methodology, analysis, and application is crucial. Programming is both an art and a science today," observes Jim Murphy, program director, WHDH-AM, Boston.

Cognizant of this, schools with programs in radio broadcasting have begun to emphasize courses in audience research, as well as other programming-related areas. An important fact for the aspiring PD to keep in mind is that more people entering broadcasting today have college backgrounds than ever before. While a college degree is not necessarily a prerequisite for the position of PD, it is clearly regarded as an asset by upper management. "It used to be that a college degree didn't mean so much. A PD came up through the ranks of programming, proved his ability, and was hired. Not that that doesn't still happen. It does. But more and more the new PD has a degree or, at the very least, several years of college," contends Joe Cortese, air personality, WROR-FM, Boston. "I majored in Communication Arts at a junior college and then transferred to a four-year school. There are many colleges offering communications courses here in the Boston area, so I'll probably take some more as a way of further preparing for the day when I'll be programming a station. That's what I eventually want to do," says Cortese.

Gary Begin, midday deejay at WSUB-FM, New London, Connecticut, has spent several years preparing for programming responsibilities. He, like Cortese, recognizes the importance of formal training, but contends that station managers are still most impressed by a programming job candidate's track record. "Granted, you can't have too much education. I'll be the first to agree that in today's marketplace college training does mean something, but the bottom line is that the person in charge of the operation wants to know what you've done in the business and where."

Begin's point is well taken. Work experience does head the list on which a station manager bases his or her selection for program director. Meanwhile, college training, at the very least, has become a criterion to the extent that if an applicant does not have any the prospective employer takes notice.

Beyond formal training and experience, Chuck Ducoty, program director of WIYY-FM, Baltimore, says a PD must possess certain innate qualities. "Common sense and a good sense of humor are necessary attributes and are in rather short supply, I think." Dick Fatherly adds sensitivity, patience, compassion, and drive to the list.

THE PROGRAM DIRECTOR'S DUTIES AND RESPONSIBILITIES

Where to begin this discussion poses no simple problem, because the PD's responsibilities and duties are so numerous and wide-ranging. Second in responsibility to the general manager, the program director is the individual responsible for everything that goes over the air. This involves working with the manager in establishing programming and format policy and overseeing their effective execution. In addition, he or she hires and supervises on-air music, and production personnel, plans various schedules, handles the programming budget, develops promotions, monitors the station and its competition, assesses research, and may even pull a daily airshift. The PD also is accountable for the presentation of news, public affairs, and sports features, although a news director often is appointed to help oversee these areas.

The program director alone does not determine a station's format. This is an upper management decision. The PD may be involved in the selection process but, more often than not, the format has been decided upon before the programmer has been hired. For example, WYYY decides it must switch from MOR to CHR in order to attract a more marketable demographic. After an in-depth examination of its own market, research on the effectiveness of CHR nationally, and advice from a program consultant and rep company, the format change is deemed appropriate. Reluctantly the station manager concludes that he must bring in a CHR specialist, which means that he must terminate the services of his present programmer, whose experience is limited to MOR. The station manager places an ad in an industry trade magazine, interviews several candidates, and hires the person he feels will take the station to the top of the ratings. When the new program director arrives, he or she is charged with the task of preparing the CHR format for its debut. Among other things, this may involve hiring new air people, the acquisition of a new music library or the updating of the existing one, preparing promos and purchasing jingles, and working in league with the sales, traffic, and engineering departments for maximum results.

Once the format is implemented, the program director must work at refining and maintaining the sound. After a short time, the programmer may feel compelled to modify air schedules

FIGURE 3.4
Programmer's monitor sheet designed for side-by-side comparison. Here the PD will check for the following: (a) the strength of each song compared to the competition's, (b) where spot set group of announcements falls in comparison to competitor's, (c) length of spot set and unit count compared to competitor's, (d) is the station into spots when competitors are playing music?, (e) is the station into news when competitors are playing music?, (f) is the LP giveaway on #1 effective?, (g) is the station's contest promo more effective than competitor's?

MONITOR SHEET

STATION WXXX-WXX1-WXX2

DATE 1-1-86

TIME 11am-12n

COMPETITOR 1	COMPETITOR #2	YOU
1. Station Id	Station Id	Station Id
2. Foreigner-I Want to	Wham-Careless	Prince-Purple
3. Elton-Sad Songs	Genesis-Taking It	ZZ Top-Legs
4. Stop Set 11:10	Pointer Sisters-Jump	Stop Set
5. :30 Second Spot	Tina Turner-Whats Love	:60 Second
6. :30 Second Spot	Huey Lewis-Heart of	:30 Second
7. :60 Second Spot	Chicago-Inspiration	JINGLE
8. Weather	Culture Club-Mistake	Led Zepp-
9. Jingle	Cars-Drive	Scandal-Loves
10. Glen Frey-Boys of	STOP SET	Contest Promo
11. John Waite-Missing You	:30 Second Spot	Lauper-She
12. Liner-Lp Giveaway	News 2:00	PSA&Jingle

either by shifting deejays around or replacing those who do not enhance the format. The PD prepares weekend and holiday schedules as well, and this generally requires the hiring of part-time announcers. A station may employ as few as one or two part-timers or fill-in people, or as many as eight to ten. This largely depends on whether deejays are on a five- or six-day schedule. At most stations, air people are hired to work a six-day week. The objective of scheduling is not merely to fill slots but to maintain continuity and consistency of sound. A PD prefers to tamper with shifts as little as possible and fervently hopes that he has filled weekend slots with people who are reliable. "The importance of dependable, trustworthy air people cannot be overemphasized. It's great to have talented deejays, but if they don't show up when they are supposed to because of one reason or another they don't do you a lot of good. You need people who are cooperative. I have no patience with individuals who try to deceive me or fail to live up to their responsibilities," says WQIK's Mitchell. A station that is constantly introducing new air personnel has a difficult time establishing listener habit. The PD knows that in order to succeed he or she must present a stable and dependable sound, and this is a significant programming challenge.

Production schedules also are prepared by the programmer. Deejays are usually tapped for production duties before or after their airshifts. For example, the morning person who is on the air from six to ten may be assigned production and copy chores from ten until noon. Meanwhile, the midday deejay who is on the air from ten until three is given production assignments from three to five, and so on. Large radio stations frequently employ a full-time production person. If so, this individual handles all production responsibilities and is supervised by the program director.

A program director traditionally handles the department's budget, which generally constitutes 30 to 40 percent of the station's operating budget. Working with the station manager, the PD ascertains the financial needs of the programming area. The size and scope of the budget varies from station to station. Most programming budgets include funds for the acquisition of program materials, such as albums, features, and contest paraphernalia. A separate promotional budget usually exists, and it too may be managed by the PD. The programmer's budgetary responsibilities range from monumental at some outlets to minuscule at others. Personnel salaries and even equipment purchases may fall within the province of the program department's budget. Brian Mitchell believes that "an understanding of the total financial structure of the company or corporation and how programming fits into the scheme of things is a real asset to a programmer."

Devising station promotions and contests also places demands on the PD's time. Large stations often appoint a promotion director. When this is the case, the PD and promotion director work together in the planning, development, and execution of the promotional campaign. The PD, however, retains final veto power should he or she feel that the promotion or contest fails to complement the station's format. When the PD alone handles promotions and contests, he or she may involve other members of the programming or sales department in brainstorming sessions designed to come up with original and interesting concepts. The programmer is aware that the right promotion or contest can have a major impact on ratings. Thus he or she is constantly on the lookout for an appropriate vehicle. In the quest to find the promotion that will launch the station on the path to a larger audience, the PD may seek assistance from one of dozens of companies that offer promotional services.

The program director's major objective is to program for results. If the station's programming fails to attract a sufficient following, the ratings will reflect that unhappy fact. All medium and larger markets are surveyed by ratings companies, primarily Arbitron and Birch. Very few small rural markets, with perhaps one or two stations, are surveyed. If a small market station is poorly programmed, the results will be apparent in the negative reactions of the local retailers. Simply put, the station will not be bought by enough advertisers to make the operation a profitable venture. In the bigger markets, where several stations compete for advertising dollars, the ratings are used to determine which is the most effective or cost-efficient station to buy. "In order to make it to the top of the ratings in your particular market, you have to be doing the best job around. It's the PD who is going to get the station the numbers it needs to make a buck. If he doesn't turn the trick, he's back in the job market," observes Dick Fatherly.

SMALL MARKETS
- Program Director
- Airstaff

MEDIUM MARKETS
- Program Director
- Music Director
- Production Director
- News Director
- Engineering
- Promotion
- Airstaff

LARGE MARKETS
- Program Director
- Assistant Program Director
- Music Director
- Music Librarian
- Programming Secretary
- Research Director
- Production Director
- Promotion Director
- News Director
- Tech Assistant
- Marketing Director
- Airstaff

FIGURE 3.5 Personnel structure of the programming department in different size markets.

Program directors constantly monitor the competition by analyzing the ratings and by listening. A radio station's programming is often constructed in reaction to a direct competitor's. For example, rock stations in the same market often counterprogram newscasts by airing them at different times in order to grab up their competitor's tune-outs. However, rather than contrast with each other, pop stations tend to reflect one another. This, in fact, has been the basis of arguments by critics who object to the so-called "mirroring" effect. What happens is easily understood. If a station does well by presenting a particular format, other stations are going to exploit the sound in the hopes of doing well also. WYYY promotes commercial-free sweeps of music, and captures big ratings, and soon its competitor programs likewise. "Program directors use what has proven to be effective. It is more a matter of survival than anything. I think most of us try to be original to the degree that we can be, but there is very little new under the sun. Programming moves cyclically. Today we're all doing this. Tomorrow we'll all be doing that. The medium reacts to trends or fads. It's the nature of the beast," notes WQIK's Mitchell. Keeping in step with, or rather one step ahead of, the competition requires that the PD know what is happening around him or her at all times.

Nearly 85 percent of the nation's PDs pull an airshift (go on the air themselves) on either a full-time or part-time basis. A difference of opinion exists among programmers concerning their on-air participation. Many feel that being on the air gives them a true sense of the station's sound, which aids them in their programming efforts. Others contend that the three or four hours that they spend on the air take them away from important programming duties. Major market PDs are less likely to be heard on the air than their peers in smaller markets because of additional duties created by the size and status of the station. Meanwhile, small and medium market stations often expect their PDs to be seasoned air people capable of filling a key shift. "It has been my experience when applying for programming jobs that managers are looking for PDs with excellent announcing skills. It is pretty rare to find a small market PD who does not have a daily airshift. It comes with the territory," says WSUB's Gary Begin.

Whether PDs are involved in actual airshifts or not, almost all participate in the production of commercials, public service announcements, and promos. In lieu of an airshift, a PD may spend several hours each day in the station's production facilities. The programmer may, in fact, serve as the primary copywriter and spot producer. This is especially true at non–major market outlets that do not employ a full-time production person.

In summation, the program director must possess an imposing list of skills to effectively perform the countless tasks confronting him or her daily. There is no one person, other than the general manager, whose responsibilities outweigh the programmer's. The program director can either make or break the radio station.

ELEMENTS OF PROGRAMMING

Few programmers entrust the selection and scheduling of music and other sound elements to deejays and announcers. There is too much at stake and too many variables, both internal and external, which must be considered in order to achieve maximum results within a chosen format. In most cases, the PD determines how much music is to be programmed hourly and in what rotation, and when news, public affairs features, and commercials are to be slotted. Program wheels, also variously known as sound hours, hot clocks, and format disks, are carefully designed by the PD to ensure the effective presentation of on-air ingredients. Program wheels are posted in the control studio to inform and guide air people as to what is to be broadcast and at what point in the hour. Although not every station provides deejays with such specific programming schemata, today very few stations leave things up to chance since the inappropriate scheduling and sequencing of sound elements may drive listeners to a competitor. Radio programming has become that much of

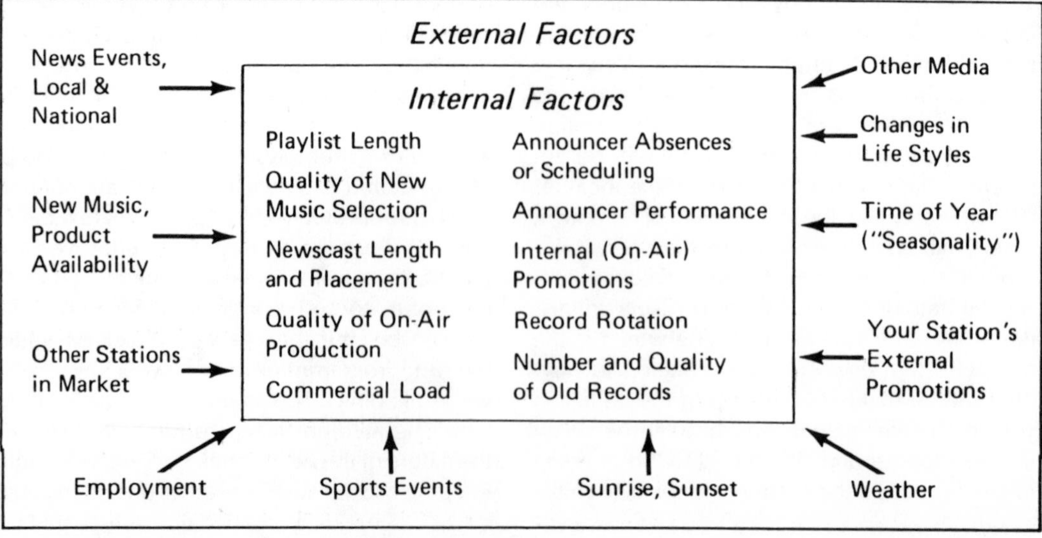

FIGURE 3.6
A model of a station's competitive environment as conceived by Arbitron. Courtesy Arbitron.

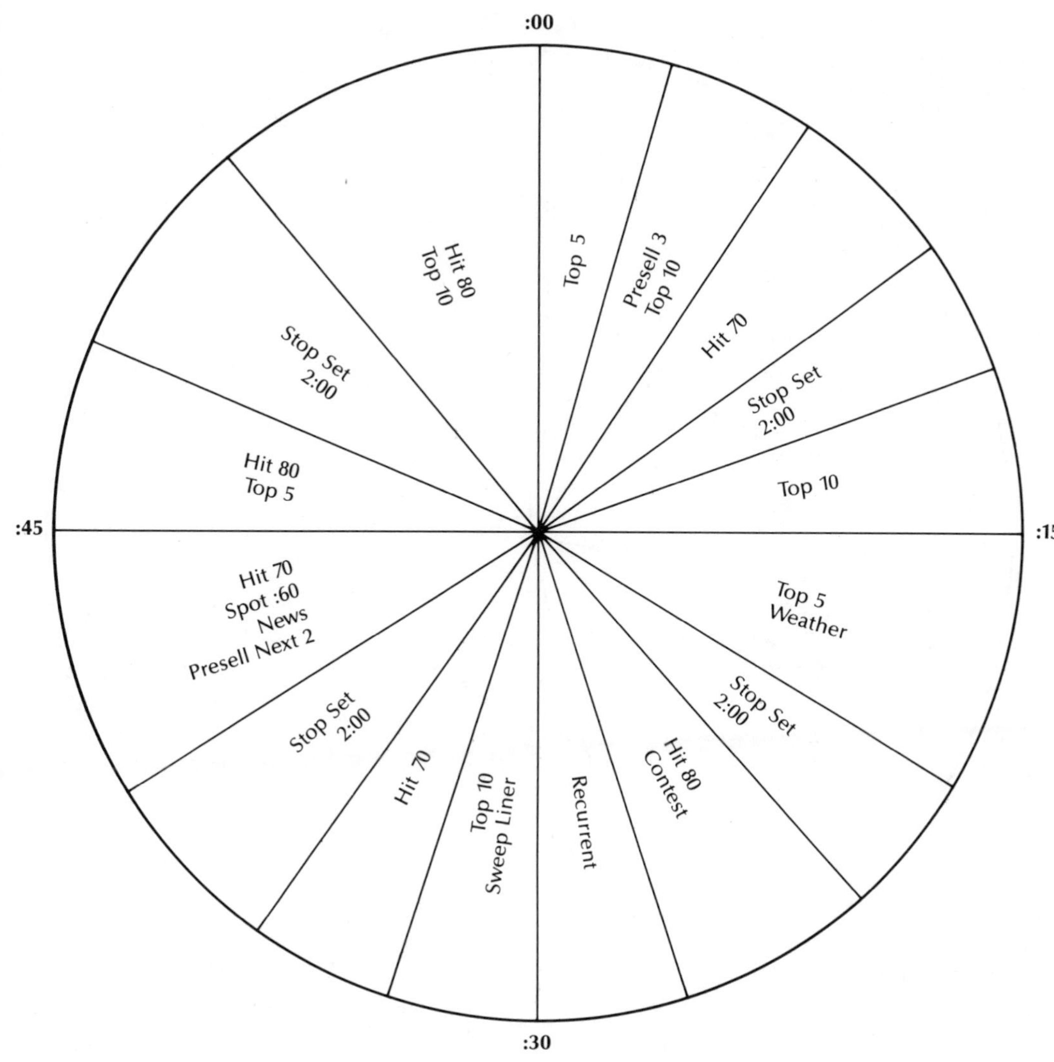

FIGURE 3.7
This is a typical morning drivetime CHR clock. It reflects a nine-minute commercial load maximum per hour. Notice that the stop sets occur away from the quarter-hour. A Top 5 record is often aired on the quarter-hour to give the station the best ratings advantage. This is called "quarter-hour maintenance."

an exacting science. With few exceptions, stations use some kind of formula in the conveyance of their programming material.

At one time, Top 40 stations were the unrivaled leaders of formula programming. Today, however, even MOR and AOR outlets, which once were the least formulaic, have become more sensitive to form. The age of free-form commercial radio has long since passed, and it is doubtful, given the state of the marketplace, that it will return. Of course, stranger things have happened in radio. Depending on the extent to which a PD prescribes the content of a sound hour, programming clocks may be elaborate in their detail or quite rudimentary. Music clocks are used to plot out elements. Clocks reflect the minutes of the standard hour, and the PD places elements where they actually are to occur during the hour. Many programmers use a set of clocks, or clocks that change with each hour.

Program clocks are set up with the competition and market factors in mind. For example, programmers will devise a clock that reflects morning and afternoon drive periods in their market. Not all markets have identical commuter hours. In some cities morning drive may start as early as 5:30 A.M., while in others it may begin at 7:00 A.M. The programmer sets up clocks accordingly. A clock parallels the activities of the community in which the station operates.

Music stations are not the only ones that use program wheels. News and Talk stations do so as well. News stations, like music outlets, use key format elements in order to maintain ratings

FIGURE 3.8
Easy Listening morning drivetime program wheel. Weather forecasts, time checks, sports scores, and traffic reports will be scheduled in and around spot sets.

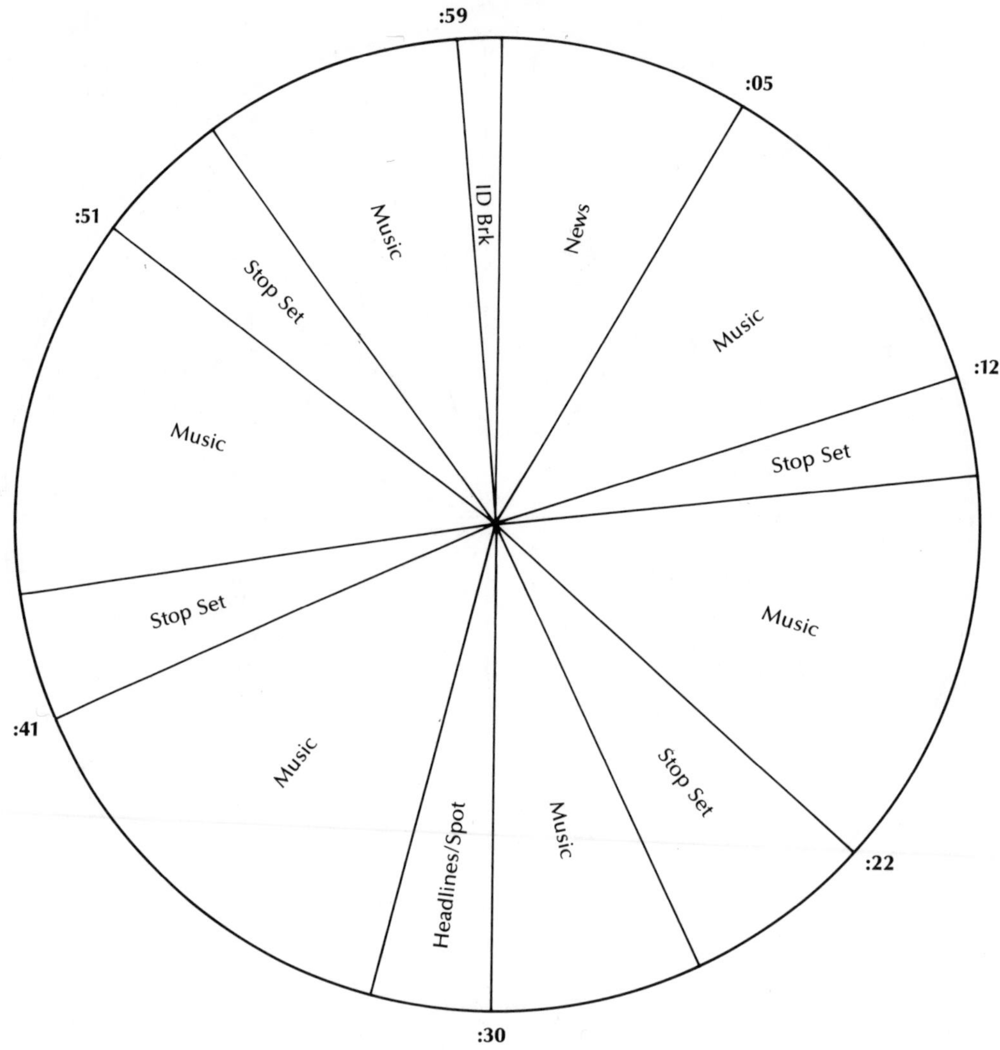

through the hour. Many news stations work their clocks in twenty-minute cycles. During this segment news is arranged according to its degree of importance and geographic relevance, such as local, regional, national, and international. Most News stations lead with their top local stories. News stories of particular interest are repeated during the segment. Sports, weather, and other news-related information, such as traffic and stock market reports, constitute a part of the segment. Elements may be juggled around or different ones inserted during successive twenty-minute blocks to keep things from sounding repetitious.

In the Talk format two-way conversation and interviews fill the space generally allotted to songs in the music format. Therefore, Talk wheels often resemble music wheels in their structure. For example, news is offered at the top of the hour followed by a talk sweep that precedes a spot set. This is done in a fashion that is reminiscent of Easy Listening.

Of course, not all stations arrange their sound hours as depicted in these pages. Many variations exist, but these examples are fairly representative of some of the program schematics used in today's radio marketplace.

Program wheels keep a station on a preordained path and prevent wandering. As stated, each programming element—commercial, news, promo, weather, etc.—is strategically located in the sound hour to enhance flow and optimize impact. Balance is imperative. Too much deejay patter on a station promoting more music–less talk and listeners become disenchanted. Too little news and information on a

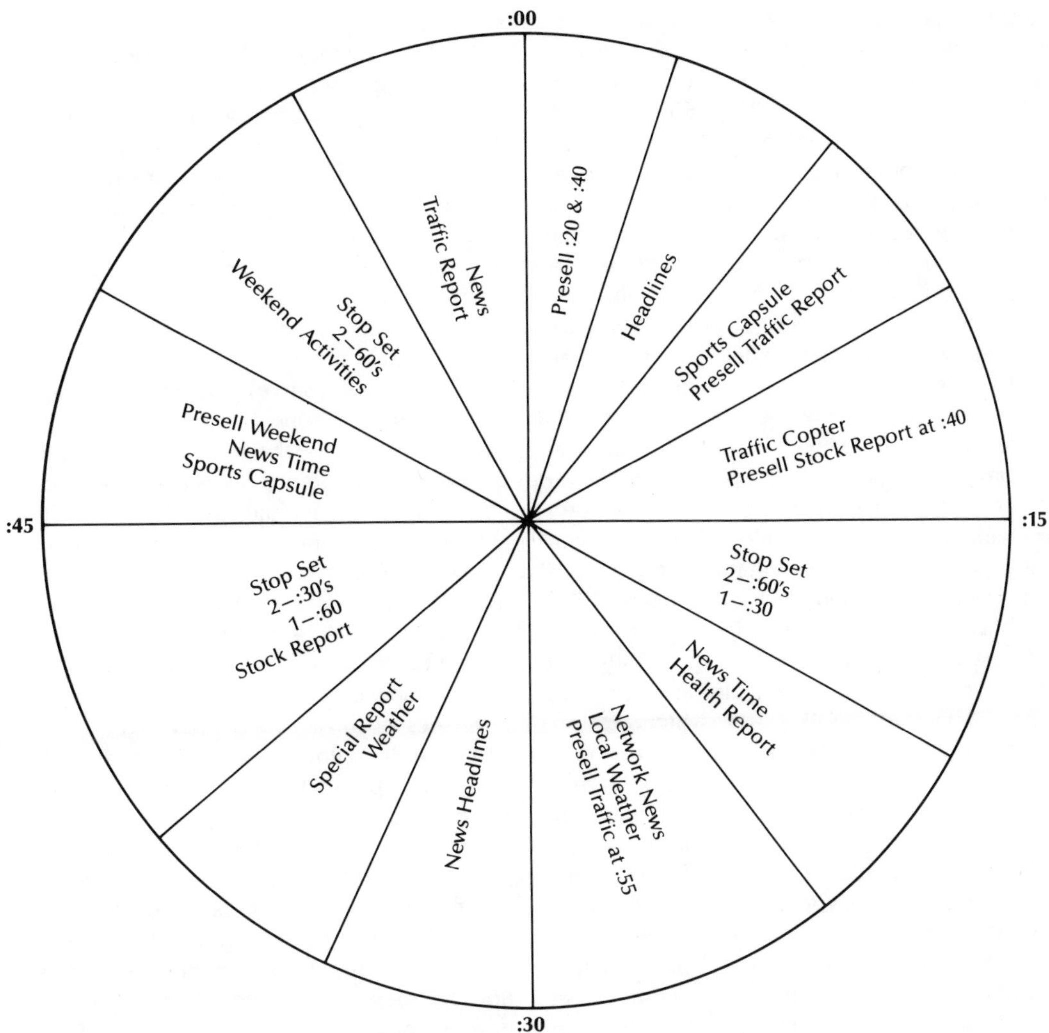

FIGURE 3.9
One approach to a News station sound hour.

station targeting the over-thirty male commuter and the competition benefits. "When constructing or arranging the program clock you have to work forward and backward to make sure that everything fits and is positioned correctly. One element out of place can become that proverbial hole in the dam. Spots, jock breaks, music—it all must be weighed before clocking. A lot of experimentation, not to mention research, goes into this," observes Lorna Ozman, program director, WROR-FM, Boston.

It was previously pointed out that a station with a more-music slant will limit announcer discourse in order to schedule additional tunes. Some formats, in particular Easy Listening, have reduced the role of the announcer to not much more than occasional live promos and IDs, which are written on liner or flip cards. Nothing is left to chance. This also is true of stations airing the supertight hit music format. Deejays say what is written and move the music. At stations where deejays are given more control, wheels play a less crucial function. Outlets where a particular personality has ruled the ratings for years often let that person have more input as to what music is aired. However, even in these cases, playlists generally are provided and followed.

The radio personality has enjoyed varying degrees of popularity since the 1950s. Over the years, Top 40, more than any other format, has toyed with the extent of deejay involvement on the air. The pendulum has swung from heavy personality presence in the 1950s and early 1960s to a drastically reduced role in the mid and later 1960s. This dramatic shift came as the result of programmer Bill Drake's attempt to streamline Top 40. In the 1970s, the air personality regained some of his status, but in the 1980s the narrowing of hit station playlists brought about a new leanness and austerity which again diminished the jock's presence. In the mid-1980s, some pop music stations began to give the deejay more to do. "There's sort of a pattern to it all. For a while, deejays are the gems in the crown, and then they're just the metal holding the precious stones in place for another period of time. What went on in the mid-1970s with personality is beginning to recur in the latter part of the 1980s. Of course, there are a few new twists in the tiara, but what it comes down to is the temporary restoration of the hit radio personality. It's a back and forth movement, kind of like a tide. It comes in, then retreats, but each time something new washes up. Deejays screamed at the teens in the 1950s, mellowed out some in the 1960s and 1970s, and have gone hyper again in the 1980s, but conservatism seems to be gaining favor once again," observes WQIK's Brian Mitchell.

In addition to concentrating on the role deejays play in the sound hour, the PD pays careful attention to the general nature and quality of other ingredients. Music is, of course, of paramount importance. Songs must fit the format to begin with, but beyond the obvious, the quality of the artistry and the audio mix must meet certain criteria. A substandard musical arrangement or a disk with poor fidelity detracts from the station's sound. Jingles and promos must effectively establish the tone and tenor of the format, or they have the reverse effect of their intended purpose, which is to attract and hold listeners. Commercials, too, must be compatible with the program elements that surround them.

In all, the program director scrutinizes every component of the program wheel to keep the station true to form. The wheel helps maintain consistency, without which a station cannot hope to cultivate a following. Erratic programming in today's highly competitive marketplace is tantamount to directing listeners to other stations.

THE PROGRAM DIRECTOR AND THE AUDIENCE

The programmer must possess a clear perception of the type of listener the station management wants to attract. Initially, a station decides on a given format because it is convinced that it will make money with the new-found audience, meaning that the people who tune in to the station will look good to prospective advertisers. The purpose of any format is to win a desirable segment of the radio audience. Just who these people are and what makes them tick are questions that the PD must constantly address in order to achieve reach and retention. An informed programmer is aware that different types of music appeal to different types of people. For example, surveys have long concluded that heavy rock appeals more to men than it does to women, and that rock music, in general, is more popular among teens and young adults than it is with individuals over forty. This is no guarded secret, and certainly the programmer who is out to gain the over-forty crowd is doing himself and his station a disservice by programming even an occasional hard rock tune. This should be obvious.

A station's demographics refers to the characteristics of those tuned—sex, age, income, and so forth. Within its demographics a station may exhibit particular strength in specific areas, or "cells" as they have come to be termed. For example, an Adult Contemporary station targeting the twenty-four- to thirty-nine-year-old group may have a prominent cell in females over thirty. The general information provided by the major ratings surveys indicate to the station the age and sex of those listening, but little beyond that. To find out more, the PD may conduct an in-house survey or employ the services of a research firm.

Since radio accompanies listeners practically everywhere, broadcasters pay particular attention to the lifestyle activities of their target audience. A station's geographic locale often dictates its program offerings. For example, hoping to capture the attention of the thirty-five-year-old male, a radio outlet located in a small coastal city along the Gulf of Mexico might decide to air a series of one-minute informational tips on outdoor activities, such as tennis, golf, and deep-sea fishing, that are exceptionally popular in the area. Stations have always catered to the interests of their listeners, but in the 1970s audience research became much more lifestyle oriented.

In the 1980s, broadcasters have delved further into audience behavior through psychographic research, which, by examining motivational factors, provides programmers with

FIGURE 3.10
Radio reflects the lifestyles of its users. Courtesy Westinghouse Electric.

information beyond the purely quantitative. Perhaps one of the best examples of a station's efforts to conform to the lifestyle of its listeners is "dayparting," a topic briefly touched upon in the discussion of program wheels. For the sake of illustration, let us discuss how an MOR station may daypart (segmentalize) its broadcast day. To begin with, the station is targeting an over-forty audience, somewhat skewed toward males. The PD concludes that the station's biggest listening hours are mornings between seven and nine and afternoons between four and six, and that most of those tuned during these periods are in their cars commuting to or from work. It is evident to the programmer that the station's programming approach must be modified during drivetime to reflect the needs of the audience. Obviously traffic reports, news and sports updates, weather forecasts, and frequent time checks are suitable fare for the station's morning audience. The interests of homebound commuters contrast slightly with those of workbound commuters. Weather and time are less important, and most sports information from the previous night is old hat by the time the listener heads for home. Stock-market reports and information about upcoming games and activities pick up the slack. Midday hours call for further modification, since the lifestyle of the station's audience is different. Aware that the majority of those listening are homemakers (in a less enlightened age this daypart was referred to as "housewife" time), the PD reduces the amount of news and information, replacing them with music and deejay conversation designed specifically to complement the activities of those tuned. In the evening, the station redirects its programming and schedules sports and talk features, going exclusively talk after midnight. All of these adjustments are made to attract and retain audience interest.

The program director relies on survey information and research data to better gauge and understand the station's audience. However, as a member of the community that the station serves, the programmer knows that not everything is contained in formal documentation. He or she gains unique insight into the mood and mentality of the area within the station's signal simply by taking part in the activities of day-to-day life. A programmer with a real feel for the area in which the station is located, as well as a fundamental grasp of research methodology and its application, is in the best possible position to direct the on-air efforts of a radio station. Concerning the role of audience research, WGAN's Peter Falconi says, "You can't run a station on research alone. Yes, research helps to an extent, but it can't replace your own observations and instincts." WQIK's Brian Mitchell agrees with Falconi. "I feel research is important, but how you react to research is more important. A PD also has to heed his gut feelings. Gaps exist in research, too. If I can't figure out what to do without data to point the way every time I make a move, I should get out of radio. Success comes from taking chances once in a while, too. Sometimes it's wiser to turn your back on the tried and tested. Of course, you had better know who's out there before you try anything. A PD who doesn't study his audience and community is like a race car driver who doesn't familiarize himself with the track. Both can end up off the road and out of the race."

THE PROGRAM DIRECTOR AND THE MUSIC

Not all radio stations have a music director. The larger the station, the more likely it is to have such a person. In any case, it is the PD who is ultimately responsible for the music that goes over the air, even when the position of music director exists. The duties of the music director vary from station to station. Although the title suggests that the individual performing this function would supervise the station's music programming from the selection and acquisition of records to the preparation of playlists, this is not always the case. At some stations, the position is primarily administrative or clerical in nature, leaving the PD to make the major decisions concerning airplay. In this instance, one of the primary duties of the music director might be to improve service from record distributors to keep the station well supplied with the latest releases. A radio station with poor record service may actually be forced to purchase music. This can be prevented to a great extent by maintaining close ties with the various record distributor reps.

When albums arrive at the station, the music director (sometimes more appropriately called the music librarian or music assistant) processes them through the system. This may take place after the program director has screened them. Records are categorized, indexed, and eventually added to the library if they suit the station's format. Each station approaches cataloguing in its own fashion. Here is a simple example: an Adult Contemporary outlet receives an album by a popular female vocalist whose last name begins with an *L*. The program director auditions the album and decides to place three cuts into regular on-air rotation. The music director then assigns the cuts the following catalogue numbers: L106/U/F, L106/D/F, and L106/M/F. L106 indicates where the album may be located in the library. In this case, the library is set up alphabetically, then numerically within the given letter that represents the artist's last name. In other words, this would be the 106th album found in the section reserved for female vocalists whose names begin with an *L*. The next symbol indicates the tempo of the cut: U(p) tempo, D(own) tempo, and M(edium) tempo. The *F* that follows the tempo symbol indicates the artist's gender: Female.

When a station is computerized, this information, including the frequency or rotation of airplay as determined by the PD, will be entered accordingly. Playlists are then assembled and printed by the computer. The music director sees that these lists are placed in the control room for use by the deejays. This last step is eliminated when the on-air studio is equipped with a computer terminal. Deejays then simply punch up the playlists designed for their particular airshifts.

The use of computers in music programming has become widespread, especially in larger markets where the cost of computerization is absorbed more easily. The number of computer companies selling both hardware and software designed for use by programmers has soared.

****
* MORN ZOO *
* 7 AM FRIDAY 8/17/84 *
****

CART	TITLE	ARTIST	INTRO/E	RUNTM	
	WRCS RADIO - PROGRAMMED EXCLUSIVELY BY SELECTOR				
607-	**GOING MOBILE**	WHO	/14/C	3:40	M
664-	**SHE BOP**	CYNDI LAUPER	8/14/C	3:45	A
	BREAK NO. 1, WEATHER, LINER PROMO UPCOMING BIT			4:30	
62-	**I THINK I M IN LOVE**	EDDIE MONEY	/18/F	3:10	E
746-	**DANCE HALL DAYS**	WANG CHUNG	7/14/S	3:34	D
	SWEEPER				
789-	**ROCK N ROLL WOMAN**	BUFFALO SPRINGFIELD	/ 7/S	2:44	G
	BREAK NO. 2 -B.A., LINER OR PROMO, NEWS & TRAF			6:00	
77-	**BACK ON THE CHAIN GANG**	PRETENDERS	/25/F	3:44	I
771-	**IF THIS IS IT**	HUEY LEWIS	/ /F	3:46	A
	BREAK NO. 3 -MAJOR BIT --WRCS ROCKIN SPECIAL			6:00	
777-	**ARE WE OURSELVES**	FIXX	/16/F	2:27	C
266-	**HOTEL CALIFORNIA**	EAGLES	/52/F	6:30	F
55:00	BREAK NO. 4 -NEWS & TRAFFIC			5:00	

TOTAL TIME IS 60:00

FIGURE 3.11
Computerized music logging has made station programming an even more exacting science. This is just one of the possible "logs" you can print. Note the song titles in bold type as well as the "intro times," duration, and how the records end (outcue). Courtesy RCS.

Among those providing computerized music systems are Computer Concepts Corporation, Station Research Systems, Jefferson Pilot Data Systems, Radio Computing Systems, Rah Rah Productions, Columbine Systems. Also, some companies, such as Billboard, provide up-to-date computerized music research on new singles and albums to assist station programmers.

At those radio stations where the music director's job is less administrative and more directorial, this individual will actually audition and select what songs are to be designated for airplay. However, the music director makes decisions based upon criteria established by the station's programmer. Obviously, a music director must work within the station's prescribed format. If the PD feels that a particular song does not fit the station's sound, he will direct the music director to remove the cut from rotation. Since the PD and music director work closely together, this seldom occurs.

A song's rotation usually is relative to its position on national and local record sales charts. For instance, songs that enjoy top ranking, say those in the top ten, will get the most airplay on hit-oriented stations. When songs descend the charts, their rotation decreases proportionately. Former chart toppers are then assigned another rotation configuration that initially may result in one-tenth of their former airplay and eventually even less. PDs and music directors derive information pertaining to a record's popularity from various trade journals, such as *Billboard* and *Radio and Records,* listener surveys, area record store sales, and numerous other sources. Stations that do not program from the current charts compose their playlists of songs that were popular in years gone by. In addition,

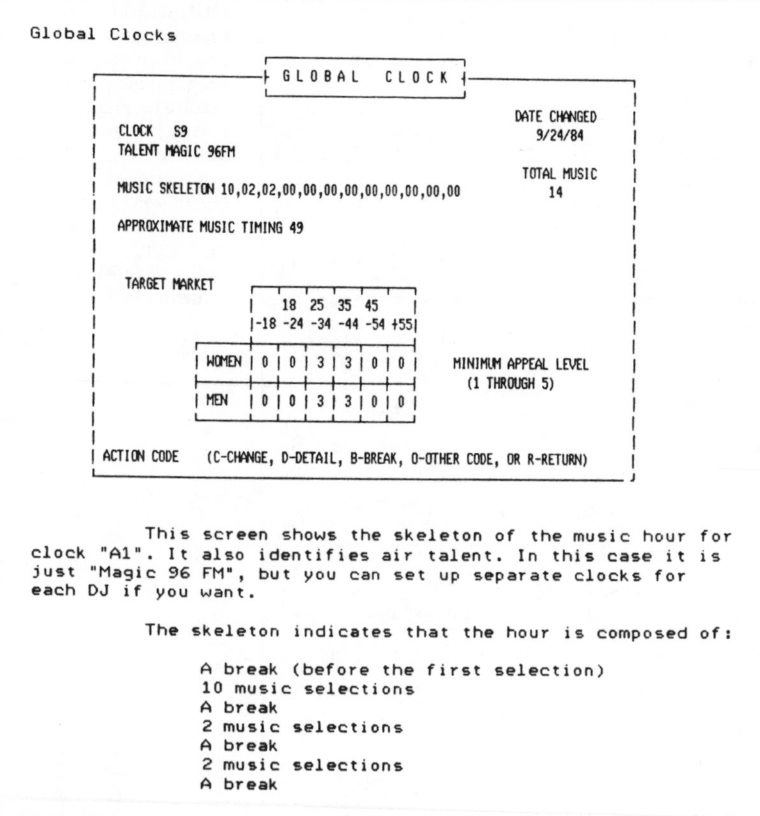

```
Global Clocks

┌──────────────┤ G L O B A L   C L O C K ├──────────────┐
|                                                        |
|                                          DATE CHANGED  |
| CLOCK  S9                                  9/24/84    |
| TALENT MAGIC 96FM                                      |
|                                          TOTAL MUSIC   |
| MUSIC SKELETON 10,02,02,00,00,00,00,00,00,00,00,00     14 |
|                                                        |
| APPROXIMATE MUSIC TIMING 49                            |
|                                                        |
|  TARGET MARKET                                         |
|              |  18  25  35  45     |                   |
|              |-18 -24 -34 -44 -54 +55|                 |
|                                                        |
|        | WOMEN | 0 | 0 | 3 | 3 | 0 | 0 |  MINIMUM APPEAL LEVEL |
|                                           (1 THROUGH 5) |
|        | MEN   | 0 | 0 | 3 | 3 | 0 | 0 |               |
|                                                        |
| ACTION CODE   (C-CHANGE, D-DETAIL, B-BREAK, O-OTHER CODE, OR R-RETURN) |
└────────────────────────────────────────────────────────┘

        This screen shows the skeleton of the music hour for
clock "A1". It also identifies air talent. In this case it is
just "Magic 96 FM", but you can set up separate clocks for
each DJ if you want.

        The skeleton indicates that the hour is composed of:

            A break (before the first selection)
            10 music selections
            A break
            2 music selections
            A break
            2 music selections
            A break
```

FIGURE 3.12 (above right) All information about a given record can be called up on the screen, as on the display shown. Courtesy RCS.

FIGURE 3.13 (above) In computer terms here is how an hourly clock might look. Courtesy RCS.

these stations often remix current hit songs to make them adaptable to their more conservative sound. While Easy Listening stations do not air popular rock songs, they do air softened versions by other artists, usually large orchestras. Critics of this technique accuse the producers of lobotomizing songs to bring them into the fold. In nonhit formats there are no "power-rotation" categories or hit positioning schemata; a song's rotation tends to be more random, although program wheels are used.

Constructing a station playlist is the single most important duty of the music programmer. What to play, when to play it, and how often are some of the key questions confronting this individual. The music director relies on a number of sources, both internal and external, to provide the answers, but also must cultivate an ear for the kind of sound the station is after. Some people are blessed with an almost innate capacity to detect a hit, while most must develop this skill over a period of time.

```
                         L I B R A R Y   I N F O R M A T I O N
-------------------------------------------------------------------------------
  CART     CATEGORY  LEVEL  PACKET                 TITLE                 LICENSE
   664        A        1       0       SHE BOP                       655     A
-------------------------------------------------------------------------------
     ARTIST ONE                          ROLE              ARTIST TWO
  CYNDI LAUPER                   113      FV
-------------------------------------------------------------------------------
MOOD ........  3 |  APPEAL      |  C H A R T  |       D A Y P A R T
TEMPO  ......  MM |             |             |   R E S T R I C T I O N S
TIMBRE ......    |  AGE   M F   |             |   1          111         11
SOUND CODES FR I |  >18         | CURRENT     |   212345678901212345678901
OPENER ....... Y | 18-24        | LAST ...        MAAAAAAAAAAANPPPPPPPPPPP
TYPE .........   | 24-34        | PEAK ...    | M-F         NNNNN
KEY .....     -  | 34-44        |             | SAT         NNNNN
RUNTIME ... 3:45 | 45-54        |  PEAKED IN  | SUN         NNNNN
INTROTIME   8/14 |  55>         |    6/84     ----------------------------
ENDING ....... C |              |             |    DAYPART ROTATION 5252
-------------------------------------------------------------------------------
            DATES          PLAYS IN  |     M O V E   I N F O R M A T I O N
ENTERED    7/20/84         CATEGORY  |    DATE          PLAYS  CAT  LEVEL  PK
LAST PLAY  8/23/84 10 PM     25      |    0/ 0/ 0   OR    0            1     0
-------------------------------------------------------------------------------
```

FIGURE 3.14 Economical, easy-to-use, in-house systems eliminate old-fashioned, time-consuming index card file method. Courtesy Jefferson Pilot Data Systems.

WRCS 10/ 1/84 — C A T E G O R Y D I R E C T O R Y — PAGE 1

CT	PK	LV	CART-ST	T I T L E	A R T I S T	RL MO	OP	TIM TEM	SCSC	TYP CUR	LAST PEAK / PKDAT	RNTM	INTRO/E	DATE ENT.
A	1	0	660	DRIVE	CARS	1		SS	R	I	1/84	3:55	11/24/S	7/26/84
A	1	0	714	I SEND A MESSAGE	INXS	2	Y	MM	BR		7/84	3:23	7/12/C	8/ 9/84
A	1	0	664	SHE BOP	CYNDI LAUPER	3	Y	MM	FR	I	6/84	3:45	8/14/C	7/20/84
A	1	0	771	IF THIS IS IT	HUEY LEWIS	2	Y	MM	WT	I	7/84	3:46	/ /F	7/26/84
A	1	0	747	WHEN YOU CLOSE YOUR EYES	NIGHTRANGER	2	Y	SS	WR		6/84	4:07	/26/F	7/26/84
A	1	0	680	ROUND AND ROUND	RATT	2	Y	MM	R		7/84	4:22	0/16/F	7/11/84
A	1	0	650	DISTANT EARLY WARNING	RUSH	2	Y	MM	RP	I	1/84	4:56	0/34/S	7/20/84
A	1	0	715	THE WARRIOR	SCANDAL	2	Y	MM	FR		7/84	4:03	5/16/F	7/26/84
A	1	0	706	COVER ME	BRUCE SPRINGSTEEN	3	Y	MM	R	I	6/84	3:26	/ 9/F	7/26/84
A	1	0	713	MISSING YOU	JOHN WAITE	1		SS	WT		6/84	4:31	0/19/F	7/20/84

COUNT FOR CATEGORY AND LEVEL IS 10

END OF REPORT

WRCS 9/24/84 — A R T I S T D I R E C T O R Y — PAGE 1

CART	T I T L E		C	PK	LV	RL M	O	TE TY	SCSC TM		RUNTM	INTRO/E
CARS		(8)										
660	DRIVE		A	1	0	1		SS	R	I	3:55	11/24/S
INXS		(90)										
714	I SEND A MESSAGE		A	1	0	2	Y	MM	BR		3:23	7/12/C
CYNDI LAUPER		(113)										
664	SHE BOP		A	1	0	3	Y	MM	FR	I	3:45	8/14/C
HUEY LEWIS		(72)										
771	IF THIS IS IT		A	1	0	2	Y	MM	WT	I	3:46	/ /F
NIGHTRANGER		(163)										
747	WHEN YOU CLOSE YOUR EYES		A	1	0	2	Y	SS	WR		4:07	/26/F
RATT		(157)										
680	ROUND AND ROUND		A	1	0	2	Y	MM	R		4:22	0/16/F
RUSH		(120)										
650	DISTANT EARLY WARNING		A	1	0	2	Y	MM	RP	I	4:56	0/34/S
SCANDAL		(167)										
715	THE WARRIOR		A	1	0	2	Y	MM	FR		4:03	5/16/F
BRUCE SPRINGSTEEN		(43)										
706	COVER ME		A	1	0	3	Y	MM	R	I	3:26	/ 9/F
JOHN WAITE		(165)										
713	MISSING YOU		A	1	0	1		SS	WT		4:31	0/19/F

END OF REPORT

FIGURE 3.15
Computers can program music and keep music library records as well. Many types of music directories can be established, such as category, artist, title, and cart (cartridge number). Courtesy RCS.

FIGURE 3.15
continued

```
WRCS 10/ 1/84            T I T L E   D I R E C T O R Y          PAGE    1

CART-ST       T I T L E                    C  PK   OP  TE    SC     LCH    PKDT
                                            LV  MD  TY   TI      CCH   PCH
    A R T I S T                                         RNTM INTRO/E
------------------------------------------------------------------------------

 765     BABY PLS DONT LET ME GO           B 1 0 3 Y   MM     R                8/84
PETER WOLF                    J GEILS.BAND               4:02  5/16/F

 772     CAGE OF FREEDOM                   B 1 0 1     MM     P                1/84
JON ANDERSON                  YES                        4:04   /14/F

 748     CRACK ME UP                       B 1 0 3 Y   MM     R   I            6/84
HUEY LEWIS                                               3:39   /15/F

 659     I LOVE YOU SUZANNE                B 1 0 2 Y   MM     R                6/84
LOU REED                                                 3:15   / 0/F

 754     JERSEY GIRL (LIVE)                B 1 0 1     SS     Y   I            8/84
BRUCE SPRINGSTEEN                                        6:40   /33/F

 774     LAYIN IT ON THE LINE              B 1 0 2 Y   MF     R   I            1/84
JEFFERSON.STARSHIP                                       4:09  8/15/F

 750     LEAVE MY KITTY ALONE              B 1 0 2     MM     T                1/64
BEATLES                                                  2:45   / 5/F

 762     NEW ROMEO                         B 1 0 3 Y   MM     R                7/84
SOUTHSIDE.JOHNNY                                         3:20   /14/F

 709     NO SURRENDER                      B 1 0 3 Y   MF     R   I            6/84
BRUCE SPRINGSTEEN                                        4:00   / 3/F

 760     RESTLESS                          B 1 0 2 Y   MM     R                6/84
ELTON JOHN                                               5:14   /32/F
         END OF REPORT
```

THE PROGRAM DIRECTOR AND THE FCC

The government is especially interested in the way a station conducts itself on the air. For instance, the program director makes certain that his station is properly identified once an hour, as close to the top of the hour as possible. The ID must include the station's call letters and the town in which it has been authorized to broadcast. Failure to properly identify the station is a violation of FCC rules.

Other on-air rules that the PD must address have to do with program content and certain types of features. For example, profane language, obscenity, sex- and drug-related statements, and even innuendos in announcements, conversations, or music lyrics jeopardize the station's license. Political messages and station editorials are carefully scrutinized by the programmer and equal time is afforded opponents. On-air contests and promotions must not resemble lotteries in which the audience must invest to win. A station that gets something in return for awarding prizes is subject to punitive actions. Neither the deejays, PD, music director, nor anyone associated with the station may receive payment for plugging a song or album on the air. This constitutes "payola" or "plugola" and was the cause of great industry upheaval in the 1950s. Today, PDs and station managers continue to be particularly careful to guard against any recurrence, although there have been charges that such practices still exist.

The program director must monitor both commercial and noncommercial messages to ensure that no false, misleading, or deceptive statements are aired, something the FCC staunchly opposes. This includes any distortion of the station's ratings survey results. A station that is not number one and claims to be is lying to the public as far as the FCC is concerned, and such behavior is not condoned.

License renewal programming promises must

WRCS 10/ 1/84 CART DIRECTORY PAGE 1

CART	TITLE	CAT	PK	LEV	ARTIST	RL MD	OP TEM	TIM	SCSC	T Y	CR	LS	PK	CHART MM/YY	RUNTM	INTRO/E	DATE ENT CATEGORY
1	BACK IN BLACK	X	1	0	AC.DC	2	Y	MM	H					1/81	4:13	/26/F	8/17/84
2	THIS TIME	E	1	0	BRYAN ADAMS	3	Y	MM	T					1/83	3:18	/18/S	5/27/84
3	DER KOMISSAR	I	1	0	AFTER.THE.FIRE	2	Y	MM	P					1/83	5:43	/64/F	7/20/84
4	ONLY TIME WILL TELL	E	1	0	ASIA/YES	1		MM	WT					1/82	4:44	/31/F	5/30/84
5	BRINGING THE HEARTBREAK	E	1	0	DEF.LEPPARD	1	Y	MM	H	I				1/81	4:33	/26/S	7/13/84
6	HIT ME WITH YR BEST SHOT	E	1	0	PAT BENATAR	3	Y	MM	F	I				1/81	2:51	/16/C	5/30/84
7	LOVE IS A BATTLEFIELD	E	1	0	PAT BENATAR	3	Y	MM	F	I				1/83	4:00	/19/F	5/30/84
8	FIRE AND ICE	E	1	0	PAT BENATAR	2	Y	MM	F	I				1/81	3:20	/14/F	5/30/84
9	MODERN LOVE	I	1	0	DAVID BOWIE	3	Y	MM		I				1/83	4:46	0/16/F	7/25/84
10	LETS DANCE	I	1	0	DAVID BOWIE	3	Y	MM		I				1/83	7:38	0/24/F	7/26/84
11	BOULEVARD	I	1	0	JACKSON BROWNE	2	Y	MM		I				1/81	3:15	/13/F	8/17/84
12	FOR A ROCKER	X	1	0	JACKSON BROWNE	0	Y	MM		I				1/83	4:05	/ 7/F	7/20/84
13	I GET AROUND	G	1	0	BEACH BOYS	3	Y	MM	T	I				1/64	2:20	/ 1/F	6/19/84
14	ROCK THE CASBAH	I	2	0	CLASH	3	Y	MM		I				1/83	3:42	0/15/F	8/17/84
15	CRUMBLING DOWN	E	1	0	JOHN COUGAR	3	Y	MM		I				1/83	3:33	/ 7/F	5/30/84
16	HURTS SO GOOD	I	1	0	JOHN COUGAR	2	Y	MM		I				1/82	3:35	/28/F	5/30/84
17	JACK AND DIANE	E	1	0	JOHN COUGAR	2	Y	MM		I				1/82	4:16	/28/F	5/30/84
18	SOUTHERN CROSS	E	1	0	CSN	1		SS	YW	I				1/82	4:40	/11/S	5/30/84
19	PHOTOGRAPH	E	1	0	DEF.LEPPARD	2	Y	MM	H	I				1/83	4:12	/15/F	5/30/84
20	FOOLIN	E	1	0	DEF.LEPPARD	1	Y	SM	H	I				1/83	4:32	/13/S	5/30/84
21	ROCK OF AGES	I	1	0	DEF.LEPPARD	2	Y	MM	H	I				1/83	4:05	/ 0/S	8/17/84
22	THE WEIGHT	G	2	0	BAND	2		SS	Y					1/72	4:40	/10/S	8/ 8/84
23	HUNGRY LIKE THE WOLF	X	1	0	DURAN DURAN	2	Y	MM	P	I				1/82	3:04	/30/F	7/26/84
24	UNION OF THE SNAKE	X	1	0	DURAN DURAN	2	Y	MM	BT	I				1/83	4:50	/19/F	7/16/84
25	1 THING LEADS TO ANOTHER	I	3	0	FIXX	3	Y	MM	T	I				1/83	3:18	/14/F	7/26/84
26	I RAN	X	1	0	FLOCK.OF.SEAGULLS	3	Y	FF	P					1/82	3:58	/26/S	7/16/84
27	WAITING 4 A GIRL LIKE U	E	1	0	FOREIGNER	1		SS	WT	I				1/81	4:44	/28/F	5/30/84

FIGURE 3.15
continued

be addressed by the PD. Time allocated hourly to commercial material should not be exceeded unless the station notifies the FCC that it is doing so for a specific period of time. The proportion of nonentertainment programming, such as news and public affairs features, pledged in the station's renewal application must be adhered to, even though such requirements have been all but eliminated. A promise is a promise. If a station claims that it will do something, it must abide by its word.

The PD helps maintain the station's Emergency Broadcast System (EBS), making certain that proper announcements are made on the air and that the EBS checklist containing an authenticator card is placed in the control room

FIGURE 3.16 Restricted Radiotelephone Operator Permit required by the FCC for those operating a broadcast facility. Permit is good for an indefinite period.

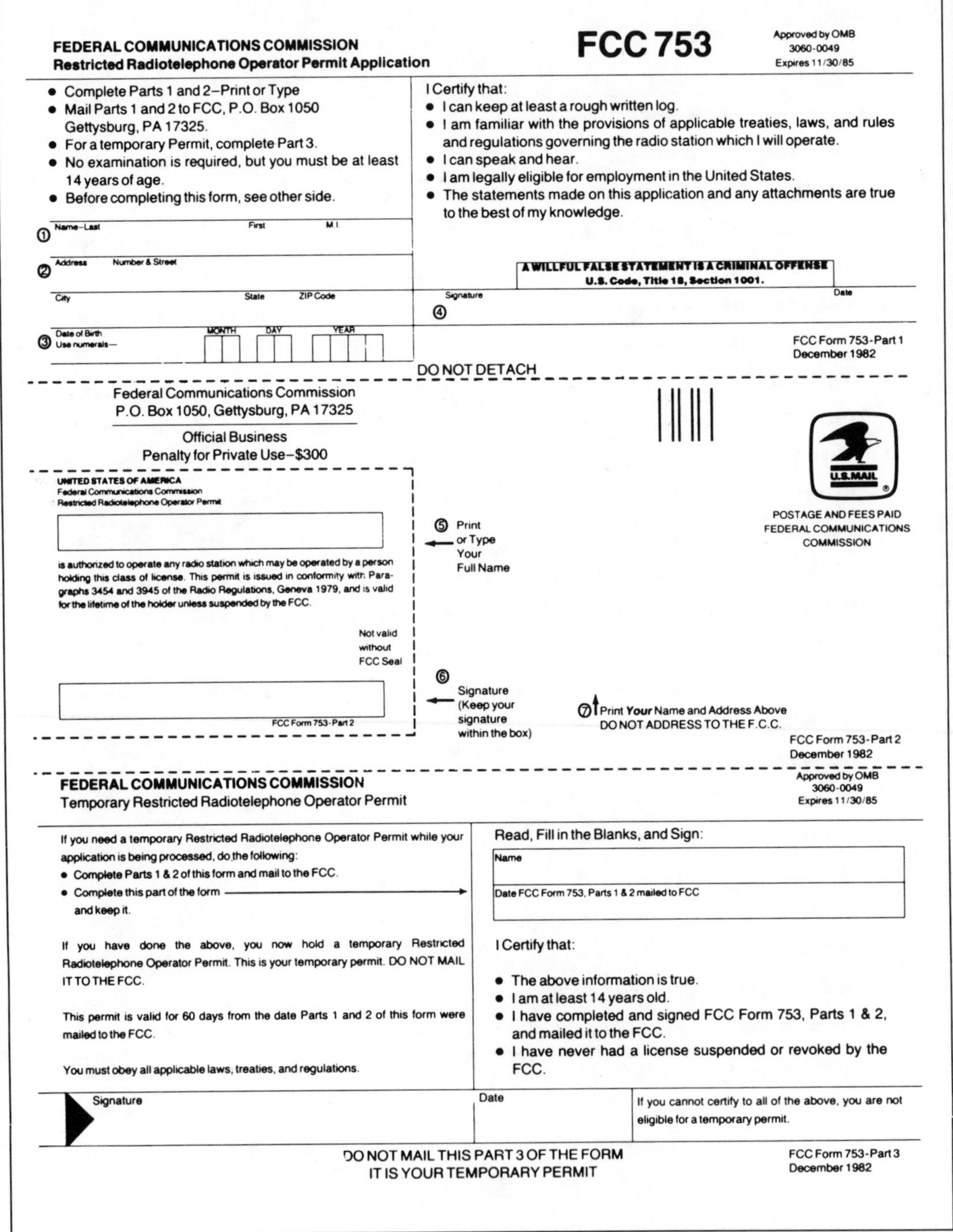

FEDERAL COMMUNICATIONS COMMISSION
Restricted Radiotelephone Operator Permit Application

FCC 753

Approved by OMB 3060-0049 Expires 11/30/85

- Complete Parts 1 and 2–Print or Type
- Mail Parts 1 and 2 to FCC, P.O. Box 1050 Gettysburg, PA 17325.
- For a temporary Permit, complete Part 3.
- No examination is required, but you must be at least 14 years of age.
- Before completing this form, see other side.

I Certify that:
- I can keep at least a rough written log.
- I am familiar with the provisions of applicable treaties, laws, and rules and regulations governing the radio station which I will operate.
- I can speak and hear.
- I am legally eligible for employment in the United States.
- The statements made on this application and any attachments are true to the best of my knowledge.

① Name–Last First M.I.

② Address Number & Street

City State ZIP Code

③ Date of Birth Use numerals— MONTH DAY YEAR

A WILLFUL FALSE STATEMENT IS A CRIMINAL OFFENSE
U.S. Code, Title 18, Section 1001.

Signature ④ Date

FCC Form 753-Part 1
December 1982

DO NOT DETACH

Federal Communications Commission
P.O. Box 1050, Gettysburg, PA 17325

Official Business
Penalty for Private Use–$300

POSTAGE AND FEES PAID
FEDERAL COMMUNICATIONS COMMISSION

U.S. MAIL

UNITED STATES OF AMERICA
Federal Communications Commission
Restricted Radiotelephone Operator Permit

is authorized to operate any radio station which may be operated by a person holding this class of license. This permit is issued in conformity with Paragraphs 3454 and 3945 of the Radio Regulations, Geneva 1979, and is valid for the lifetime of the holder unless suspended by the FCC.

Not valid without FCC Seal

FCC Form 753-Part 2

⑤ Print or Type Your Full Name

⑥ Signature (Keep your signature within the box)

⑦ Print Your Name and Address Above
DO NOT ADDRESS TO THE F.C.C.

FCC Form 753-Part 2
December 1982

FEDERAL COMMUNICATIONS COMMISSION
Temporary Restricted Radiotelephone Operator Permit

Approved by OMB 3060-0049 Expires 11/30/85

If you need a temporary Restricted Radiotelephone Operator Permit while your application is being processed, do the following:
- Complete Parts 1 & 2 of this form and mail to the FCC.
- Complete this part of the form and keep it.

If you have done the above, you now hold a temporary Restricted Radiotelephone Operator Permit. This is your temporary permit. DO NOT MAIL IT TO THE FCC.

This permit is valid for 60 days from the date Parts 1 and 2 of this form were mailed to the FCC.

You must obey all applicable laws, treaties, and regulations.

Read, Fill in the Blanks, and Sign:

Name

Date FCC Form 753, Parts 1 & 2 mailed to FCC

I Certify that:
- The above information is true.
- I am at least 14 years old.
- I have completed and signed FCC Form 753, Parts 1 & 2, and mailed it to the FCC.
- I have never had a license suspended or revoked by the FCC.

Signature | Date | If you cannot certify to all of the above, you are not eligible for a temporary permit.

DO NOT MAIL THIS PART 3 OF THE FORM
IT IS YOUR TEMPORARY PERMIT

FCC Form 753-Part 3
December 1982

area. PDs also instruct personnel in the proper procedures used when conducting on-air telephone conversations to guarantee that the rights of callers are not violated.

The station log is examined by the PD for accuracy, and he also must see to it that operators have permits and that they are posted in the on-air studio. In addition, the station manager may assign the PD the responsibility for maintaining the station's public file. If so, the PD must be fully aware of what the file is required to contain. The FCC and many broadcast associations will provide station operators with a public file checklist upon request. This

FIGURE 3.16
continued

Note: All U.S. citizens are considered, for the purposes of this application, to be legally eligible for employment in the United States.

If you are not a U.S. citizen, ***and*** you are not eligible for employment in the United States, but you need an operator permit because—

1. You hold an FAA pilot certificate and need to operate aircraft radio stations—

OR

2. You hold an FCC radio station license issued in your name, and will use the permit for operation of that particular station—

THEN use FCC Form 755 to apply for the permit, instead of this form.

Notice to Individuals Required by the Privacy Act and the Paperwork Reduction Act

The Communications Act of 1934 authorizes solicitation of personal information requested in this application. The information is to be used principally to determine if the benefits requested are consistent with the public interest, convenience and necessity.

Commission staff will routinely use the information to evaluate and render a judgement as to whether to grant or deny this application.

If all of the requested information is not provided, the application may be returned without action or processing may be delayed while a request is made to provide the missing information. Therefore, extreme care should be exercised in making certain that all the information requested is provided.

Limited file material may be included in the Commission Computer Facility. Where a possible violation of law is indicated, the records may, as a matter of routine use, be referred to the Commission's General Counsel and forwarded to other appropriate agencies charged with responsibilities of investigating or prosecuting such violations.

The Foregoing Notice is Required by the Privacy Act of 1974, p.l. 93-579, December 31, 1974, 5 U.S.C. 552a (e) (3), and the Paperwork Reduction Act of 1980, P.L. 96-511, December 11, 1980, 44 U.S.C. 3504 (C) (3) (C).

(cut on this line)

Terms and Conditions

It is your responsibility to know the laws, treaties, rules, and regulations which currently govern any station you operate. Do not operate any radio transmitter unless such operation is authorized by a valid radio station license. Operation of an unlicensed radio transmitter is a violation of Section 301 of the Communications Act of 1934, as amended, and is punishable by fine and/or imprisonment.

As a licensed radio operator, it is illegal for you to:

- willfully interfere with any radio communication or signal.
- transmit false or deceptive signals or communications by radio.
- falsely identify a radio station by transmitting a call sign which has not been assigned by proper authority to that station.
- transmit unidentified radio communications or signals.
- without authorization, divulge, publish, or use for your benefit or the benefit of another not entitled thereto, the existence, contents, substance, purport, effect, or meaning of any interstate or foreign communications by radio, other than transmissions intended for the use of the general public, transmissions relating to ships, aircraft, vehicles, or persons in distress, or transmissions by an amateur or citizens band radio operator.

Terms and Conditions

It is your responsibility to know the laws, treaties, rules, and regulations which currently govern any station you operate. Do not operate any radio transmitter unless such operation is authorized by a valid radio station license. Operation of an unlicensed radio transmitter is a violation of Section 301 of the Communications Act of 1934, as amended, and is punishable by fine and/or imprisonment.

As a licensed radio operator, it is illegal for you to:

- willfully interfere with any radio communication or signal.
- transmit false or deceptive signals or communications by radio.
- falsely identify a radio station by transmitting a call sign which has not been assigned by proper authority to that station.
- transmit unidentified radio communications or signals.
- without authorization, divulge, publish, or use for your benefit or the benefit of another not entitled thereto, the existence, contents, substance, purport, effect, or meaning of any interstate or foreign communications by radio, other than transmissions intended for the use of the general public, transmissions relating to ships, aircraft, vehicles, or persons in distress, or transmissions by an amateur or citizens band radio operator.

FCC Form 753
December 1982

information is available in the *Code of Federal Regulations* as well.

Additional programming areas of interest to the FCC include procedures governing rebroadcasts, simulcasts, and subcarrier activities. The program director also must be aware that the government is keenly interested in employment practices. The programmer, station manager, and other department heads are under obligation to familiarize themselves with equal employment opportunity (EEO) and affirmative action rules. An annual employment report must be sent to the FCC.

The preceding is only a partial listing of the

FIGURE 3.17 Employment application with EEO policy statement. Courtesy WHDH-AM.

RADIO 85

FOUR FORTY-ONE STUART STREET
BOSTON, MASSACHUSETTS 02116
TELEPHONE (617)267-3302

APPLICATION FOR EMPLOYMENT

"Employment discrimination because of race, color, religion, national origin, sex or age is prohibited. Anyone who believes that he or she has been a victim of discrimination in seeking employment at this station is urged to report the matter promptly to the President of WHDH and may notify the Federal Communications Commission or other appropriate agency."

This application will be given every consideration. Its acceptance, however, does not imply that the applicant will be employed. If an applicant is not employed, it will be retained in our active file for one year.

All appointments to positions are on trial and subject to satisfactory work. If it is found that an employee is not adapted to radio work or is likely to prove useful, he or she may be dismissed.

WHDH CORPORATION
EQUAL EMPLOYMENT OPPORTUNITY POLICY

Equal employment opportunity is the policy of WHDH Corporation (WHDH). Prejudice or discrimination, based upon race, color, religion, national origin, or sex, is strictly prohibited by WHDH in recruitment, selection and hiring, placement and promotion, pay, working conditions, demotion, termination, and in all employment practices generally. All personnel are required diligently and affirmatively to observe this policy and are absolutely forbidden to violate, frustrate or diminish that policy. Any applicant for employment, any employee or any other person having reason to believe that there exists any prejudice or discrimination based upon race, color, religion, national origin, or sex, in any area of WHDH's employment practices, is urged to report the matter promptly to the President of WHDH or any other officer, executive or supervisor employed by WHDH Corporation. Any person believing himself to have been a victim of prohibited discrimination in any area of WHDW's employment practices also has the right to notify the Federal Communications Commission or other appropriate federal or state agencies, or all of these, of such incident or incidents. It is requested that all employees and applicants for employment make this policy of equal employment opportunity known to others, whenever appropriate.

concerns set forth by the government relative to the program director's position.

THE PROGRAM DIRECTOR AND UPPER MANAGEMENT

The pressures of the program director's position should be apparent by now. The station programmer knows well that his or her job entails satisfying the desires of many—the audience, government, air staff, and, of course, management. The relationship between the PD and the station's upper echelon is not always serene or without incident. Although usually a mutually fulfilling and productive alliance, difficulties can and do occur when philosophies or practices clash. "Most inhibiting and detrimental to the

FIGURE 3.17
continued

EDUCATIONAL RECORD	IN THE UNITED STATES (Names of Schools attended)	CITY/TOWN	No. of Years Completed	If Grad. give Mo. and Yr.	If not Grad give date of leaving
High or Preparatory School					
Commercial School					
College					
Graduate School					

If you did not graduate from school or college, state reason for leaving: ______

Were you ever suspended or expelled from any of the institutions above? ______ If so, state the reason fully:

What educational courses are you now taking? ______

PERSONAL RECORD

Date of Birth (OPTIONAL) ______ Are you a citizen of the United States? ______

Have you ever been arrested or convicted for an offense other than a minor traffic violation? ______ If yes, explain

Give number of persons in immediate family ______ Does this include (if living) both father and mother? ______

If not living at home, with whom do you live? Relatives ______ Friends ______ Alone ______ Board ______

Number of persons financially dependent upon you? Fully ______ Partially ______

What is your present selective service classification? ______

Have you served in the armed forces of the United States? ______ Branch of service ______

Month and year of induction ______ Date of discharge or transfer to reserve ______

Rank at time of discharge ______

HEALTH RECORD

How much time have you lost from work or school during the last three years on account of illness? ______

What was the nature of the illness? ______

Are you related to any director, officer or employee of WHDH? ______ Relatives name ______

Describe your outside interests or hobbies ______

Please describe any special assets you bring to WHDH ______

List below, five personal references (not relatives, former employers or fellow employees) who have known you well during the past five or more years.

Print Name in Full	Address (Street / City)	No of Years Known	Occupation

THIS SECTION TO BE FILLED OUT ONLY BY APPLICANTS FOR CLERICAL POSITIONS:

What shorthand method do you use? ______ Words per minute ______ Do you type? ______

Words per minute ______ List office machines that you have operated ______

I hereby affirm that my answers to the foregoing questions are true and correct, and that I have not knowingly withheld any fact or circumstance that would, if disclosed, affect my application unfavorably.

Please sign HERE ______

On a separate sheet of paper you may give any further information you desire to explain your qualifications.

FIGURE 3.17
continued

Please answer each and every question completely, except you are not required to give age or date of birth. It is preferred that the information be supplied in your normal handwriting.

Date ______________ 19 ____

Name in Full ______________ Social Security Number ______________

Address ______________ Tel. No. ______________
STREET CITY STATE ZIP

Referred By ______________

Name ______________ Tel. No. ______________

Address ______________
STREET CITY STATE ZIP

Position Desired ______________ Salary Desired ______________

EMPLOYMENT RECORD (This record *must* account for *all* previous employment including summer and part time employment, if any, while attending school.)

Firm **Name** of **Present** or
1—**Most Recent** Employer ______________

	YOUR POSITION		EMPLOYED	
	Salary at Start	When Leaving	FROM	TO
Address			Mo.	Mo.
City and State			Yr.	Yr.
Are you now working for this employer?				
Reason for leaving:				

Firm Name of Next
2—Previous Employer ______________

	YOUR POSITION		EMPLOYED	
	Salary at Start	When Leaving	FROM	TO
Address			Mo.	Mo.
City and State			Yr.	Yr.
Reason for Leaving:				

Firm Name of Next
3—Previous Employer ______________

	YOUR POSITION		EMPLOYED	
	Salary at Start	When Leaving	FROM	TO
Address			Mo.	Mo.
City and State			Yr.	Yr.
Reason for leaving:				

Firm Name of Next
4—Previous Employer ______________

	YOUR POSITION		EMPLOYED	
	Salary at Start	When Leaving	FROM	TO
Address			Mo.	Mo.
City and State			Yr.	Yr.
Reason for leaving:				

If a complete record of past employment has not been given above, please continue on a separate sheet of paper.

	From		To	
5—Please describe your activities during any period of unemployment in excess of three (3) months from the month you left school to the present time.	Mo.	Yr.	Mo.	Yr.

Have you ever been discharged or requested to resign from any position? ________ If so, give full particulars: ________

Has your application for a surety or fidelity bond ever been refused? ________ If so, state the circumstances: ________

PD is the GM who lacks a broad base of experience but imposes his opinions on you anyway. The guy who has come up through sales and has never spent a minute in the studio can be a real thorn in the side. Without a thorough knowledge of programming, management should rely on the expertise of that person hired who does. I don't mean, 'Hey, GM, get out of

FOR OFFICE USE ONLY

_____ Application

_____ New employee authorization

_____ Health insurance brochures issued

_____ ID card issued

_____ Withhold information

_____ Employment contract (account executives only)

PAYROLL CHANGES:

Eff. Date	New Rate	New Position, if any	New Distribution, if any	Change Per	Entered By

* * *

Employment terminated on ______________________ per ______________________

Recommended for re-employment? ______________________________

____

Keys and ID returned? ______________________________

Letter of Resignation on file? ______________________________

FIGURE 3.17
continued

the way!' What I'm saying is, don't impose programming ideas and policies without at least conferring with that individual who ends up taking the heat if the air product fails to bring in the listeners," says KTOP's Dick Fatherly.

WIYY's Chuck Ducoty contends that managers can enhance as well as inhibit the style of programmers. "I've worked for some managers who give their PDs a great deal of space and others who attempt to control every aspect of programming. From the station manager's perspective, I think the key to a good experience with those who work for you is to find excellent people from the start and then have enough confidence in your judgment to let them do their job with minimal interference. Breathing down the neck of the PD is just going to create tension and resentment."

WGAN's programmer, Peter Falconi, believes that a sincere effort to get to know and understand one another should be exerted by both the PD and manager. "You have to be on the same wavelength, and there has to be an excellent line of communication. When a manager has confidence and trust in his PD, he'll generally let him run with the ball. It's a two-way street. Most problems can be resolved when there is honesty and openness."

An adversarial situation between the station's PD and upper management does not have to exist. The station that cultivates an atmosphere of cooperation and mutual respect seldom becomes embroiled in skirmishes that deplete energy—energy better spent raising revenues and ratings.

CHAPTER HIGHLIGHTS

1. The Adult Contemporary (A/C) format features current (since the 1970s) pop standards. It appeals to the twenty-five to forty-nine age group, which attracts advertisers. It often utilizes music sweeps and clustered commercials.

2. Contemporary Hit Radio (CHR) features current, fast-selling hits from the Top 40. It targets teens, broadcasts minimal news, and is very promotion/contest oriented.

3. Country is the fastest growing format since the 1960s. More prevalent in the South and Midwest, it attracts a broad age group. More Country stations are AM.

4. Easy Listening stations evolved from the Beautiful Music stations of the 1960s and 1970s. Featuring mostly instrumentals and minimal talk, many stations have become automated and use prepackaged programming. The primary audience is over age fifty.

5. Album-Oriented Rock (AOR) stations began in the mid-1960s to counter Top 40 stations. Featuring music sweeps with a large airplay library, they play rock album cuts. News is minimal. The format attracts a predominantly male audience aged eighteen to thirty-four.

6. All-News stations rotate time blocks of local, regional, and national news and features to avoid repetition. The format requires three to four times the staff and budget of most music operations.

7. All-Talk combines discussion and call-in shows. It is primarily a medium and major market format. Like All-News, All-Talk is mostly found on AM.

8. The Nostalgia playlist emphasizes popular tunes from the 1940s and pre-rock 1950s, presenting its music in sweeps with a relatively low deejay profile.

9. The Oldies playlist includes hits from the 1950s and 1960s, relying on fine air personalities. Commercials are randomly placed and songs are spaced to allow deejay patter.

10. Urban Contemporary (UC) is the "melting pot" format, attracting a heterogeneous audience. Its upbeat, danceable sound and hip, friendly deejays attract the eighteen to thirty-four age group. Contests and promotions are important.

11. Classical commercial outlets are few, but they have a loyal audience. Primarily an FM format appealing to a higher-income, college-educated (upscale, twenty-five to forty-nine) audience, Classical features a conservative, straightforward air sound.

12. Religious stations are most prevalent on the AM band. Religious broadcasters usually approach programming in one of two ways. One includes music as a primary part of its presentation, while the other does not.

13. Ethnic stations serve the listening needs of minority groups. Black and Hispanic listeners constitute the largest ethnic audiences.

14. Middle-of-the-Road (MOR) stations rely on the strength of air personalities and features. Mostly an AM format, MOR attempts to be all things to all people, attracting an over-forty audience.

15. Program directors (PDs) are hired to fit whatever format the station management has selected. They are chosen for their experience, primarily, although education level is important.

16. The PD is responsible for everything that is aired. Second in responsibility to the general manager, the PD establishes programming and format policy; hires and supervises on-air, music, and production personnel; handles the programming budget; develops promotions; monitors the stations and its competitors and assesses research; is accountable for news, public affairs, and sports features; and may even pull an airshift.

17. The PD's effectiveness is measured by ratings in large markets and by sales in smaller markets.

18. The PD determines the content of each

sound hour, utilizing program clocks to ensure that each element—commercial, news, promo, weather, music, etc.—is strategically located to enhance flow and optimize impact. Even News/Talk stations need program clocks.

19. PDs must adjust programming to the life-style activities of the target audience. They must develop a feel for the area in which the station is located, as well as an understanding of survey information and research data.

20. Finally, the PD must ensure that the station adheres to all FCC regulations pertaining to programming practices, anticipating problems before they occur.

SUGGESTED FURTHER READING

Arbitron Company. *Research Guidelines for Programming Decision Makers*. Beltsville, Md.: Arbitron Company, 1977.

Armstrong, Ben. *The Electronic Church*. Nashville: J. Nelson, 1979.

Baines, Rey L. "Program Decision Making in Small Market AM Radio Stations." Ph.D. diss., University of Iowa, 1970.

Busby, Linda, and Parker, Donald. *The Art and Science of Radio*. Boston: Allyn and Bacon, 1984.

Chapple, Steve, and Garofalo, R. *Rock 'n' Roll Is Here to Pay*. Chicago: Nelson-Hall, 1977.

Cliff, Charles, and Greer, A. *Broadcasting Programming: The Current Perspective*. Washington, D.C.: University Press of America, 1974 to date, revised annually.

Coddington, R.H. *Modern Radio Programming*. Blue Ridge Summit, Pa.: Tab Books, 1970.

DeLong, Thomas, A. *The Mighty Music Box*. Los Angeles: Amber Crest Books, 1980.

Denisoff, R.S. *Solid Gold: The Popular Record Industry*. New York: Transaction Books, 1976.

Dolan, Robert E. *Music in Modern Media*. New York: G. Schirmer and Company, 1967.

Eastman, Susan T. *Broadcast Programming: Strategies for Winning Television and Radio Audiences*. Belmont, Calif.: Wadsworth Publishing, 1981.

Hall, Claude, and Hall, Barbara. *This Business of Radio Programming*. New York: Billboard Publishing, 1977.

Keirstead, Phillip A. *All-News Radio*. Blue Ridge Summit, Pa.: Tab Books, 1980.

Lujack, Larry, and Jedlicka, D.A. *Superjock: The Loud, Frantic, Non-Stop World of Rock Radio Deejays*. Chicago: Regnery, 1975.

MacFarland, David T. *The Development of the Top 40 Format*. New York: Arno Press, 1979.

Passman, Arnold. *The Deejays*. New York: Macmillan, 1971.

Routt, Ed., McGrath, James B., and Weiss, Frederic A. *The Radio Format Conundrum*. New York: Hastings House, 1978.

Sklar, Rick. *Rocking America: How the All-Hit Radio Stations Took Over*. New York: St. Martin's Press, 1984.

Smith, V. Jackson. *Programming for Radio and Television*. Washington, D.C.: University Press of America, 1980.

4 Sales

COMMERCIALIZATION: A RETROSPECTIVE

Selling commercials keeps the radio station on the air. It is that simple, yet not so simple. In the 1920s broadcasters realized the necessity of converting the medium into a sponsor-supported industry. It seemed to be the most viable option and the key to growth and prosperity. Not everyone approved of the method, however. Opponents of commercialization argued that advertising would decrease the medium's ability to effectively serve the public's good, and one United States senator voiced fears that advertisers would turn radio into an on-air pawnshop. These predictions would prove to be somewhat accurate. By the mid-1920s, most radio outlets sold airtime, and few restrictions existed pertaining to the substance and content of messages. Commercials promoting everything from miracle pain relievers to instant hair-growing solutions filled the broadcast day. It was not until the 1930s that the government developed regulations that addressed the issue of false advertising claims. This resulted in the gradual elimination of sponsors peddling dubious products.

Program sponsorships were the most popular form of radio advertising in the 1920s and early 1930s. Stations, networks, and advertising agencies often lured clients onto the air by naming or renaming programs after their products—"Eveready Hour," "Palmolive Hour," "Fleishman Hour," "Cliquot Club Eskimos," "The Coty Playgirl," and so on. Since formidable opposition to commercialization existed in the beginning, sponsorships, in which the only reference to a product was in a program's title, appeared the best path to take. This approach was known as "indirect" advertising.

As the outcry against advertiser-supported radio subsided, stations became more blatant or "direct" in their presentation of commercial material. Parcels of time, anywhere from one to five, even ten minutes, were sold to advertisers eager to convey the virtues of their labors. Industry advertising revenues soared throughout the 1930s, despite the broken economy. Radio salespeople were among the few who had a salable product. World War II increased the rate of sales revenues twofold. As it became the foremost source of news and information during that bleak period in American history, the value of radio's stock reached new highs, as did the incomes of salespeople. The tide would shift, however, with the advent of television in the late 1940s.

When television unseated radio as the number one source of entertainment in the early 1950s, time sellers for the deposed medium found their fortunes sagging. In the face of adversity, as well as opportunity, many radio salespeople abandoned the old in preference for the new, opting to sell for television. Radio was, indeed, in a dilemma of frightening proportions, but it soon put itself back on course by renovating its programming approach. By 1957 the medium had undergone an almost total transformation and was once again enjoying the rewards of success. Salespeople concentrated on selling airtime to advertisers interested in reaching specific segments of the population. Radio became extremely localized and, out of necessity, the networks diverted their attention to television.

Competition also became keener. Thousands of new outlets began to broadcast between 1950 and 1970. Meanwhile, FM started to spread its wings, preparing to surpass its older rival. AM music stations experienced great difficulties in the face of the mass exodus to FM, which culminated in 1979 when FM exceeded AM's listenership. FM became the medium to sell.

In the first half of the 1980s, radio programming was easily divisible. Talk was found on AM and music on FM. Today, hundreds of AM stations have become stereo in an attempt to improve their marketing potential, while some FM outlets have taken to airing nonmusic formats as a way of surviving the ratings battles.

Selling time in the 1980s is a far cry from what it was like during the medium's heyday. Station account executives no longer search for advertisers willing to sponsor half-hour sitcoms, quiz shows, or mysteries. Today, the advertising dollar generally is spent on spots scheduled for airing during specific dayparts on stations that attract a particular piece of the highly fractionalized radio audience pie. The majority of radio outlets in the 1980s program prerecorded music, and it is that which constitutes the station's product. The salesperson sells the audience, which the station's music attracts, to the advertiser or time-buyer.

SELLING AIRTIME

Airtime is intangible. You cannot see it or hold it in your hand. It is not like any other form of advertising. Newspaper and magazine ads can be cut out by the advertiser and pinned to a bulletin board or taped to a window as tangible evidence of money spent. Television commercials can be seen, but radio commercials are sounds flitting through the ether with no visual component to attest to their existence. They are ephemeral, or fleeting, to use words that are often associated with radio advertising. However, any informed account executive will respond to such terms by stating the simple fact that an effective radio commercial makes a strong and lasting impression on the mind of the listener in much the same way that a popular song tends to permeate the gray matter. "The so-called intangible nature of a radio commercial really only means you can't see it or touch it. There is little doubt, however, that a good spot is concrete in its own unique way. Few of us have gone unaffected or, better still, untouched by radio commercials. If a spot is good, it is felt, and that's a tangible," says Charles W. Friedman, sales manager, WKVT AM/FM, Brattleboro, Vermont.

Initially considered an experimental or novel way to publicize a product, it soon became apparent to advertisers that radio was far more. Early sponsors who earmarked a small portion of their advertising budgets to the new electronic "gadget," while pouring the rest into print, were surprised by the results. Encouraged by radio's performance, advertisers began to spend more heavily. By the 1930s many prominent companies were reallocating substantial portions of their print advertising budgets for radio. To these convinced advertisers, radio was, indeed, a concrete way to market their products.

Yet the feeling that radio is an unconventional mode of advertising continues to persist to some extent even today, especially among small, print-oriented retailers. Usually, the small market radio station's prime competitor for ad dollars is the local newspaper. Many retailers have used papers for years and perceive radio as a secondary or even frivolous means of advertising, contends Friedman. "Retailers who have used print since opening their doors for business are reluctant to change. The toughest factor facing a radio salesperson is the notion that the old way works the best. It is difficult to overcome inertia."

Radio is one of the most effective means of advertising when used correctly. Of course, there is a right way and a wrong way to utilize the medium, and the salesperson who knows and understands the unique character of his or her product is in the best position to succeed. To the extent that a radio commercial cannot be held or taped to a cash register, it is intangible. However, the results produced by a carefully conceived campaign can be seen in the cash register. Consistent radio users, from the giant multinational corporations to the so-called "mom and pop" shops, know that a radio commercial can capture people's attention as effectively as anything crossing their field of vision. A 1950s promotional slogan says it best: "Radio gives you more than you can see."

BECOMING AN ACCOUNT EXECUTIVE

A notion held by some sales managers is that salespeople are born and not made. This position holds that a salesperson either "has it" or does not: "it" meaning the innate gift to sell, without which all the schooling and training in the world means little. Although this theory is not embraced by all sales managers, many agree with the view that anyone attempting a career in sales should first and foremost possess an unflagging desire to make money, because without it failure is almost assured.

According to RAB figures, 70 percent of the radio salespeople hired by stations are gone within three years, a figure comparable to the turnover among new insurance salespeople. While this sounds less than encouraging, it also must be stated that to succeed in broadcast sales invariably means substantial earnings and rapid advancement. True, the battle can be a tough one and the dropout rate is high, but the rewards of success are great.

The majority of newly hired account executives have college training. An understanding of research, marketing, and finance is important. Formal instruction in these areas is particularly advisable for persons considering a career in broadcast sales. Broadcast sales has become a familiar course offering at many schools with programs in electronic media. Research and marketing courses designed for the broadcast major also have become more prevalent since the 1970s. "Young people applying for sales positions here, for the most part, have college backgrounds. A degree indicates a certain amount of tenacity and perseverance, which are important qualities in anyone wanting to sell radio. Not only that, but the candidate with a degree often is more articulate and self-assured. As in most other areas of radio, ten or fifteen years ago fewer people had college diplomas, but the business has become so much more sophisticated and complex because of the greater competition and emphasis on research that managers actually look for salespeople with college training," says WRCH General Manager Richard Bremkamp.

Whether a candidate for a sales position has extensive formal training or not, he or she must possess a knowledge of the product in order to be hired. "To begin with, an applicant must show me that she knows something about radio; after all, that is what we're selling. The individual doesn't necessarily have to have a consummate understanding of the medium, although that would be nice, but she must have some product knowledge. Most stations are willing to train to an extent. I suppose you always look for someone with some sales experience, whether in radio or in some other field, "says Bob Turley, general sales manager, WQBE AM/FM, Charleston, West Virginia.

Stations do prefer a candidate with sales exposure, be it selling vacuum cleaners door-to-door or shoes in a retail store. "Being out in front of the public, especially in a selling situation, regardless of the product, is excellent training for the prospective radio sales rep. I started in the transportation industry, first in customer service, then in sales. After that I owned and operated a restaurant. Radio sales was a whole new ball game for me when I went to work for WCFR in Springfield, Vermont, in the mid-1970s. Having dealt with the public for two decades served me well. In two years I rose to be the station's top biller, concentrating mainly on direct retail sales. In 1979, WKVT in Brattleboro hired me as sales manager," recalls Charles Friedman.

Hiring inexperienced salespeople is a gamble that a small radio station generally must take. The larger outlets almost always require radio sales credentials, a luxury that lesser stations cannot afford. Thus they must hire untested salespeople, and there are no guarantees that a person who has sold lawn mowers can sell airtime, that is, assuming that the newly hired salesperson has ever sold anything at all. In most cases, he or she comes to radio and sales without experience, and the station must provide at least a modicum of training. New salespeople commonly are given two to three months to display their wares and exhibit their potential. If they prove themselves to the sales manager by generating new business, they are asked to stay. On the other hand, if the sales manager is not convinced that the apprentice salesperson has the ability to bring in the accounts, then he or she is shown the door.

During the trial period, the salesperson is given a modest draw against sales, or a "no-strings" salary on which to subsist. In the former case, the salesperson eventually must pay back, through commissions on sales, the amount that he or she has drawn. Thus, after a few months, a new salesperson may well find himself in debt for two or three thousand dollars. As a show of confidence and to encourage the new salesperson who has shown an affinity for radio sales, management may erase the debt. If a station decides to terminate its association with the new salesperson, it must absorb the loss of time, energy, and money invested in the employee.

Characteristics that managers most often look for in prospective salespeople include, among other things, ambition, confidence, energy, determination, honesty, and intelligence. "Ambition is the cornerstone of success. It's one of the first things I look for. Without it, forget it. You have to be hungry. It's a great motivator. You have to be a quick thinker—be able to think on your feet, under pressure, too. Personal appearance and grooming also are important," says Gene Etheridge, general sales manager, KOUL-FM, Corpus Christi, Texas.

Joe Martin, general sales manager of WBQW-AM, Scranton, Pennsylvania, places a premium on energy, persistence, creativity, and organization. "Someone who is a self-starter makes life a lot easier. No manager likes having to keep after members of his sales team. A salesperson should take initiative and be adept at planning his day. Too many people make half the calls they should and could. A salesperson without organizational skills is simply not going to bring in the same number of orders as the person who does. There is a correlation between the number of pitches and the number of sales."

Peg Kelly, general sales manager of WNBC-AM, New York, values people skills. "Empathy is right up there. I need someone who possesses the ability to get quick answers, learn the client's business quickly, and relate well to clients. It's a people business, really." Charles Friedman also looks for sales personnel who are "people oriented." "My experience has proven that a prospective salesperson had better like people. In other words, be gregarious and friendly. This is something that's hard to fake. Sincerity, or the lack of it, shows. If you're selling airtime, or I guess anything for that matter, you had better enjoy talking with people as well as listening. A good listener is a good salesperson. Another thing that is absolutely essential is the ability to take rejection objectively. It usually takes several 'no's' to get to the 'yes.' "

Friedman believes that it is important for a salesperson to have insight into human nature and behavior. "You really must be adept at psychology. Selling really is a matter of anticipating what the prospect is thinking and knowing how best to address his concerns. It's not so much a matter of outthinking the prospective client, but rather being cognizant of the things that play a significant role in his life. Empathy requires the ability to appreciate the experiences of others. A salesperson who is insensitive to a client's moods or states of mind usually will come away empty handed."

In recent years, sales managers have recruited more heavily from within the radio station itself rather than immediately looking elsewhere for salespeople. For decades, it was felt that programming people were not suited for sales. An inexplicable barrier seemed to separate the two areas. Since the 1970s, however, this attitude has changed to some degree, and sales managers now give serious consideration to on-air people who desire to make the transition into sales. The major advantage of hiring programming people to sell the station is that they have a practical understanding of the product. "A lot of former deejays make good account reps because they had to sell the listener on the product. A deejay really is a salesperson, when you get right down to it," observes WBQW's Joe Martin.

Realizing, too, that sales is the most direct path into station management, programming people often are eager to make the shift. In the 1980s, there is a greater trend than ever to recruit managers from the programming ranks. However, a sales background is still preferred.

The salesperson is invariably among the best-paid members of a station. How much a salesperson earns is usually left up to the individual to determine. Contrary to popular opinion, the salesperson's salary generally exceeds the deejay's, especially in the smaller markets. In the larger markets, certain air personalities' salaries are astronomical and even surpass the general manager's income, but major market sales salaries are commonly in the five- and even six-figure range.

Entry-level sales positions are fairly abundant, and stations are always on the lookout for good people. Perhaps no other position in the radio station affords an individual the opportunities that sales does, but most salespeople will never go beyond entry level in sales. Yet for those who are successful, the payoff is worthwhile. "The climb itself can be the most exhilarating part, I think. But you've got to have a lot of reserve in your tanks, because the air can be pretty thin at times," observes KOUL's Etheridge.

THE SALES MANAGER

The general sales manager directs the marketing of the radio station's airtime. This person is responsible for moving inventory, which in the case of the radio outlet constitutes the selling of spot and feature schedules to advertisers. To achieve this end, the sales manager directs the daily efforts of the station's account executives, establishes sales department policies, develops sales plans and materials, conceives of sales and marketing campaigns and promotions, sets quotas, and also may sell as well.

The organizational structure of a station's sales department customarily includes the positions of national, regional, and local sales managers. The national responsibilities usually are handled by the general sales manager. This in-

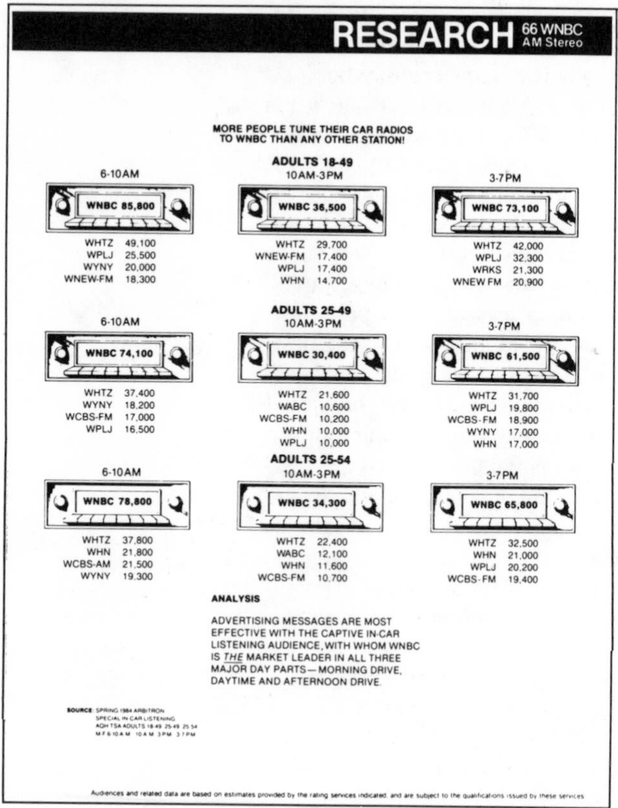

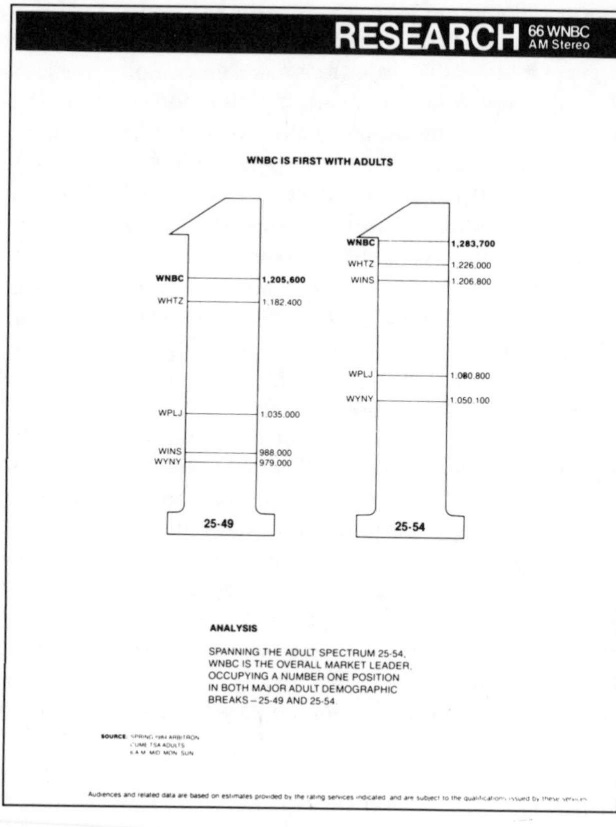

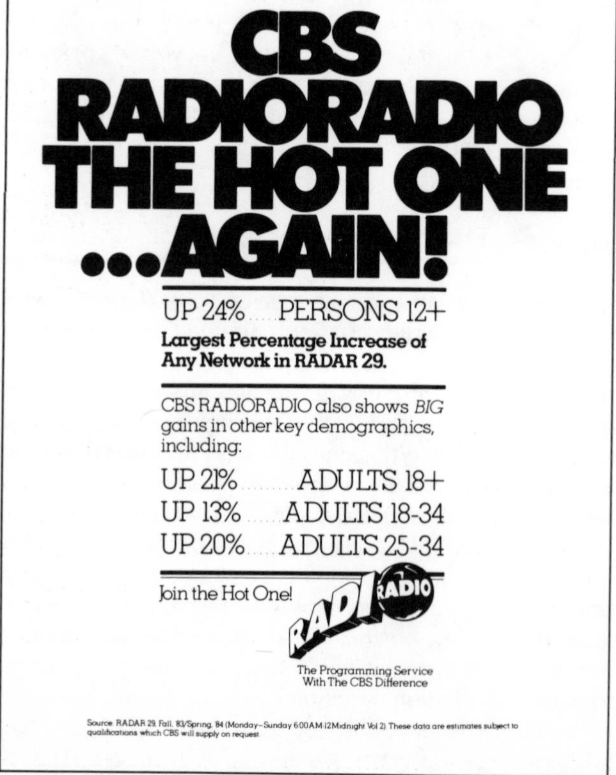

FIGURE 4.1
Marketing a station makes it a more viable product. Courtesy WNBC and CBS.

cludes working with the station's rep company to stimulate business from national advertisers. The regional sales manager is given the responsibility of exploring sales possibilities in a broad geographical area surrounding the station. For example, the regional person for an outlet in New York City may be assigned portions of Connecticut, New Jersey, and Long Island. The local sales manager at the same station would concentrate on advertisers within the city proper. The general sales manager oversees the efforts of each of these individuals.

The size of a station's sales staff varies according to its location and reach. A typical small market radio station employs between two and four account executives, while the medium market station averages about five. Large, top-ranked metropolitan outlets employ as many as eight to ten salespeople, although it is more typical for the major market station to have about a half dozen account executives.

The general sales manager reports directly to the station's general manager and works closely with the program director in developing salable features. Regular daily and weekly sales meetings are scheduled and headed by the sales manager, during which time goals are set and problems addressed. The sales manager also assigns account lists to members of his or her staff and helps coordiante trade and co-op deals.

As mentioned earlier, the head of the sales department usually is responsible for maintaining close contact with the station's rep company as a way of generating income from national advertisers who are handled by advertising agencies. The relationship of the sales manager and rep company is a particularly important one and will be discussed in greater detail later in this chapter. In addition, the sales manager must be adept at working ratings figures to the station's advantage for inclusion in sales promotional materials that are used on both the national and local level.

All sales come under the scrutiny of the sales manager, who determines if an account is appropriate for the station and whether conditions of the sale meet established standards. In addition, the sales manager may have a policy that requires credit checks to be made on every new account and that each new client pay for a portion of their spot schedule "up front" as a show of good faith. Again, policies vary from station to station.

It is up to the head of sales to keep abreast of local and national sales and marketing trends that can be used to the station's advantage. This requires that the sales manager constantly survey trade magazines, like *Advertising Age*, and attend industry seminars, such as those conducted by the Radio Advertising Bureau. No sales department can operate in a vacuum and hope to succeed in today's dynamic radio marketplace.

Statistics continue to bear out the fact that sales managers are most often recruited to fill the position of general manager. It is also becoming more commonplace for sales managers to have experience in other areas of a station's operations, such as programming and production, a factor that has become increasingly important to the person who hires the chief account executive.

FIGURE 4.2
Coverage maps are used to sell those areas within the station's signal. Courtesy WLLT-FM.

You already know that W·Lite is an excellent advertising vehicle for reaching Cincinnati adults 18-49, particularly women 25-49.

But, you may not know that because of W·Lite's unique tower location, a strong signal is broadcast into Dayton and surrounding areas. In fact, W·Lite is one of the most listened-to NON-Dayton radio stations in Dayton's Arbitron defined Metro Survey Area!

The W·Lite signal, covering both Cincinnati and Dayton, gives you "The W·Lite Advantage". Look what happens when the populations of the Cincinnati and Dayton MSA, plus the population of Butler County, are added! The combined population is greater than the nation's 13th largest radio market!

As a retailer with locations in Cincinnati and Dayton, or in between, you can benefit from "The W·Lite Advantage!"

CINCINNATI MSA 12+	1,155,600	#27 Population Rank
DAYTON MSA 12+	680,100	#48 Population Rank
BUTLER COUNTY 12+	214,878	
TOTAL POPULATION	2,050,578	Greater than #13! (St. Louis)

RADIO SALES TOOLS

The fees that a station charges for airtime are published in its rate card. Rates for airtime depend upon the size of a station's listenership, that is, the bigger the audience the higher the rates. At the same time, the unit cost for a spot or a feature is affected by the quantity or amount purchased: the bigger the "buy," the cheaper the unit price. Clients also get discounts for consecutive week purchases over a prescribed period of time, say twenty-six or fifty-two weeks.

The sales manager and station manager work together in designing the rate card, basing their decisions on ratings and what their market can support. A typical rate card will include a brief policy statement concerning terms of payment and commission: "Bills due and payable when rendered. Without prior credit approval, cash in advance required. Commission to recognized advertising agencies on net charges for station time—15%." A statement pertaining to the nature of copy and when it is due at the station also may be included in the rate card: "All programs and announcements are subject to removal without notice for any broadcast which, in our opinion, is not in the public's interest. Copy must be at the station 48 hours prior to broadcast date and before noon on days preceding weekends and holidays." A station's approach to discounting must, of practical necessity, be included in the rate card: "All programs, features, and announcements are provided a 5% discount if on the air for 26 consecutive weeks and a 10% discount if on the air for 52 weeks."

FIGURE 4.3
Rate cards are designed to convey the station's image. Courtesy WFBQ-FM.

It is important to state as emphatically and clearly as possible the station's position on all possible topics affecting a sale. Most stations provide clients rate protection for a designated period of time should fees for airtime change. This means that if a client purchases a three-month spot schedule in May, and the station raises its rates in June, the advertiser continues

BE PART OF NEW ENGLAND'S COUNTRY

NEWS...

News Sponsorships can give your product credibility.
You are part of the big stories of the day.

WBOS NEW ENGLAND'S COUNTRY

FEATURES

- National and Local Features
- Traffic Reports by WBOS Chopper and Captain Todd
- Weather — With Shane Hollett
- Sports Updates

Drive Time Availability's

5:50 A.M. — 6:20 A.M. — 6:50 A.M. — 7:20 A.M.
7:50 A.M. — 8:20 A.M. — 8:50 A.M. — 9:20 A.M.
and 9:50 A.M.

3:20 P.M. — 4:50 P.M. — 5:20 P.M. — 5:50 P.M.
6:20 P.M. — 6:50 P.M.

You can get 10 Exclusive Sponsorships per week at Guaranteed times every week.

	Weekly	**13 Weeks**
Retail Value:	$400.00	$5200.00
The New England's Country Special Tie In Package	$350.00	$3550.00
A Savings of		$1700.00

BOOK IT NOW FOR AVAILABILITIES

160 NORTH WASHINGTON STREET • BOSTON, MA 02114 • 617-367-9003

FIGURE 4.4
Features are major income generators for stations. Courtesy WBOS.

FIGURE 4.5
The WBOS and KNEW rate cards show use of the grid structure. Courtesy WBOS and KNEW.

RATE CARD II

★ ★

60 SECONDS	AAA	AA	A	BTA
GRID I	$75.00	65.00	60.00	50.00
GRID II	$70.00	60.00	55.00	40.00
GRID III	$60.00	50.00	40.00	35.00

TOTAL AUDIENCE PACKAGE

1/3 AAA, 1/3 AA, 1/3 A (THIRTY SECONDS OR LESS = 80% OF ONE MINUTE RATE

TIMES PER WEEK (:60)	PER ANNOUNCEMENT	COST PER WEEK
6X	$60.00	$360.00
12X	50.00	600.00
18X	40.00	720.00
24X	35.00	840.00

BULK RATE — :60 520 @ $40.00 AAA-BULK
T.A.P. — :60 1040 @ $35.00 55.00

AAA = Monday - Saturday: 6:00 A.M. — 10:00 A.M. and 3:00 P.M. - 7:00 P.M.
AA = Monday - Saturday 10:00 A.M. — 3:00 P.M. and Sunday
A = Monday - Saturday 7:00 P.M. — 12:00 P.M. and Overnight

GENERAL INFORMATION

*News, Sports, Features including: 10 Opening and Closing = 1.5X Spot Rate

30 sec. 80% of 60 sec. rate 10 sec. 50% of 60 sec. rate

★ ★

160 NORTH WASHINGTON STREET • BOSTON, MA 02114 • 617-367-9003

FIGURE 4.5
continued

KNEW 91 AM/KSAN 95 FM

Vice President &
General Manager
Steve Edwards

General Sales Manager
Joel K. Schwartz

Represented nationally by
Katz Radio

rate card

Weekly 60 Second Spot Rate

GRID	AAA	AA	A	B
I	500	460	450	200
II	450	410	400	150
III	400	360	350	120
IV	350	310	300	100

Time Classifications

AAA	Monday - Saturday Saturday	5:00 AM-10:00 AM 10:00 AM- 3:00 PM
AA	Monday-Saturday	3:00 PM- 8:00 PM
A	Monday-Friday Sunday	10:00 AM- 3:00 PM
B	Monday-Saturday Sunday	8:00 PM- 1:00 AM 8:00 AM-10:00 AM 3:00 PM-10:30 PM

The cost per announcement quoted is as if for one station. The schedule is also duplicated on the other station.

Notes:
1. Grid IV subject to immediate preemption without notice.
2. Limitations: Maximum ¼ of all spots in AM drive Monday - Friday.
   Maximum of 2 spots 10:00 AM - 3:00 PM Saturday.
   Any spots over these limits are subject to availability and will only be sold at Grid I.
3. News Sponsorships: Applicable rate plus $20.
4. 30 second rates are 90% of the weekly 60 second rate.
5. 1:00 AM - 5:00 AM Monday - Sunday: No charge with every B time spot purchased.
6. For billing purposes only: AM/FM breakdown is 50/50.
7. KNEW or KSAN only: Deduct $1 from applicable combo rate.

Conditions

Contracts are accepted for a maximum period of one year. Rates quoted herein are guaranteed for a period of 30 days from the effective date of any increase. Rates quoted represent station time only. There is an agency commission of 15% to recognized agencies: no cash discounts. Bills due and payable upon receipt.

All contracts are governed by the conditions detailed on the KNEW/KSAN Broadcast Contract.

All advertising material must be received at least 48 hours before each broadcast. Advertising material must conform to the standards of the station and the station reserves the right to refuse or discontinue any advertising for reasons satisfactory to itself.

Unfulfilled contracts are subject to appropriate short rate.

This rate card is for information purposes and does not constitute an offer on the part of KNEW/KSAN.

● MALRITE COMMUNICATIONS GROUP ● 66 JACK LONDON SQUARE ● OAKLAND, CA 94607 ● (415) 836-0910
YE IEF 1084

to pay the original rates until the expiration of its current contract.

The rate card also contains its feature and spot rates. Among the most prevalant features that stations offer are traffic, sports, weather, and business reports. Newscasts also are available to advertisers. Features generally include an open (introduction) and a thirty- or sixty-second announcement. They are particularly effective advertising vehicles because listeners tend to pay greater attention. Conditions pertaining to feature buys usually appear in the rate card: "All feature sales are subject to four weeks' notice for renewal and cancellation." A station wants to establish credibility with its features and therefore prefers to maintain continuity among its sponsors. A feature with a regular sponsor conveys stability, and that is what a station seeks.

Since the size of a radio station's audience generally varies depending on the time of day, rates for spots (commercials) or features must reflect that fact. Thus, the broadcast day is dividing into time classifications. Six to ten A.M. weekdays is typically a station's prime selling period and therefore may be designated "AAA," while afternoon drivetime, usually three to seven P.M., may be called "AA" because of its secondary drawing power. Under this system, the midday segment, ten to three P.M., would be given a single "A" designation, and evenings, seven to midnight, a "B." Overnights, midnight to six A.M., may be classified as "C" time. Obviously, the fees charged for spots are established on an ascending scale from "C" to "AAA." A station may charge three hundred dollars for an announcement aired at eight A.M. and forty-five dollars for one aired at two A.M. The difference in the size of the station's audience at those hours warrants the contrast.

As previously mentioned, the more airtime a client purchases, the less expensive the cost for an individual commercial or unit. For instance, if an advertiser buys ten spots a week during "AAA" time, the cost of each spot would be slightly less than if the sponsor purchased two spots a week. A client must buy a specified number of spots in order to benefit from the frequency discount. A 6X rate, meaning six spots per week, for "AAA" 60's (sixty-second announcements) may be $75, while the 12X rate may be $71 and the 18X rate $68, and so forth. Thirty-second spots are usually priced at two-thirds the cost of a sixty. Should a client desire that a spot be aired at a fixed time, say 7:10 A.M. daily, the station will tack on an additional charge, possibly 20 percent. Fixed position drivetime spots are among the most expensive in a station's inventory. Grid structures frequently are established in rate cards to reflect premium and nonpremium calendar periods. For example, because of Christmas, a time when radio sales peak, a station may sell from its Grid I level. Rates will be higher because of the demand for airtime. During lighter advertising periods, the lower rates on the Grid III or IV levels will be offered sponsors.

Announcements are rotated or "orbited" within time classifications to maximize the number of different listeners reached. If a client buys three drivetime spots per week to be aired on a Monday, Wednesday, and Friday, over a four-week period, the time they are scheduled will be different each day. Here is a possible rotation set-up:

	MON	WED	FRI
Week I	7:15	6:25	9:10
Week II	8:22	7:36	8:05
Week III	6:11	9:12	7:46
Week IV	9:20	8:34	6:52

Clients are offered several spot schedule plans suited for their advertising and budgetary needs. For advertisers with limited funds, run-of-station (ROS) or best-time-available (BTA) plans are usually an option. Rates are lower under these plans since no guarantee is given as to what times the spots will be aired. However, most stations make a concerted effort to rotate ROS and BTA spots as equitably as possible, and during periods when commercial loads are light they frequently are scheduled during premium times. Of course, when a station is loaded down with spot schedules, especially around holidays or elections, ROS and BTA spots may find themselves "buried." In the long run, advertisers using these plans receive a more than fair amount of choice times and at rates considerably lower than those clients who buy specific dayparts.

Total audience plan (TAP) is another popular package offered clients by many stations. It is designed to distribute a client's spots among the various dayparts for maximum audience penetration, while costing less than an exclusive prime-time schedule. The rate for a TAP spot is arrived at by averaging out the cost for spots in several time classifications. For example, AAA = $80, AA = $70, A = $58, B = $31, thus the TAP rate per spot is $59. The advantages are obvious. The advertiser is getting a significant discount on the spots scheduled during morning and afternoon drive periods. At the same time, the advertiser is paying more for airtime during evenings. However, TAP is very attractive because it does expose a client's message to every possible segment of a station's listening audience with a measure of cost effectiveness.

Bulk or annual discounts are available to advertisers who buy a heavy schedule of commercials over the course of a year. Large companies in particular take advantage of volume discounts because the savings are significant.

Rather than purchase a consecutive week schedule, advertisers may choose to purchase time in flights, an alternating pattern—on one week and off the next. For instance, a client with a seasonal business or one that is geared toward holiday sales may set up a plan in which spots are scheduled at specific times throughout the year. Thus, an annual flight schedule may look something like this:

Feb. 13–19 Washington's Birthday Sale. 10 A 60's.
Mar. 14–17 St. Patrick's Day Celebration. 8 ROS 30's.
Apr. 16–21 Easter Parade Days. 20 TAP 30's.
May 7–12 Mother's Day Sale. 6 AAA and 6 AA 30's.
June 1–15 Summer Sale Days. 30 ROS 60's.
Aug. 20–30 Back-to-School Sale. 15 A 60's.
Sept. 24–Oct. 6 Fall Sale Bonanza. 25 ROS 60's.
Nov. 25–Dec. 19 Christmas Sale. 25 AAA 60's and 20 A 30's.

A rate card is used by salespeople to plan and compute buys. It is generally perceived as a poor idea to simply leave it with a prospective client to figure out, even if such a request is made. First off, few laypersons are really adept at reading rate cards and, secondly, a station does not like to publicize its rates to its competition, which is what happens when too many station rate cards are in circulation. Granted, it is quite easy for any station to obtain a competitor's sales portfolio, but stations prefer to keep a low profile as a means of retaining a competitive edge.

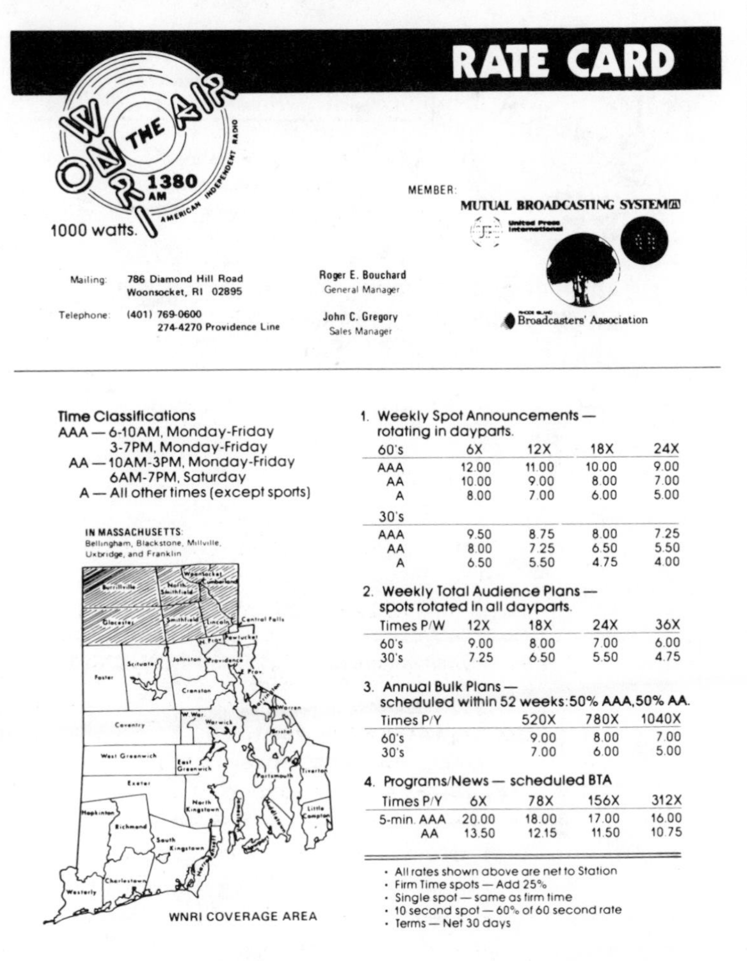

RATE CARD

WNRI ON THE AIR 1380 AM AMERICAN INDEPENDENT RADIO
1000 watts.

MEMBER:
MUTUAL BROADCASTING SYSTEM
United Press International
Broadcasters' Association

Mailing: 786 Diamond Hill Road
Woonsocket, RI 02895
Telephone: (401) 769-0600
274-4270 Providence Line

Roger E. Bouchard
General Manager

John C. Gregory
Sales Manager

Time Classifications
AAA — 6-10AM, Monday-Friday
3-7PM, Monday-Friday
AA — 10AM-3PM, Monday-Friday
6AM-7PM, Saturday
A — All other times (except sports)

IN MASSACHUSETTS: Bellingham, Blackstone, Millville, Uxbridge, and Franklin

WNRI COVERAGE AREA

1. Weekly Spot Announcements — rotating in dayparts.

60's	6X	12X	18X	24X
AAA	12.00	11.00	10.00	9.00
AA	10.00	9.00	8.00	7.00
A	8.00	7.00	6.00	5.00
30's				
AAA	9.50	8.75	8.00	7.25
AA	8.00	7.25	6.50	5.50
A	6.50	5.50	4.75	4.00

2. Weekly Total Audience Plans — spots rotated in all dayparts.

Times P/W	12X	18X	24X	36X
60's	9.00	8.00	7.00	6.00
30's	7.25	6.50	5.50	4.75

3. Annual Bulk Plans — scheduled within 52 weeks: 50% AAA, 50% AA.

Times P/Y	520X	780X	1040X
60's	9.00	8.00	7.00
30's	7.00	6.00	5.00

4. Programs/News — scheduled BTA

Times P/Y	6X	78X	156X	312X
5-min. AAA	20.00	18.00	17.00	16.00
AA	13.50	12.15	11.50	10.75

- All rates shown above are net to Station
- Firm Time spots — Add 25%
- Single spot — same as firm time
- 10 second spot — 60% of 60 second rate
- Terms — Net 30 days

FIGURE 4.6
Rate card featuring TAP, Bulk, and BTA plans. Courtesy WNRI-AM.

POINTS OF THE PITCH

Not all sales are made on the first call; nonetheless, the salesperson does go in with the hopes of "closing" an account. The first call generally is designed to introduce the station to the prospective sponsor and to determine their needs. However, the salesperson should always be prepared to propose a buy that is suitable for the account. This means that some homework must be done relative to the business before an approach is made. "First determine the client's needs, as best as possible. Then address those needs with a schedule built to reach the client's customers. Don't walk into a business cold or without some sense of what the place is about," advises WKVT's Friedman.

Should all go smoothly during the initial call, the salesperson may opt to go for an order there and then. If the account obliges, fine. In the event that the prospective advertiser is not prepared to make an immediate decision, a follow-up appointment must be made. The call-back should be accomplished as close to the initial presentation as possible to prevent the impression made then from fading or growing cold. The primary objective of the return call is to close the deal and get the order. To strengthen the odds, the salesperson must review and assess any objections or reservations that may have arisen during the first call and devise a plan to overcome them. Meanwhile, the initial proposal may be beefed up to appear even more attractive to the client, and a "spec" tape (see the later section in this chapter) for the business can be prepared as further enticement.

Should the salesperson's efforts fail the second time out, a third and even fourth call are made. Perseverance does pay off, and many salespeople admit that just when they figured a

FIGURE 4.7
Flyer offering special rates on a spot schedule. Courtesy WNRI-AM.

situation was hopeless an account said yes. "Of course, beating your head against the wall accomplishes nothing. You have to know when your time is being wasted. Never give up entirely on an account; just approach it more sensibly. A phone call or a drop-in every so often keeps you in their thoughts," says Ron Piro, general sales manager, WHTT-FM, Boston.

What follows are two checklists. The "DO" list contains some suggestions conducive to a positive sales experience, while the "DON'T" list contains things that will have a negative or counterproductive effect:

DO

- Research advertiser. Be prepared. Have a relevant plan in mind.
- Be enthusiastic. Think positive.
- Display self-confidence. Believe in self and product.
- Smile. Exude friendliness, warmth, and sincerity.
- Listen. Be polite, sympathetic, and interested.
- Tell of station's successes. Provide testimonial material.
- Think creatively.
- Know your competition.
- Maintain integrity and poise.
- Look your best. Check your appearance.
- Be objective and keep proper perspective.
- Pitch the decision maker.
- Ask for the order that will do the job.
- Service account after the sale.

DON'T

- Pitch without a plan.
- Criticize or demean client's previous advertising efforts.
- Argue with client. This just creates greater resistance.
- Badmouth competition.
- Talk too much.
- Brag or be overly aggressive.
- Lie, exaggerate, or make unrealistic promises.
- Smoke or chew gum in front of client.
- Procrastinate or put things off.
- Be intimidated or kept waiting an unreasonable amount of time.
- Make a presentation unless you have client's undivided attention.
- Lose your temper.
- Ask for too little. Never undersell a client.
- Fail to follow up.
- Accept a "no" as final.

Checklists like the preceding ones can only serve as basic guidelines. Anyone who has spent time on the street as a station account executive can expand on this or any other such checklist. For the positive-thinking radio salesperson, every call gives something back, whether a sale is made or not.

Overcoming common objections is a necessary step toward achieving the sale. Here are some typical "put-offs" presented to radio sales reps:

1. Nobody listens to radio commercials.
2. Newspaper ads are more effective.
3. Radio costs too much.
4. Nobody listens to your station.
5. We tried radio, and it didn't work.
6. We don't need any more business.

WACE — RADIO 730

serving over three million in six states with 5000 watts

P.O. BOX 1 • SPRINGFIELD, MASSACHUSETTS 01101 • (413) 594-6654/(413) 781-2240

WACE RADIO
IS SOMETHING ELSE
AND THE DIFFERENCE
SPELLS RESULTS

WACE IS A DIFFERENT "BREED" OF METRO RADIO.....SO DIFFERENT, IN FACT, THAT SOME OF THE "HEADS UP" AD AGENCIES AND ADVERTISERS MISS THE SUBTLE VALUE OF THIS DISTINCT DIFFERENCE. CONSIDER THESE DECIDED ADVANTAGES IN USING THIS DIFFERENT "BREED OF RADIO".

- WACE airs features close to the hearts of its listeners: thus, an amazing audience loyalty to the advertiser whom the listener supports for sustaining their brand of radio.
- WACE does not need to pace the 'ratings race' to deliver big results. Our religious oriented and conservative minded listeners as well as our liberal friends constitute the financial cornerstone of this great locale.
- WACE'S vast radio family listens basically for inspiration and information: thus they do not use WACE for subdued background entertainment but quite to the contrary...they actually hear what is being programmed...commercial messages included.
- Finally, WACE is also a DIFFERENT BREED OF METRO RADIO in that the basic format is derived from programs and a minor share from commercial spots. Conventional stations receive almost all of their revenue from spot advertising. Most conventional radio facilities overpopulate their programming with commercials and in doing so, cut the audience "pie" into very small portions. WACE obviously has only limited participating openings for commercial advertisers: consequently we are able to deliver our great and responsive audience to a few clients on an almost exclusive basis.

FIGURE 4.8
Radio station sales packets contain material specifically designed to sell the prospective advertiser. Courtesy WACE.

7. We've already allocated our advertising budget.
8. We can get another station for less.
9. Business is off, and we haven't got the money.
10. My partner doesn't like radio.

And so on. There are countless rebuttals for each of these statements, and a knowledgeable and skilled radio salesperson can turn such objections into positives.

LEVELS OF SALES

There are three levels from which the medium draws its sales—retail, local, and national. Retail accounts for the biggest percentage of the

FIGURE 4.8
continued

WACE/730
Chicopee/Springfield

Dear Sponsor;

As a business person serving the greater Chicopee/Springfield/Holyoke area, I'm sure you are keenly aware of **WACE**'s long-running Sunday Polish program.

This program was aired on **WACE** every Sunday for thirty-four years with Andy Szuberla. We, like you, miss the program and more importantly, we miss this fine gentleman.

With Andy's passing, **WACE** was confronted with a decision regarding the program. Should we continue the Polish music program or should we cancel it? We knew we could never replace Andy. After an extensive survey within the Polish community, we found that we should—by all means—continue the tradition begun by Andy and run successfully for so many years.

I'd like to take this opportunity to thank those of you whom I had the pleasure of meeting personally for your interest and suggestions.

We will continue to entertain and inform our large, loyal Polish audience with quality programming. With the implementation of new programming ideas, we will gain an even larger audience—with new listeners and supporters. The overall intent, of course, is to retain our present audience while welcoming new listeners all the time.

Together we will reach the entire Polish community with our message.

We know you will enjoy what we have to offer, and we welcome your participation.

Sincerely,

Ken Carter
President/**WACE**

P.O. Box 1 • Springfield, Massachusetts 01101 • (413) 594-6654 or (413) 781-2240

industry's income, over 70 percent. Retail, also referred to as "direct," sales involve the radio station on a one-to-one basis with advertisers within its signal area. In this case, a station's account executive works directly with the client and earns a commission of approximately 15 percent on the airtime he or she sells. An advertiser who spends one thousand dollars would benefit the salesperson to the tune of one hundred fifty dollars. A newly hired salesperson without previous experience generally will work on a direct retail basis and will not be assigned advertising agencies until he or she has become more seasoned and has displayed some ability. Generally speaking, the smaller the radio station, the more dependent it is on retail sales,

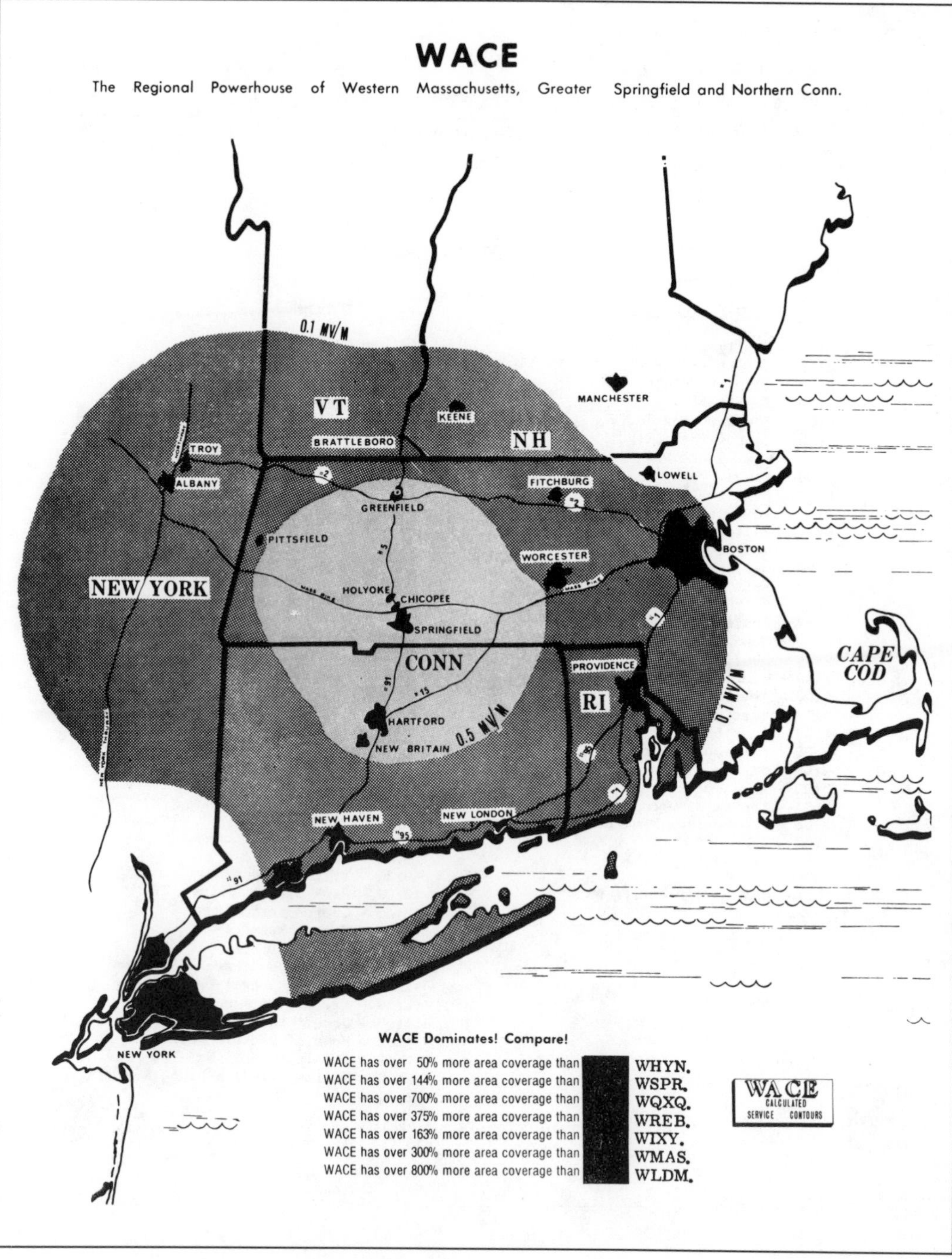

FIGURE 4.8
continued

although most medium and metro market stations would be in trouble without strong business on this level.

All stations, regardless of size, have some contact with advertising agencies. Here again, however, the larger a market, the more a station will derive its business from ad agencies. This level of station sales generally is classified as "local." The number of advertising agencies in a market will vary depending on its size. A sales manager will divide the market's agencies among his reps as equitably as possible, sometimes using a merit system. In this way, an account executive who has worked hard and produced results will be rewarded for his efforts by being given an agency to work. The top bill-

FIGURE 4.8
continued

WACE — RADIO 730

serving over three million in six states with 5000 watts

P.O. BOX 1 • SPRINGFIELD, MASSACHUSETTS 01101 • (413) 594-6654/(413) 781-2240

WACE PROGRAM GUIDE

MONDAY - FRIDAY

6:00 - Family Altar
6:15 - The King's Hour
6:30 - Thru the Bible
7:00 - Songtime
8:00 - The Word for Today
8:30 - Impact for Living
8:45 - The Lighthouse Ministries
9:00 - Peace, Poise & Power
9:15 - Mission to Children
9:30 - Voice of Prophecy
9:45 - Believer's Voice/Copeland
10:00 - The Bible Speaks
11:00 - News/Gospel Music
11:15 - Faith Messenger/Popoff
11:30 - Faith Alive
11:45 - Insight/Gospel Music
12:00 - News/Gospel Music
12:15 - Healing & Restoration Hour
12:30 - Old Time Gospel Hour
1:00 - News/Gospel Music
1:15 - Word of Life
1:30 - Swap Shop
2:00 - News on the Hour
2:05 - Yankee Kitchen/Saunders
3:00 - News on the Hour
3:05 - Yankee Kitchen/Saunders
4:00 - Roy Mastars Program
4:30 - Sign Off

December Sign Off - 4:15 PM

SATURDAY

6:00 - Family Altar
6:15 - Gospel Music
6:45 - American Indian Hour
7:00 - News/Gospel Music
8:00 - God's News Behind the News
8:15 - House of Prayer
8:30 - People's Gospel Hour
9:00 - Life Can Be Beautiful
9:30 - Radio Bible Class
10:00 - The Bible Speaks
11:00 - Irish Hours w/Jim Sullivan
1:00 - News on the Hour
1:05 -Irish Melodies
2:00 - News on the Hour
2:05 - Irish Melodies
3:00 - News on the Hour
3:05 - Irish Melodies
3:30 - Tom Griffin Music Hall
4:30 - Sign Off

SUNDAY

6:00 - Little White Church
6:30 - Gospel Music
7:15 - Sar Shalom
7:30 - Wings of Healing
8:00 - Oral Roberts
8:30 - Radio Bible Class
9:00 - Word of Life
9:30 - Voice of Prophecy
10:00 - Park Street Chuch
10:30 - Church in the Home
10:45 - Gospel Music
11:00 - Gates Unbarred
11:30 - The Bible Standard Program
12:00 - Healing & Restoration Hour
12:30 - Fr. Justin Rosary Hour
1:00 - Jesus' Love Unveiled
1:05 - Hour of Decision/Graham
1:30 - The Truth Will Set You Free
2:00 - Polish Melodies w/Joe Koziol
4:30 - Sign Off

NATIONAL ASSOCIATION OF BROADCASTERS — NAB

ers, that is, those salespeople who bring in the most business, often possess the greatest number of agencies, or at least the most active. Although the percentage of commission a salesperson is accorded, typically 6 to 8 percent, is less than that derived from retail sales, the size of the agency buys usually are far more substantial.

The third category of station sales comes from the "national" level. In most cases, it is the general sales manager who works with the station's rep company to secure buys from advertising agencies that handle national accounts. Again, national business is greater for the metro station than it is for the rural. Agencies justify a buy on numbers and little else, although it is not uncommon for small market stations, which do not even appear in ratings surveys, to be bought by major accounts interested in maintaining a strong local or community image.

Each level of sales—retail, local, or national—must be sufficiently cultivated if a station is to enjoy maximum prosperity. To neglect any one would result in a loss of station revenue.

WQGN Q105

WQGN 105.5 FM Stereo Rock

1064 Poquonnock Road
Groton, Connecticut 06340
Tel: (203)446-1980

RADIO CONTRACT

ADVERTISER ____
PRODUCT ____
CO/OP NAME ____
ACCOUNT NO. ____
ACCOUNT EXEC. ____
DATE ____

STARTING DATE ____ ENDING DATE ____

____ (number) ____ (length) @ $ ____ EACH = $ ____ PLAN # ____
____ (number) ____ (length) @ $ ____ EACH = $ ____ PLAN # ____
____ (number) ____ (length) @ $ ____ EACH = $ ____ PLAN # ____
____ (number) ____ (length) @ $ ____ EACH = $ ____ PLAN # ____

NUMBER ____ PER WEEK TOTAL COST ____ PER WEEK

TOTAL ANNOUNCEMENTS PER SCHEDULE ____

SPECIAL INSTRUCTIONS:
AFFIDAVITS ☐
EXACT TIMES ☐
CO/OP COPY ☐

For above services advertiser agrees to pay Lightfoot Broadcasting Corporation as follows:

$ .... TOTAL COST OF THIS SCHEDULE; .... WEEKS. •COMMERCIAL PRODUCTION = $ ....

TO BE BILLED MONTHLY.

By ____ Station Representative
By ____ Advertiser Representative

ADVERTISER/AGENCY ____
ADDRESS ____
CITY ____ STATE ____ ZIP ____

FIGURE 4.9
Station contract for a spot schedule. Courtesy WQGN.

SPEC TAPES

One of the most effective ways to convince an advertiser to use a station is to provide a fully produced sample commercial, or "spec tape." If prepared properly and imaginatively, a client will find it difficult to deny its potential. Spec tapes often are used in call-backs when a salesperson needs to break down a client's resistance. More than once a clever spec tape has converted an adamant "no" into an "okay, let's give it a shot." Spec tapes also are used to reactivate the interest of former accounts who may not have spent money on the station for a while and who need some justification to do so.

Specs also are effective tools for motivating clients to "heavy-up" or increase their current spot schedules. A good idea can move a mountain, and salespeople are encouraged by the sales manager to develop spec tape ideas. Many sales managers require that account executives make at least one spec tape presentation each week. The sales manager may even choose to critique spec spots during regularly scheduled meetings.

The information needed to prepare a spec spot is acquired in several ways. If a salesperson already has called on a prospective client, he should have a very good idea of what the business is about as well as the attitude of the retailer toward the enterprise. The station sales rep is then in a very good position to prepare a spot that directly appeals to the needs and perceptions of the would-be advertiser. If a salesperson decides that the first call on a client warrants preparing a spec tape, then he or she may collect information on the business by actually browsing through the store as a customer might. This gives the salesperson an accurate, firsthand impression of the store's environment and merchandise. An idea of how the store perceives itself and specific information, such as address and hours, can be derived by checking its display ad in the Yellow Pages, if it has one, or by examining any ads it may have run in the local newspaper. Flyers that the business may have distributed also provide useful information for the formulation of the copy used in the spec spot. Listening to commercials the advertiser may be running on another station also gives the salesperson an idea of the direction in which to move.

Again, the primary purpose of a spec tape is to motivate a possible advertiser to buy time. A spec that fails to capture the interest and appreciation of the individual for which it has been prepared may be lacking in the necessary ingredients. It is generally a good rule of thumb to avoid humor in a spec, unless the salesperson has had some firsthand experience with the advertiser. Nothing fails as abysmally as a com-

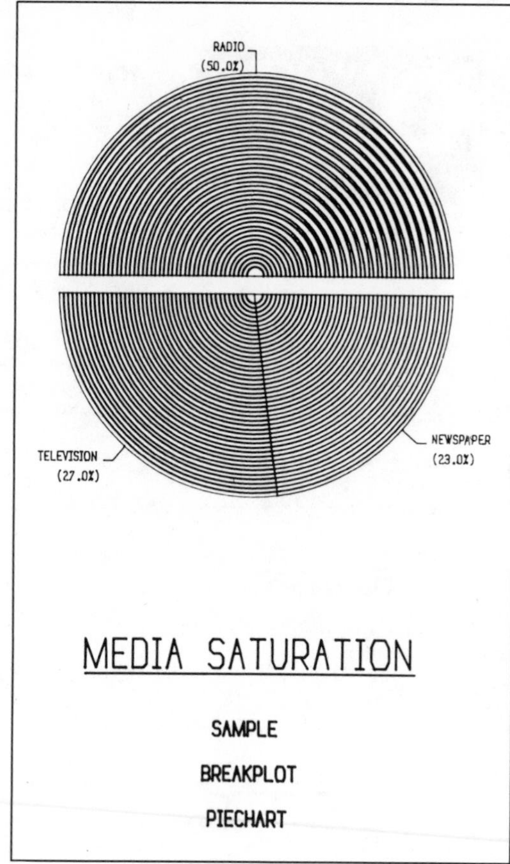

FIGURE 4.10
Sales visuals are designed to convince advertisers of medium's effectiveness. Courtesy Jefferson Pilot Data Systems.

mercial that attempts to be funny and does not come across as such to the client. Thus the saying, "What is funny to one person may be silly or offensive to another."

Although spec spots are, to some extent, a gamble, they should be prepared in such a way that the odds are not too great. Of course, a salesperson who believes in an idea must have the gumption to go with it. Great sales are often inspired by unconventional concepts.

OBJECTIVES OF THE BUY

A single spot on a radio station seldom brings instant riches to an advertiser. However, a thoughtfully devised plan based upon a formula of frequency and consistency will achieve impressive results, contends John Gregory, general sales manager, WNRI-AM Woonsocket, Rhode Island. "It has to be made clear from the start what a client hopes to accomplish by advertising on your station. Then a schedule that realistically corresponds with the client's goals must be put together. This means selling the advertiser a sufficient number of commercials spread over a specific period of time. An occasional spot here and there doesn't do much in this medium. There's a right way to sell radio, and that isn't it."

Our lists of "dos" and "don'ts" of selling suggested that the salesperson "ask for the order that will do the job." It also said not to undersell an account. Implicit in the first point is the idea that the salesperson has determined what kind of schedule the advertiser should buy to get the results expected. Too often salespeople fail to ask for what they need for fear the client will balk. Thus they settle for what they can get without much resistance. This, in fact, may be doing the advertiser a disservice since the buy that the salesperson settles for may not fulfill declared objectives. "It takes a little courage to persist until you get what you think will do the job. There is the temptation just to take what the client hands you and run, but that technique usually backfires when the client doesn't get what he expected. As a radio sales rep you should know how best to sell the medium. Don't be apologetic or easily compromised. Sell the medium the way it should be sold. Write enough of an order to get the job done," says WHTT's Piro.

Inflated claims and unrealistic promises should never be a part of a sales presentation. Avoid "If you buy spots on my station you'll have to hire additional salespeople to handle the huge crowds." Salespeople must be honest in their projections and in what a client may expect from the spot schedule he purchases. "You will notice a gradual increase in store traffic over the next few weeks as the audience is exposed to your commercial over WXXX" is the better approach. Unfulfilled promises ruin any chances of future buys. Too often salespeople caught up in the enthusiasm of the pitch make claims that cannot be achieved. Radio is a phenomenally effective advertising medium. This is a proven fact. Those who have successfully used the medium can attest to the importance of placing an adequate order. "An advertiser has to buy a decent schedule to get strong results. Frequency is essential in radio," notes Piro. A radio sales axiom says it best: "The more spots aired the more impressions made, and the more impressions made the more impressed the client."

== BreakOut ==

MULTI-STATION CPM GRID

< PROJECTED RATES BASED ON: ADULTS 25-54, MON-FRI 6A-7P >

[DALLAS/FT WORTH ARB METRO: SPRNG-83]

#	STATION	AQH PER	CPMs 3.50	4.00	4.50	5.00	5.50	6.00	6.50
			RESULTING RATES						
1	KVIL-FM	39,400	137.90	157.60	177.30	197.00	216.70	236.40	256.10
2	KSCS-FM	26,900	94.15	107.60	121.05	134.50	147.95	161.40	174.85
3	KPLX-FM	25,700	89.95	102.80	115.65	128.50	141.35	154.20	167.05
4	KRLD-AM	21,300	74.55	85.20	95.85	106.50	117.15	127.80	138.45
5	WBAP-AM	19,600	68.60	78.40	88.20	98.00	107.80	117.60	127.40
6	KMEZ-FM	16,900	59.15	67.60	76.05	84.50	92.95	101.40	109.85
7	KKDA-FM	13,000	45.50	52.00	58.50	65.00	71.50	78.00	84.50
8	KMGC-FM	12,700	44.45	50.80	57.15	63.50	69.85	76.20	82.55
9	KOAX-FM	12,000	42.00	48.00	54.00	60.00	66.00	72.00	78.00
10	KZEW-FM	11,800	41.30	47.20	53.10	59.00	64.90	70.80	76.70
11	KNOK-FM	10,300	36.05	41.20	46.35	51.50	56.65	61.80	66.95
12	KEGL-FM	9,800	34.30	39.20	44.10	49.00	53.90	58.80	63.70
13	KLVU-FM	8,300	29.05	33.20	37.35	41.50	45.65	49.80	53.95
14	KTXQ-FM	7,500	26.25	30.00	33.75	37.50	41.25	45.00	48.75
15	KAAM-AM	6,900	24.15	27.60	31.05	34.50	37.95	41.40	44.85
16	KAFM-FM	6,800	23.80	27.20	30.60	34.00	37.40	40.80	44.20
17	KKDA-AM	6,400	22.40	25.60	28.80	32.00	35.20	38.40	41.60
18	WFAA-AM	6,400	22.40	25.60	28.80	32.00	35.20	38.40	41.60
19	KESS-FM	5,700	19.95	22.80	25.65	28.50	31.35	34.20	37.05
20	KIXK-FM	5,200	18.20	20.80	23.40	26.00	28.60	31.20	33.80

NOTE: Population for "ADULTS 25-54" is 1,337,900

[BreakOut Report copyright 1983 Jefferson-Pilot Data Systems Charlotte NC]

IF YOU KNOW YOUR CPM AND WANT TO SEE EQUIVALENT RATES FOR THE REST OF THE MARKET, THIS REPORT WILL SHOW THEM TO YOU. SOME PEOPLE CALL THIS A "REVERSE CPM". YOU PICK THE CPM LEVELS AND INTERVALS AND ANY OR ALL STATIONS AND THIS REPORT WILL DO THE REST. THIS WILL HELP YOU JUSTIFY YOUR RATES WHENEVER YOU NEED TO. YOU MAY ALSO DISPLAY RATES BASED ON COST-PER-POINT VALUES.

FIGURE 4.11 Computerized breakouts show a client how well a station performs. Computers have become an integral part of radio sales. Courtesy Jefferson Pilot Data Systems.

PROSPECTING AND LIST BUILDING

When a salesperson is hired by a radio station, he or she is customarily provided with a list of accounts to which airtime may be sold. For an inexperienced salesperson, this list may consist of essentially inactive or dormant accounts, that is, businesses that either have been on the air in the past or those that have never purchased airtime on the station. The new sales rep is expected to breathe life into the list by selling spot schedules to those accounts listed, as well as by adding to the list by bringing in new business. This is called list building, and it is the primary challenge facing the new account executive.

A more active list, one that generates commissions, will be given to the more experienced radio salesperson. A salesperson may be persuaded to leave one station in favor of another based upon the contents of a list, which may include large accounts and prominent advertising agencies. Lists held by a station's top billers

FIGURE 4.11
continued

```
==============        ---------------------------------------
== BreakOut ==        SCHEDULE FOR Dallas Metro Ford Dealers
==============        ---------------------------------------

                          < ADULTS 25-54, MON-FRI  6A-7P >

                      [ DALLAS/FT WORTH ARB METRO: SPRNG-83 ]

Fall New Car Introduction

Campaign

Population:  1,337,900   --  Desired frequency: 3.0 over 2 week(s)

Radio's reach is 1,277,500 (95.5%)

Campaign length: 2 WEEKS -- Campaign Budget: $    12,000
```

	SPOTS /WEEK	CUME (00)	AQH (00)	CPM	COST/ POINT	CMPGN GRPS	-- REACH -- (00)	%DEMO	CMPGN FREQ	CAMPAIGN COST
KVIL-FM	6	267,7	39,4	5.33	$71	32.4	159,7	11.9%	2.7	$2,310
KSCS-FM	6	194,6	26,9	7.43	$99	22.1	113,2	8.5%	2.6	$2,200
KPLX-FM	5	169,4	25,7	7.00	$93	19.2	98,4	7.4%	2.6	$1,800
KMEZ-FM	6	129,1	16,9	9.47	$126	15.2	76,0	5.7%	2.7	$1,920
		Summary -->		6.89	$92	88.9	447,4	33.4%	3.0	$8,230

```
                    NET REACH is    394,1 29.5
                    UNDER BUDGET by       $3,770

NOTE: "%DEMO" column represents the percent of ADULTS 25-54 reached.
      Reach & Frequency is derived from the Group W reach curves.
```

THIS IS THE "SCHEDULE GENERATOR". TAKING THE STATIONS YOU SELECT, THE BUDGET LIMIT YOU SET AND YOUR REACH & FREQUENCY GOALS, THIS PLANNER WILL CREATE A SCHEDULE YOU CAN USE OR MODIFY TO SUIT YOUR NEEDS.

invariably contain the most enthusiastic radio users. Salespeople cultivate their lists as a farmer does his fields. The more the account list yields, the more commissions in the salesperson's pocket.

New accounts are added to a sales rep's list in several ways. Once the status of the list's existing accounts is determined, and this is accomplished through a series of in-person calls and presentations, a salesperson must begin prospecting for additional business. Area newspapers are a common source. When a salesperson finds an account that he wishes to add to his list, the account must be "declared." This involves consulting the sales manager for approval to add the account to the salesperson's existing list. In some cases the account declared may already belong to another salesperson. If it is an "open" account, the individual who comes forward first is usually allowed to add it to his list.

Other sources for new accounts include the Yellow Pages, television stations, and competing radio outlets. In the first case, every business in the area is listed in this directory, and many have display ads that provide useful informa-

```
==============
== BreakOut ==                 ----------------------------
==============                 SCHEDULE ANALYZER -- PART I
                               ----------------------------

                               < SPOT PLACEMENT -- ADULTS >

                        [ DALLAS/FT WORTH ARB METRO: SPRNG-83 ]


========================
= Schedule for KVIL-FM =
*-----------------------------------+-----------------------------------*
* daypart          spots     rate   daypart            spots     rate   *
*-----------------------------------+-----------------------------------*
* MON-FRI 6A-10A                    MON-SUN 6A-MID                      *
* MON-FRI 10A-3P                                                        *
* MON-FRI  3P-7P                    MON-FRI 6A-MID                      *
* MON-FRI 7P-MID                                                        *
*                                   SAT-SUN 6A-MID                      *
*   SAT   6A-10A                                                        *
*   SAT   10A-3P                                                        *
*   SAT    3P-7P                    MON-FRI DRIVES                      *
*   SAT   7P-MID                    MON-FRI  6A-7P      12  @ $ 210     *
*                                                                       *
*   SUN   6A-10A                    MON-SAT 6A-10A                      *
*   SUN   10A-3P                    MON-SAT 10A-3P                      *
*   SUN    3P-7P                    MON-SAT  3P-7P                      *
*   SUN   7P-MID                    MON-SAT 7P-MID                      *
*-----------------------------------+-----------------------------------*
*  12 spots run /// Total cost: $  2520.00 /// Average cost: $  210.00  *
*-----------------------------------+-----------------------------------*
```

THE "SCHEDULE ANALYZER" IS A 2-PART REPORT. IT LETS YOU SEE EXACTLY HOW A PLANNED SCHEDULE WILL LOOK. PART I SHOWS THE SPOT PLACEMENT/COST DETAILS. PART II SHOWS THE ACTUAL ANALYSIS. THE NEXT FEW PAGES SHOW YOU A SINGLE STATION SCHEDULE ANALYZER AND A "BUY" INVOLVING SEVERAL STATIONS.

FIGURE 4.11
continued

tion. Local television stations are viewed with an eye toward its advertisers. Television can be an expensive proposition, even in smaller markets, and businesses that spend money on it may find radio's rates more palatable. On the other hand, if a business can afford to buy television, it often can afford to embellish its advertising campaign with radio spots. Many advertisers place money in several media—newspaper, radio, television—simultaneously. This is called a "mixed media" buy and is a proven advertising formula for the obvious reason that the client is reaching all possible audiences. Finally, accounts currently on other stations constitute good prospects since they obviously already have been sold on the medium.

In the course of an average workday, a salesperson will pass hundreds of businesses, some of which may have just opened their doors, or are about to do so. Sales reps must keep their eyes open and be prepared to make an impromptu call. The old saying "The early bird gets the worm" is particularly relevant in radio sales. The first account executive into a newly launched business often is the one who gets the buy.

A list containing dozens of accounts does not necessarily assure a good income. If those businesses listed are small spenders or inactive, little in the way of commissions will be generated and billing will be low. The objective of list building is not merely to increase the number of accounts, but rather to raise the level of commissions it produces. In other words, a list that contains thirty accounts, of which twenty-two are active, is preferable to one with fifty accounts containing only twelve that are doing business with the station. A salesperson does not get points for having a lot of names on his list.

FIGURE 4.12
Computer-generated customized sales presentations have become commonplace, especially in larger markets. Courtesy Jefferson Pilot Data Systems.

It is the sales manager's prerogative to shift an account from one salesperson's list to another's if he believes the account is being neglected or handled incorrectly. At the same time, certain in-house accounts, those handled by the sales manager, may be added to a sales rep's list as a reward for performing well. A salesperson's account list also may be pared down if the sales manager concludes that it is disproportional with the others at the station. The attempt to more equitably distribute the wealth may cause a brouhaha with the account person whose list is being trimmed. The sales manager attempting this feat may find himself losing a top biller; thus, he must consider the ramifications of such a move and proceed accordingly. This may even mean letting things remain as they are. The top biller often is responsible for as much as 30 to 40 percent of the station's earnings.

PLANNING THE SALES DAY

A radio salesperson makes between seventy-five and one hundred in-person calls a week, or on the average of fifteen to twenty each day. This requires careful planning and organization. WHTT's Ron Piro advises preparing a day's itinerary the night before. "There's nothing worse than facing the day without an idea of where to go. A salesperson can spare himself that dreaded sensation and a lot of lost time by preparing a complete schedule of calls the night before."

When preparing a daily call sheet a salesperson, especially one whose station covers a vast area, attempts to centralize, as much as possible, the businesses to be contacted. Time, energy, and gas are needlessly expended through poor planning. A sales rep who is traveling ten miles between each presentation can only get to half as many clients as the person with a consolidated call sheet. Of course, there are days when a salesperson must spend more time traveling. Not every day can be ideally plotted. It may be necessary to make a call in one part of the city at nine A.M. and be in another part at ten A.M. A salesperson must be where he or she feels the buys are going to be made. "Go first to those businesses likeliest to buy. The tone of the day will be sweetened by an early sale," contends Piro.

Sales managers advise their reps to list more prospects than they expect to contact. In so doing, they are not likely to run out of places to go should those prospects they had planned to see be unavailable. "You have to make the calls to make the sales. The more calls you make, the more the odds favor a sale," points out KOUL's Gene Etheridge.

Itineraries should be adhered to regardless of whether a sale is made early in the day, says WNRI's John Gregory. "You can't pack it in at ten in the morning because you've closed an account. A salesperson who is easily satisfied is one who will never make much money. You must stay true to your day's game plan and follow through. No all-day coffee klatch at the local Ho-Jo's or movie matinee because you nailed an order after two calls."

The telephone is one of the salesperson's best tools. While it is true that a client cannot sign a contract over the phone, much time and energy can be saved through its effective use. Appointments can be made and a client can be qualified via the telephone. That is to say, a salesperson can ascertain when the decision maker will be available. "Rather than travel twenty miles without knowing if the person who has the authority to make a buy will be around, take a couple of minutes and make a phone call. As they say, 'Time is money.' In the time spent finding out that the store manager or owner is not on the premises when you get there, other, more productive calls can be made," says WKVT's Friedman.

If a client is not available when the salesperson appears, a call-back should be arranged for either later the same day or soon thereafter. The prospective advertiser should never be forgotten or relegated to a call three months hence. If the sales rep is able to rearrange his schedule to accommodate a return visit the same day, given that the person to see is available, then he should do so. However, it is futile to make a presentation to someone who cannot give full attention. The sales rep who arrives at a business only to find the decision maker overwhelmed by distractions is wise to ask for another appointment. In fact, the client will perceive this as an act of kindness and consideration. Timing is important.

A record of each call should be kept for follow-up purposes. When calling on a myriad of accounts, it is easy to lose track of what transpired during a particular call. Maintaining a record of a call requires little more than a brief notation after it is made. Notes may then be periodically reviewed to help determine what action should be taken on the account. Follow-ups are crucial. There is nothing more embarrassing and disheartening than to discover a client, who was pitched and then forgotten, advertising on another station. Sales managers usually require that salespeople turn in copies of their call sheets on a daily or weekly basis for review purposes.

GAIL MITCHELL

A FRESH START

Rebutting Sales Objections

Welcome to 1985! It's a fresh start on a new programming year: a new lease on attracting listeners, ratings, and those all-important sales. And speaking of sales, TM Beautiful Music Director/Programming Steve Hibbard sent in a timely piece I'd like to share. It offers suggested rebuttals against common format sales objections:

Here are some frequently-heard arguments related to the Easy Listening format, and a number of responses that salespeople should have in reserve.

CLIENT: "Your listeners are too old."

ANSWER: "Who is it you want to reach? We know we're not the best buy for skating rinks, rock concerts, or motorcycles. But we *do* reach a lot of the educated, affluent adults over the age of 35. Those are the people with the most discretionary income."

(Obviously, your approach here depends on the prospect's product or service, and whether you are a rating service subscriber with numbers to pull out if you have to.)

CLIENT: "Your radio station is background. No one will hear my commercial."

ANSWER: "All radio is background. With the exception of News and Talk formats, true foreground radio went out with 'Amos 'n' Andy.' When did you last see people gathered around the radio for any length of time? People use radio as a companion to their activities, matching a present mood or changing one. WXXX is the only station that provides a relaxing mood."

CLIENT: "That's what I mean; relaxing music is too background."

ANSWER: "Actually a relaxing mood is the *best* environment for your commercial. Haven't you noticed how everything on those other stations is grabbing for the listener's attention – the DJs, the jingles, the contests, the features, and *lots* of commercials? We know why many advertisers run hard-sell spots on those stations. It's an attempt to be heard over everything else. We solve that by creating an uncluttered environment for your spot . . . 12 or 13 minutes of music and then only two or three messages."

CLIENT: "But those long periods of relaxing music; they put the whole station into the background."

ANSWER: "Have you listened to our station lately? I'll bet it's brighter and more contemporary than you think! But the fact remains that WXXX *is* the station for relaxing. Won't you agree that your prospective customer is going to be more receptive to your message if he or she is in a comfortable, relaxed frame of mind after enjoying a quarter-hour of uninterrupted music?"

CLIENT: "That makes sense, and I don't have a problem with your ratings, but I'm still not convinced that people will hear and respond to the commercials."

ANSWER: (At this point you will be glad if your station has conducted the "Listen While You Work" promotion. You should be prepared to produce evidence of all the people who wrote in.)

"What if I told you that people took time to write us just for a *chance* at a drawing for a coffee cake for their morning break? Furthermore, all these people heard and responded to our promotion while listening *on the job!*"

CLIENT: "You're too expensive. I can get another station for less."

ANSWER: "Yes, but will it be effective for you? Not only do we deliver upscale adults with discretionary income, but we very strictly limit clutter on our station so that your message actually *stands out.*"

FIGURE 4.13
Trade magazine article on rebutting objections related to the Easy Listening format. Reprinted with permission from *Radio and Records*.

SELLING WITH AND WITHOUT NUMBERS

Not all stations can claim to be number one or two in the ratings. In fact, not all stations appear in any formal ratings survey. Very small markets are not visited by Arbitron or Birch for the simple reason that there may only be one station broadcasting in the area. An outlet in a nonsurvey area relies on its good reputation in the community to attract advertisers. In small markets salespeople do not work out of a ratings book, and clients are not concerned with cumes and shares. In the truest sense of the word, an account person must sell the station. Local businesses often account for more than 95 percent of a small market station's revenue. Thus, the stronger the ties with the community the better. Broadcasters in rural markets must foster an image of good citizenship in order to make a living.

Civic-mindedness is not as marketable a commodity in the larger markets as are ratings points. In the sophisticated multistation urban market, the ratings book is the bible. A station without numbers in the highly competitive environment finds the task of earning an income a difficult

WXXX-FM Call Sheet

AE: Sue Hart   Date: July 16, 1988

Berk's Jewelry

Mr. Berk not interested now. Wants to do something in the Fall. Will return in late August with a plan.

Hart Housewares

Sold 20 ROS 60's for week of July 24. Will do more in August.

B and R Travel

Still not ready to budge. Sold on newspapers. Will return with a spec tape and special package plan next week. Contact: Rose Fitzgerald.

Sum La's Garden

Interested in drive-time. On WZZZ three years ago. Not happy with results. Likes our format. Is warming. Will follow-up end of week for the close. Contact: Mr. Sum Lee.

Furniturama

Business slow. Won't move on proposal. Maybe in September. Checking for co-op dollars. Call back appointment August 23. Contact: Bill Meyers.

FIGURE 4.14
Excerpt from a salesperson's call sheet.

one, although there are numerous examples of low-rated stations that do very well. However, "no-numbers" pretty much puts a metro area station out of the running for agency business. Agencies almost invariably "buy by the book." A station without numbers "works the street," to use the popular phrase, focusing its sales efforts on direct business.

An obvious difference in approaches exists between selling the station with ratings and the one without. In the first case, a station centers its entire presentation around its high ratings. "According to the latest Arbitron, WXXX-FM is number one with adults 24 to 39." Never out of the conversation for very long are the station's numbers, and at advertising agencies the station's standing speaks for itself. "We'll buy WXXX because the book shows that they have the largest audience in the demos we're after."

The station without ratings numbers sells itself on a more personal level, perhaps focusing on its unique features and special blend of music and personalities, and so forth. In an effort to attract advertisers, nonrated outlets often develop programs with a targeted retail market in mind; for example, a home "how-to" show designed to interest hardware and interior decor stores, or a cooking feature aimed at food and appliance stores.

The salesperson working for the station with the cherished "good book" must be especially adept at talking numbers, since they are the key subject of the presentation in most situations. "Selling a top-rated metro station like WHTT requires more than a pedestrian knowledge of numbers, especially when dealing with agencies. In big cities, retailers have plenty of book savvy, too," contends Piro.

Selling without numbers demands its own unique set of skills, notes WNRI's Gregory. "There are really two different types of radio selling—with numbers and without. In the former instance, you'd better know your math, whereas in the latter, you've got to be really effective at molding your station to suit the desires of the individual advertiser. Without the numbers to speak for you, you have to do all the selling yourself. Flexibility and ingenuity are the keys to the sale."

ADVERTISING AGENCIES

Advertising agencies came into existence more than a century ago and have played an integral role in broadcasting since its inception. During radio's famed heyday, advertising agencies were omnipotent. Not only did they handle the advertising budgets of some of the nation's largest businesses, but they also provided the networks with fully produced programs. The programs were designed by the agencies for the specific satisfaction of their clients. If the networks and certain independent stations wanted a company's business, they had little choice but to air the agency's program. This practice in the 1920s and 1930s gave ad agencies unprecedented power. At one point, advertising agencies were the biggest supplier of network radio programming. By the 1940s, agencies were forced to abandon their direct programming involvement, and the industry was left to its own devices, or almost. Agencies continued to influence the content of what was aired. Their presence continues to be felt today, but not to the extent that it did prior to the advent of television.

Agencies annually account for hundreds of millions in radio ad dollars. The long, and at times turbulent, marriage of radio and advertising agencies was and continues to be based on the need of national companies to convey their messages on the local level and the need of the local broadcaster for national business. It is a two-way street.

Today nearly two thousand advertising agencies use the radio medium. They range in size from mammoth to minute. Agencies such as Young and Rubicam, J. Walter Thompson, Dancer Fitzgerald Sample, and Leo Burnett bill in the hundreds of millions annually and employ hundreds. More typical, however, are the agencies scattered throughout the country that bill between one-half and two-and-a-half million dollars each year and employ anywhere from a half dozen to twenty people. Agencies come in all shapes and sizes and provide various services, depending on their scope and dimensions.

The process of getting national business onto a local station is an involved one. The major agencies must compete against dozens of others to win the right to handle the advertising of large companies. This usually involves elaborate presentations and substantial investments by agencies. When and if the account is secured, the agency must then prepare the materials—audio, video, print—for the campaign and see to it that the advertiser's money is spent in the most effective way possible. Little is done without extensive marketing research and planning. The agency's media buyer oversees the placement of dollars in the various media. Media buyers at national agencies deal with station and network reps and not directly with the stations themselves. It would be impossible for an agency placing a buy on four hundred stations to personally transact with each.

There basically are three types of agencies: *full-service agencies,* which provide clients with a complete range of services, including research, marketing, and production; *modular agencies,* which provide specific services to advertisers; and *in-house agencies,* which handle the advertising needs of their own business.

The standard commission that an agency receives for its service is 15 percent on billing. For example, if an agency places one hundred thousand dollars on radio it earns fifteen thousand dollars for its efforts. Agencies often charge clients additional fees to cover production costs, and some agencies receive a retainer from clients.

The business generated by agencies constitutes an important percentage of radio's revenues, especially for medium and large market stations. However, compared to other media, such as television for example, radio's allocation is diminutive. The nation's top three agencies invest over 80 percent of their broadcast budgets in television. Nonetheless, hundreds of millions of dollars are channeled into radio by agencies that recognize the effectiveness of the medium.

REP COMPANIES

Rep companies are the industry's middlemen. Rep companies are given the task of convincing national agency media buyers to place money on the stations they represent. Without their existence, radio stations would have to find a way to reach the myriad of agencies on their own—an impossible feat.

With few exceptions, radio outlets contract the services of a station rep company. Even the smallest station wants to be included in buys on the national level. The rep company basically is an extension of a station's sales department. The rep and the station's sales manager work closely. Information about a station and its market are crucial to the rep. The burden of keeping the rep fully aware of what is happening back at the station rests on the sales manager's shoulders. Since a rep company based in New York or Chicago would have no way of knowing that its client-station in Arkansas has decided to carry the local college's basketball games, it is the station's responsibility to make the information available. A rep cannot sell what it does not know exists. Of course, a good rep will keep in contact with a station on a regular basis simply to keep up on station changes.

There are far fewer radio station reps than there are ad agencies. Approximately 150 reps handle the nine thousand plus commercial stations around the country. Major rep firms, such as Katz, Eastman, Blair, and Torbet, pitch agencies on behalf of hundreds of client stations. The large and very successful reps often refuse to act as the envoy for small market stations because of their lack of earning potential. A rep company typically receives a commission of between 5 and 12 percent on the spot buys made by agencies, and since the national advertising money usually is directed first to the medium and large markets, the bigger commissions are not to be made from handling small market outlets. Many rep companies specialize in small market stations, however.

While a rep company may work the agencies on behalf of numerous stations, it will not handle two radio outlets in the same market. Doing so could result in a rep company being placed in the untenable position of competing with itself for a buy, thus creating an obvious conflict of interest.

The majority of station reps provide additional services. In recent years many have expanded into the area of programming and management consultancy, while almost all offer clients audience research data, as well as aid in developing station promotions and designing sales materials such as rate cards.

CO-OP SALES

It is estimated that over six hundred million dollars in radio revenue comes from co-op advertising—no small piece of change, indeed. Co-op advertising involves the cooperation of three parties: the retailer whose business is being promoted, the manufacturer whose product is being promoted, and the medium used for the promotion. In other words, a retailer and manufacturer get together to share advertising expenses. For example, Smith's Sporting Goods is informed by the Converse Running Shoes representative that the company will match, dollar for dollar up to five thousand, the money that the retailer invests in radio advertising. The only stipulation of the deal is that Converse be promoted in the commercials on which the money is spent. This means that no competitive product can be mentioned. Converse demands exclusivity for its contribution.

Manufacturers of practically every conceivable type of product, from lawn mowers to mobile homes, establish co-op advertising budgets. A radio salesperson can use co-op to great advantage. First the station account executive must determine the extent of co-op subsidy a client is entitled to receive. Most of the time the retailer knows the answer to this. Frequently, however, retailers do not take full advantage of the co-op funds that manufacturers make available. In some instances, retailers are not aware that a particular manufacturer will share radio advertising expenses. Many potential advertisers have been motivated to go on the air after discovering the existence of co-op dollars. Midsized retailers account for the biggest chunk of the industry's co-op revenues. However, even the smallest retailer likely is eligible for some subsidy, and a salesperson can make this fact known for everyone's mutual advantage.

The sales manager generally directs a station's co-op efforts. Large stations often employ a full-time co-op specialist. The individual responsible for stimulating co-op revenue will survey retail trade journals for pertinent information about available dollars. Retail associations also are a good source of information, since they generally possess manufacturer co-op advertising lists. The importance of taking advantage of co-op opportunities cannot be overstressed. Some stations, especially metro market outlets, earn hundreds of thousands of dollars in additional ad revenue through their co-op efforts.

From the retailer's perspective, co-op advertising is not always a great bargain. This usually stems from copy constraints imposed by certain manufacturers, which give the retailer a ten-second tag-out in a thirty- or sixty-second commercial. Obviously, this does not please the retailer who has split the cost of advertising fifty-fifty. In recent years, this type of copy domination by the manufacturer has decreased some and a more equitable approach, whereby both parties share evenly the exposure and the expense, is more commonplace.

Co-op also is appealing to radio stations since they do not have to modify their billing practices to accommodate the third party. Stations simply bill the retailer and provide an affidavit attesting to the time commercials aired. The retailer, in turn, bills the manufacturer for its share of the airtime. For its part, the manufacturer requires receipt of an affidavit before making payment. In certain cases, the station is asked to mail affidavits directly to the manufacturer. Some manufacturer's stipulate that bills be sent to audit houses, which inspect the materials before authorizing payment.

TRADE-OUTS

Stations commonly exchange airtime for goods, although top-rated outlets, whose time is sold at a premium, are less likely to swap spots for anything other than cash. Rather than pay for needed items, such as office supplies and furnishings, studio equipment, meals for clients and listeners, new cars, and so forth, a station may choose to strike a deal with merchants in which airtime is traded for merchandise. There are advertisers who only use radio on a trade basis. A station may start out in an exclusively trade relationship with a client in the hope of eventually converting him to cash. Split contracts also are written when a client agrees to provide both money and merchandise. For example, WXXX-FM needs two new office desks. The total cost of the desks is eight hundred dollars. An agreement is made whereby the client receives a fourteen-hundred-dollar ROS spot schedule in exchange for the desks and six hundred dollars in cash. Trade-outs are not always this equitable. Stations often provide trade clients with airtime worth two or three times the merchandise value in order to get what is needed. Thus the saying "Need inspires deals."

Many sales managers also feel that it makes good business sense to write radio trade contracts to fill available and unsold airtime, rather than let it pass unused. Once airtime is gone, it cannot be retrieved, and yesterday's unfilled availability is a lost opportunity.

CHAPTER HIGHLIGHTS

1. Selling commercials keeps radio stations on the air. Between 1920 and today, advertising revenues and forms reflected the ebb and flow of radio's popularity. Today, advertising dollars are selectively spent on spots aired during times of the day and on stations that attract the type of audience the advertiser wants to reach.

2. An effective radio commercial makes a strong and lasting impression on the mind of the listener.

3. A successful account executive needs an understanding of research methods, marketing, finance; some form of sales experience; and such personal traits as ambition, confidence, honesty, energy, determination, intelligence, and good grooming.

4. Since the 1970s programming people have made successful job transitions to sales because they have a practical understanding of the product they are selling.

5. Although an increasing number of station managers are being drawn from programming people, a sales background is still preferred.

6. The sales manager, who reports directly to the station's general manager, oversees the account executives, establishes departmental policies, develops sales plans and materials, conceives campaigns and promotions, sets quotas, works closely with the program director to develop salable features, and sometimes sells.

7. Rates for selling airtime vary according to listenership and are published on the station's rate card. The card lists terms of payment and commission, nature of copy and due dates, station's approach to discounting, rate protection policy, as well as feature and spot rates.

8. Station listenership varies according to time of day, so rate card daypart classifications range from the highest-costing AAA (typically 6–10 A.M. weekdays) to C (usually midnight–6 A.M.). Fixed position drivetime spots are usually among the most expensive to purchase.

9. For advertisers with limited funds, run-of-station (ROS), best time available (BTA), or total audience plan (TAP) are cost-effective alternatives.

10. Since few accounts are "closed" on the first call, it is used to introduce the station to the client and to determine its needs. Follow-up calls are made to offset reservations and, if necessary, to improve the proposal. Perseverance is essential.

11. Radio sales are drawn from three levels: retail, local, and national. Retail sales are direct sales to advertisers within the station's signal area. Local sales are obtained from advertising agencies representing businesses in the market area. National sales are obtained by the station's rep company from agencies representing national accounts.

12. A fully produced sample commercial (spec tape) is an effective selling tool. It is used to break down client resistance on call-backs, to interest former clients who have not bought time recently, and to encourage clients to increase their schedules.

13. The salesperson should commit the advertiser to sufficient commercials, placed properly, to ensure achieving the advertiser's objectives. Underselling is as self-defeating as overselling.

14. New accounts are added to a salesperson's list by "prospecting": searching newspapers, Yellow Pages, television ads, competing radio station ads, and new store openings. Only "open" accounts may be added (those not already "declared" by another salesperson at the same station).

15. Because a salesperson must average fifteen to twenty in-person calls each day, when preparing a daily call sheet it is important to logically sequence and centralize the businesses to be contacted. Also, advance telephone contacts can eliminate much wasted time.

16. A salesperson at a station with a high rating has a decided advantage when contacting advertisers. Stations with low or no "numbers" must focus on retail sales (work the street), developing programs and programming to attract targeted clients. Stations in nonsurvey areas must rely on their image of good citizenship and strong community ties.

17. Ad agencies annually supply hundreds of millions of dollars in advertising revenue to stations with good ratings. Media buyers at the agencies deal directly with station and network reps.

18. A station's rep company must convince national agency media buyers to select their station as their advertising outlet for the area. Therefore, the station's sales manager and the rep must work together closely.

19. Co-op advertising involves the sharing of advertising expenses by the retailer of the business being promoted and the manufacturer of the product being promoted.

20. Rather than pay for needed items or to obtain something of value for unsold time, a station may trade (trade-out) advertising airtime with a merchant in exchange for specific merchandise.

SUGGESTED FURTHER READING

Aaker, David A., and Myers, John G. *Advertising Management,* 2nd ed. Englewood Cliffs, N.J.: Prentice-Hall, 1982.

Barnouw, Erik. *The Sponsor: Notes on a Modern Potentate*. New York: Oxford University Press, 1978.

Bergendorf, Fred. *Broadcast Advertising*. New York: Hastings House, 1983.

Bovee, Courtland, and Arena, William F. *Contemporary Advertising*. Homewood, Ill.: Irwin, 1982.

Burton, Philip Ward, and Sandhusen, Richard. *Cases in Advertising*. Columbus, Ohio: Grid Publishing, 1981.

Culligan, Matthew J. *Getting Back to the Basics of Selling*. New York: Crown, 1981.

Cundiff, Edward W., Still, Richard R., and Govoni, Norman A.P. *Fundamentals of Modern Marketing,* 3rd ed. Englewood Cliffs, N.J.: Prentice-Hall, 1980.

Delmar, Ken. *Winning Moves: The Body Language of Selling*. New York: Warner, 1984.

Dunn, W. Watson, and Barban, Arnold M. *Advertising: Its Role in Modern Marketing,* 4th ed. Hinsdale, Ill.: Dryden Press, 1978.

Gardner, Herbert S. Jr. *The Advertising Agency Business*. Chicago: Crain Books, 1976.

Gilson, Christopher, and Berkman, Harold W. *Advertising Concepts and Strategies*. New York: Random House, 1980.

Heighton, Elizabeth J., and Cunningham, Don R. *Advertising in the Broadcast and Cable Media,* 2nd ed. Belmont, Calif.: Wadsworth Publishing, 1984.

Hoffer, Jay, and McRae, John. *The Complete Broadcast Sales Guide for Stations, Reps, and Ad Agencies*. Blue Ridge Summit, Pa.: Tab Books, 1981.

Johnson, J. Douglas. *Advertising Today*. Chicago: SRA, 1978.

Jugenheimer, Donald W., and Turk, Peter B. *Advertising Media*. Columbus, Ohio: Grid Publishing, 1980.

Kleppner, Otto. *Advertising Procedure,* 7th ed. Englewood Cliffs, N.J.: Prentice-Hall, 1979.

McGee, William L. *Broadcast Co-op, the Untapped Goldmine*. San Francisco: Broadcast Marketing Company, 1975.

———. *Building Store Traffic with Broadcast Advertising*. San Francisco: Broadcast Marketing Company, 1978.

Montgomery, Robert Leo. *How to Sell in the 1980's*. Englewood Cliffs, N.J.: Prentice-Hall, 1980.

Murphy, Jonne. *Handbook of Radio Advertising*. Radnor, Pa.: Chilton Books, 1980.

Schultz, Don E. *Essentials of Advertising Strategy*. Chicago: Crain Books, 1981.

Sissors, Jack Z., and Surmanek, Jim. *Advertising Media Planning,* 2nd ed. Chicago: Crain Books, 1982.

Standard Rate and Data Service: Spot Radio. Skokie, Ill.: SRDS, annual.

Willing, Si. *How to Sell Radio Advertising*. Blue Ridge Summit, Pa.: Tab Books, 1977.

5 News

NEWS FROM THE START

The medium of radio was used to convey news before news of the medium had reached the majority of the general public. Ironically, it was the sinking of the *Titanic* in 1912 and the subsequent rebroadcast of the ship's coded distress message that helped launch a wider awareness and appreciation of the new-fangled gadget called the "wireless telegraphy." It was not until the early 1920s, when the "wireless" had become known as the "radio," that broadcast journalism actually began to evolve.

A historical benchmark in radio news is the broadcast of the Harding–Cox election results in 1920 by stations WWJ in Detroit and KDKA in Pittsburgh, although the first actual newscast is reported to have occurred in California a decade earlier. Despite these early ventures, news programming progressed slowly until the late 1920s. By then, two networks, NBC and CBS, were providing national audiences with certain news and information features.

Until 1932 radio depended on newspapers for its stories. That year newspapers officially perceived the electronic medium as a competitive threat. Fearing a decline in readership, they imposed a blackout. Radio was left to its own resources. The networks put forth substantial efforts to gather news and did very well without the wire services that they had come to rely on. Late in 1934, United Press (UP), International News Services (INS), and Associated Press (AP) agreed to sell their news services to radio, thus ending the boycott. However, by then the medium had demonstrated its ability to fend for itself.

Radio has served as a vital source of news and information throughout the century's most signifcant historical events. When the nation was gripped by economic turmoil in the 1930s, the incumbent head-of-state, Franklin D. Roosevelt, demonstrated the tremendous reach of the medium by using it to address the people. The majority of Americans heard and responded to the president's talks.

Radio's status as a news source reached its apex during World War II. On-the-spot reports and interviews, as well as commentaries, brought the war into the nation's living rooms. In contrast to World War I, when the fledgling wireless was exclusively used for military purposes, during World War II radio served as the primary link between those at home and the foreign battlefronts around the globe. News programming during this troubled period matured, while the public adjusted its perception of the medium, casting it in a more austere light. Radio journalism became a more credible profession.

The effects of television on radio news were wide ranging. While the medium in general reeled from the blow dealt it by the *enfant terrible,* in the late 1940s and early 1950s news programming underwent a sort of metamorphosis. Faced with drastically reduced network schedules, radio stations began to localize their news efforts. Attention was focused on area news events rather than national and international. Stations that had relied almost exclusively on network news began to hire newspeople and broadcast a regular schedule of local newscasts. By the mid-1950s the transformation was nearly complete, and radio news had become a local programming matter. Radio news had undergone a 180-degree turn, even before the medium gave up trying to directly compete on a program-for-program basis with television. By the time radio set a new and revivifying course for itself by programming for specific segments of the listening audience, local newscasts were the norm.

Since its period of reconstruction in the 1950s, radio has proven time and time again to be the nation's first source of information about major news events. The majority of Americans first heard of the assassination of President Kennedy and the subsequent shootings of Martin Luther King, Jr., and Senator Robert Kennedy over radio. In 1965 when most of the Northeast was crippled by a power blackout, battery-powered radios literally became a lifeline for millions of people by providing continuous news and information until power was restored.

During the 1970s and 1980s, news on both the world and local levels reached radio listeners first. Today the public knows that radio is the first place to turn for up-to-the-minute news about occurrences halfway around the world, as well as in its own backyard.

NEWS AND TODAY'S RADIO

More people claim to listen to radio for music than for any other reason. Somewhat surprising, however, is that most of these same people admit to relying on the medium for the news they receive. A study by the National Association of Broadcasters found that while 97 percent of those surveyed tuned in to radio for entertainment, 73 percent considered news programming important. The NAB also ascertained that radio is the first morning news source for nearly 70 percent of all full-time working women.

According to the Radio Advertising Bureau, over 60 percent of young adults get their first news of the day from radio. In comparison only 16 percent of adults rely on newspapers as the first source of daily news, while 21 percent tune in to television. Practically all of the nation's nine thousand commercial stations program news to some extent. Radio's tremendous mobility and pervasiveness has made it an instant and reliable news source for over 150 million Americans.

FIGURE 5.1
News is a prominent feature on most radio stations. Courtesy RAB.

RADIO: THE IMMEDIATE MEDIUM

56% GET THEIR FIRST NEWS IN THE MORNING FROM RADIO

Where People Get the News First, 6 A.M.–10 A.M.

	Radio	TV	Newspapers	Other	None
Persons 12+	56%	21%	16%	1%	7%
Teens 12–17	56	23	13	—	7
Adults 18+	56	20	16	1	7
Adults 18–34	62	19	12	1	6
Adults 35–54	57	18	18	1	6
College Grads	56	15	23	—	6
Prof./Mgr. Males	56	16	24	1	2
Working Women	66	20	10	1	4

RADIO LEADS OTHER MEDIA AS FIRST-NEWS SOURCE IN MIDDAY

Where People Get the News First, 10 A.M.–3 P.M.

	Radio	TV	Newspapers	Other	None
Persons 12+	38%	23%	13%	2%	25%
Teens 12–17	24	22	22	2	30
Adults 18+	40	23	12	2	24
Adults 18–34	50	20	12	2	17
Adults 35–54	41	18	13	2	27
College Grads	43	9	16	1	32
Prof./Mgr. Males	41	5	16	4	34
Working Women	40	18	13	2	27

THE NEWSROOM

The number of individuals working in a radio station newsroom will vary depending on the size of the station and its format. On the average, a station in a small market employs one or two full-time newspeople. Of course, some outlets find it financially unfeasible to hire newspeople. These stations do not necessarily ignore news, rather they delegate responsibilities to their deejays to deliver brief newscasts at specified times, often at the top of the hour. Stations approaching news in this manner make it necessary for the on-air person to collect news from the wire service during record cuts and broadcast it nearly verbatim. Little, if any, rewrite is done, because the deejay simply does not have the time to do it. About the only thing that persons at "rip 'n' read" outlets can and must do is examine wire copy before going on the air. This eliminates the likelihood of mistakes. Again, all this is accomplished while the records are spinning.

Music-oriented stations in larger markets rarely allow their deejays to do news. Occasionally the person jocking the overnight shift will be expected to give a brief newscast every hour or two, but in metro markets this is fairly uncommon. There is generally a newsperson on duty around the clock. A top-rated station in a medium market typically employs four full-time newspeople; again, this varies depending on the status of the outlet and the type of programming it airs. For example, Easy Listening stations that stress music and deemphasize talk may employ only one or two newspeople. Meanwhile, an MOR station in the same market may have five people on its news staff in an attempt to promote itself as a heavy news and information outlet, even though its primary product is music. Certain music stations in major markets hire as many as a dozen news employees. This figure may include not only on-air newscasters, but writers, street reporters, and technical people as well. Stringers and interns also swell the figure.

During the prime listening periods when a station's audience is at its maximum, newscasts are programmed with greater frequency, sometimes twice as often as during other dayparts. The newsroom is a hub of activity as newspeople prepare for newscasts scheduled every twenty to thirty minutes. Half a dozen people may be involved in assembling news, but only two may actually enter the broadcast booth. A prime-time newscast schedule may look something like this:

A.M. Drive Coverage		**P.M. Drive Coverage**	
Smith	6:25 A.M.	Lopez	3:25 P.M.
Bernard	7:00 A.M.	Gardner	4:00 P.M.
Smith	7:25 A.M.	Lopez	4:25 P.M.
Bernard	8:00 A.M.	Gardner	5:00 P.M.
Smith	8:25 A.M.	Lopez	5:25 P.M.
Bernard	9:00 A.M.	Gardner	6:00 P.M.
Smith	9:25 A.M.	Lopez	6:25 P.M.

Midday and evening are far less frenetic in the newsroom, and one person per shift may be considered sufficient.

A standard-size newsroom in a medium market will contain several pieces of audio equipment, not to mention office furniture such as desks, typewriters, file cabinets, and so on. Reel-to-reel recorders and cassette and cartridge machines are important tools for the newsperson. The newsroom also will be equipped with various monitors to keep newspeople on top of what is happening at the local police and fire departments and weather bureau. Various wire-service machines provide the latest news, sports, stock, and weather information, as well as a host of other data. Depending on the station's budget, two or more news services may be used. Stations with a genuine commitment to news create work areas that are designed for maximum efficiency and productivity.

THE ALL-NEWS STATION

Stations devoted entirely to news programming arrived on the scene in the mid-1960s. Program innovator Gordon McLendon, who had been a key figure in the development of two music formats, Beautiful Music and Top 40, implemented All-News at WNUS-AM ("NEWS") in Chicago. In 1965, Group W, Westinghouse Broadcasting, changed WINS-AM in New York to All-News, and soon did the same at more of its metro outlets—KYW-AM, Philadelphia, and KFWB-AM, Los Angeles. While Group W was converting several of its outlets to nonmusic programming, CBS decided that All-News was

FIGURE 5.2
A metro market newsroom typically contains several desks, phones, tape recorders, typewriters or computer terminals, monitoring equipment, and teletype machines. Courtesy WMJX-FM.

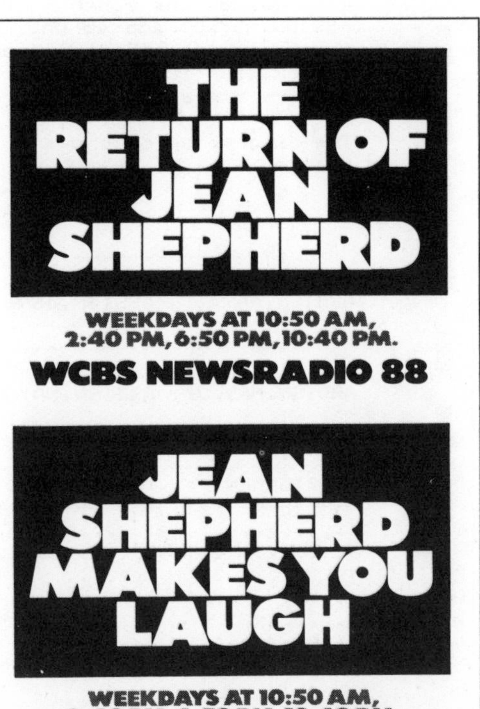

FIGURE 5.3
Promotional piece for All-News station. Courtesy WCBS.

the way to go at WCBS-AM, New York; KCBS-AM, San Francisco; and KNX-AM, Los Angeles.

Not long after KCBS in San Francisco began its All-News programming, another Bay City station, KGO-AM, introduced the hybrid News/Talk format in which news shares the microphone with conversation and interview features. Over the years, the hybrid approach has caught on and leads the pure All-News format in popularity.

Because of the exorbitant cost of running a news-only operation, it has remained primarily a metro market endeavor. It often costs several times as much to run an effective All-News station as it does one broadcasting music. This usually keeps small market outlets out of the business. Staff size in All-News far exceeds that of formats that primarily serve up music. Whereas a lone deejay is needed at an Adult Contemporary or Top 40 station, All-News requires the involvement of several people to keep the air sound credible.

While the cost of running a News station is high, the payback can more than justify expenditures. However, this is one format that requires a sizable initial investment, as well as

FILE CABINET ORGANIZATION:

Each of the "file cabinets" named above are organized just like the file cabinets you use in your newsroom today. For example, let's look at how you might arrange the ASSIGN file cabinet.

The main file cabinet is named ASSIGN. Inside are file drawers for each month. Therefore, you have one drawer named ASSIGN.MAY, another named ASSIGN.JUNE, another named ASSIGN.JULY and so on. Inside of each file drawer are the file folders. Using ASSIGN.MAY for example, there's a file folder for assignments for each day of the month -- ASSIGN.MAY.01, ASSIGN.MAY.02, ASSIGN.MAY.03 and so on. Finally inside of each file folder are the individual sheets of paper that hold the assignment information for each day. You can have as many of these as you need -- there's no limit.

The diagram below shows how this organization might appear if charted.

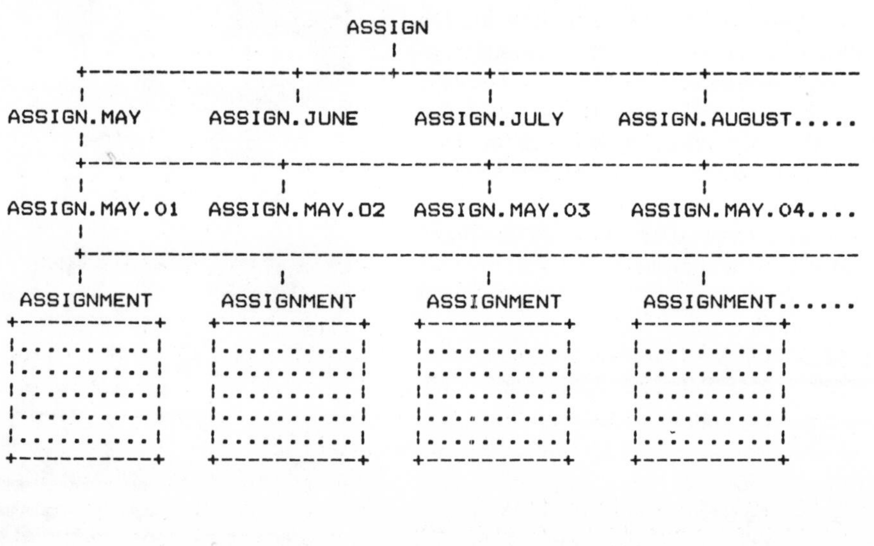

```
                              ASSIGN
                                 |
      +--------------+------+------+--------------+----------
      |              |             |              |
ASSIGN.MAY     ASSIGN.JUNE    ASSIGN.JULY    ASSIGN.AUGUST.....
      |
      +-------------+--------------+--------------+----------
      |             |              |              |
ASSIGN.MAY.01  ASSIGN.MAY.02  ASSIGN.MAY.03  ASSIGN.MAY.04....
      |
      +--------------------------------------------------------
      |             |              |              |
 ASSIGNMENT     ASSIGNMENT     ASSIGNMENT     ASSIGNMENT......
+----------+   +----------+   +----------+   +----------+
|..........|   |..........|   |..........|   |..........|
|..........|   |..........|   |..........|   |..........|
|..........|   |..........|   |..........|   |..........|
|..........|   |..........|   |..........|   |..........|
|..........|   |..........|   |..........|   |..........|
+----------+   +----------+   +----------+   +----------+
```

FIGURE 5.4 Computerized news assignment file. Computers have transformed the radio newsroom. Courtesy Jefferson Pilot Data Systems.

Tue Oct 09 12:36 page 1

SLUG	ANCHOR	WRITER	CART NO.	COPY	TAPE	TOTAL	CUME
OPENER				0:08	00:00	0:08	- 8:00
NEWS TEASER				0:18	00:00	0:18	- 7:52
COMMERCIAL ONE				0:00	01:00	1:00	- 7:34
NEWS SEGMENT				0:00	00:00	0:00	- 6:34
DEBATE REAX		cundiff	12	0:32	00:45	1:17	- 6:34
PRESIDENTIAL VISIT		cundiff	21A/B	0:34	00:30	1:04	- 5:17
I77 FATAL		cundiff	NA	0:28	00:00	0:28	- 4:13
TRANSIT MALL UPDATE		cundiff	17	0:18	00:42	1:00	- 3:45
SPORTS				0:06	01:00	1:06	- 2:45
COMMERCIAL TWO				0:00	01:00	1:00	- 1:39
WEATHER				0:30	00:00	0:30	- 0:39
CLOSE				0:09	00:00	0:09	- 0:09

FIGURE 5.5 Computerized news scheduling. Courtesy Jefferson Pilot Data Systems.

the financial wherewithal and patience to last until it becomes an established and viable entity. Considerable planning takes place before a station decides to convert to News, since it is not simply a matter of hiring new jocks and updating the music library. Switching from a music format to News is dramatic and anything but cosmetic.

AM has always been the home of the All-News station. There are only a handful of FM News outlets. The format's prevalence on AM has grown considerably since the late 1970s when FM took the lead in listeners. The percentage of News and News/Talk formats on AM continued to increase in the 1980s as the band lost more and more of its music listeners to FM. However, News stations in many metro markets keep AM at the top of the ratings charts. In the mid-1980s, it was common to find one AM outlet among the leaders, and almost invariably it programmed nonmusic. Some media observers predict that All-News will make significant inroads into FM as AM stereo outlets delve into music and more new outlets hit the airways.

THE ELECTRONIC NEWSROOM

The use of computers in the radio newsroom has increased significantly in the 1980s. Computers linked to the various wire and information services are used to access primary and background data on fast-breaking stories and features. Many stations have installed video display terminals (VDTs) in on-air studios. Instead of hand-held copy, newscasters simply broadcast off the screens. Gradually desktop computers are replacing typewriters in the newsrooms at larger stations. The speed and agility with which copy can be produced and edited makes a computer the perfect tool for broadcast journalists.

Computers are slowly appearing at non–metro market outlets. However, the majority of small stations have found the cost of computers prohibitive. Radio computer consultant Vicki Cliff of KSLY/KUNA, San Luis Obispo, California, contends that it is not cost effective for a modest-income station to purchase computers just for its newsroom. However, she does believe that a station should commit to a total system if it is financially able to justify such an investment. According to Cliff, this means a station should bill at least five hundred thousand dollars annually before computerizing. Carl Zajas, president of McZ Limited, contends that a station's income need not be as high before it considers acquiring a computer system. "I'd say a station billing in the vicinity of two hundred thousand dollars could legitimately consider computerizing. It used to be that a full-service computer system cost a small fortune, but no more. Today a station can get a multiterminal system for under twenty thousand dollars, and this means that a station eventually is going to save both time and money. As far as computers in station newsrooms are concerned, their value is inestimable."

On the whole, news directors agree with Zajas, but resistance to the use of computers does exist in some quarters. "All a newsroom really needs is phones, typewriters, and wireservice machines. A newsroom doesn't have to look like NASA's launch center. I've been doing radio news a long time, and it's my conclusion that it's good people who make things work," says Frank Titus, news director, WARA-AM, Attleboro, Massachusetts. Sherman Whitman, assistant news director of WBCN-FM, Boston, is of a similar opinion. "It's people who make the newsroom work. As far as I'm concerned, computers are not a vital necessity." News director Cecilia Mason, KGTM-AM, Wichita Falls, Texas, believes that computers can actually create some problems. "As the newspaper industry has found out, electronic does not necessarily mean better. Many papers lose stories, notes, and other valuable data when the computer breaks down. The radio medium is intangible enough. We don't need our news stories to vanish into thin air. It may sound old-fashioned, but I think it's important to maintain newsrooms as they now exist. I'm not condeming new technology, but I think it should only be incorporated into the scheme of things when it's an absolute asset and not just because it's the latest thing."

Possessing a different perspective, Judy Smith, news director at the San Antonio, Texas, station KAPE-AM, believes that the ever-increasing level of competition makes the high-tech newsroom a necessity. "Like it or not, the electronic newsroom is fast becoming a fact of life, and not just in major markets. Computers and satellites will be as much a part of a station's news operation as the telephone and tape recorder. The medium has entered a new age. There's no sense resisting it."

Despite the current debate, most industry experts predict that within the next decade computers will be standard equipment in the majority of radio station newsrooms. Larry Jewett, news director of WTOD-AM, Toledo, Ohio, suggests that today's newspeople, and individuals anticipating careers in radio news, become computer friendly. "Computers are a fact of life now. They cannot be ignored. Newspeople, and would-be newspeople, should learn all they can about the new technology because it is fast becoming an integral part of the profession."

THE NEWS DIRECTOR

News directors, like other department heads, are responsible for developing and implementing policies pertaining to their area, supervising staff members, and handling budgetary concerns. These are basic to any managerial position. However, the news department poses its own unique challenges to the individual who oversees its operation. These challenges must be met with a considerable degree of skill and know-how. Education and training are important. Surveys have concluded that station managers look for college degrees when hiring news directors. In addition, most news directors have, on the average, five years of experience in radio news before advancement to the managerial level.

The news director and program director work closely. At most stations, the PD has authority over the news department, since everything going over the air or affecting the air product is his direct concern and responsibility. Any changes in the format of the news or in the scheduling of newscasts or newscasters may, in fact, have to be approved by the station's programmer. For example, if the PD is opposed to the news director's plans to include two or more taped reports (actualities) per newscast, he may withhold approval. While the news director may feel that the reports enhance the newscasts, the PD may argue that they create congestion and clutter. In terms of establishing the on-air news schedule, the PD works with the news director to ensure that the sound of a given newsperson is suitably matched with the time slot he or she is assigned.

Getting the news out rapidly and accurately is a top priority of the news director. "People tune radio news to find out what is happening right now. That's what makes the medium such a key source for most people. While it is important to get news on the air as fast as possible, it is more important that the stories broadcast be factual and correct. You can't sacrifice accuracy for the sake of speed. As news director at KAPE my first responsibility is to inform our audience about breaking events on the local level. That's what our listeners want to hear," says Judy Smith, who functions as a one-person newsroom at the San Antonio station. "Because I'm the only newsperson on duty, I have to spend a lot of time verifying facts on the phone and recording actualities. I don't have the luxury of assigning that work to someone else, but it has to be done."

WTOD's Larry Jewett perceives his responsibilities similarly. "First and foremost, the news director's job is to keep the listener informed of what is happening in the world around him. A newsperson is a gatherer and conveyor of information. News is a serious business. A jock can be wacky and outrageous on the air and be a great success. On the other hand, a newsperson must communicate credibility or find another occupation."

Gathering local news is the most time-consuming task facing a radio news director, according to KGTM's Cecilia Mason. "To do the job well you have to keep moving. All kinds of meetings—governmental, civic, business—have to be covered if you intend on being a primary source of local news. A station with a news commitment must have the resources to be where the stories are, too. A news director has to be a logistical engineer at times. You have to be good at prioritizing and making the most out of what you have at hand. All too often there are just too many events unfolding for a news department to effectively cover, so you call the shots the best way that you can. If you know your business, your best shot is usually more than adequate."

In addition to the gathering and reporting of news, public affairs programming often is the responsibility of the news director. This generally includes the planning and preparation of local information features, such as interviews, debates, and even documentaries.

WHAT MAKES A NEWSPERSON

College training has become an increasingly important criterion to the radio news director planning to hire personnel. It is not impossible to land a news job without a degree, but formal education is a definite asset. An individual planning to enter the radio news profession should consider pursuing a broadcasting, journalism, or liberal arts degree. Courses in political science, history, economics, and literature give the aspiring newsperson the kind of well-rounded background that is most useful. "Coming into this field, especially in the 1980s, a college degree is an attractive, if not essential, credential. There's so much that a newsperson has to know. I think an education makes the kind of difference you can hear, and that's what our business is about. It's a fact that most people are more cognizant of the world and write better after attending college. Credibility is crucial in this business, and college training provides some of that. A degree is something that I would look for in prospective newspeople," says Cecilia Mason.

FIGURE 5.6
Women have found greater employment opportunities in news than in other areas of radio programming. Courtesy WMJX-FM.

While education ranks high, most news directors still look for experience first. "As far as I'm concerned, experience counts the most. I'm not suggesting that education isn't important. It is. Most news directors want the person that they are hiring to have a college background, but experience impresses them more. I believe a person should have a good understanding of the basics before attempting to make a living at something. Whereas a college education is useful, a person should not lean back and point to a degree. Mine hasn't gotten me a job yet, though I wouldn't trade it for the world," notes WTOD's Jewett. KAPE's Smith agrees, "The first thing I think most news directors really look for is experience. Although I have a bachelor of arts degree myself, I wouldn't hold out for a person with a college diploma. I think if it came down to hiring a person with a degree versus someone with solid experience, I'd go for the latter."

Gaining news experience can be somewhat difficult, at least more so than acquiring deejay experience. Small stations, where the beginner is most likely to break into the business, have slots for several deejays but seldom more than one for a newsperson. It becomes even more problematical when employers at small stations want the one person that they hire for news to bring some experience to the job. Larger stations place even greater emphasis on experience. Thus the aspiring newsperson is faced with a sort of Catch 22 situation, in which a job cannot be acquired without experience and experience cannot be acquired without a job.

News director Frank Titus says that there are ways of gaining experience that will lead to a news job. "Working in news at high school and college stations is very valid experience. That's how Dan Rather and a hundred other newsmen got started. Also working as an intern at a commercial radio station fattens out the résumé. If someone comes to me with this kind of background and a strong desire to do news, I'm interested."

Among the personal qualities that most appeal to news directors are enthusiasm, aggressiveness, energy, and inquisitiveness. "I want someone with a strong news sense and unflagging desire to get a story and get it right. A person either wants to do news or doesn't. Someone with a pedestrian interest in radio journalism is more of a hindrance to an operation than a help," contends Mason. WARA's Titus wants someone who is totally devoted to the profession. "When you get right down to it, I want someone on my staff who eats, drinks, and sleeps news."

On the practical side of the ledger, WBCN's Sherman Whitman says typing or keyboard skills are essential. "If you can't type, you can't work in a newsroom. It's an essential ability, and the more accuracy and speed the better. It's one of those skills basic to the job. A candidate for a news job can come in here with two degrees, but if that person can't type, that person won't be hired. Broadcast students should learn to type." Meanwhile, WTOD's Jewett stresses the value of possessing a firm command of the English language. "Proper punctuation, spelling, and syntax make a news story intelligible. A newsperson doesn't have to be a grammarian, but he or she had better know where to put a comma and a period and how to compose a good clean sentence. A copy of Strunk and White's *Elements of Style* is good to have around."

An individual who is knowledgeable about the area in which a station is located has a major advantage over those who are not, says Whitman. "A newsperson here at WBCN has to know Boston inside out. I'd advise anybody about to be interviewed for a news position to find out as much as possible about the station's coverage area. Read back issues of newspapers, get socioeconomic stats from the library or chamber of commerce, and study street directories and maps of the town or city in which the station is located. Go into the job interview well informed, and you'll make a strong impression."

Unlike a print journalist, a radio newsperson also must be a performer. In addition to good writing and news-gathering skills, the newsperson in radio must have announcing abilities. Again, training is usually essential. "Not only must a radio newsperson be able to write a story, but he or she has to be able to present it on the air. You have to be an announcer, too. It takes both training and experience to become a really effective newscaster. Voice performance courses can provide a foundation," says KAPE's Smith. Most colleges with broadcasting programs offer announcing and newscasting instruction.

Entry-level news positions pay modestly, while newspeople at metro market stations earn impressive incomes. With experience come the better paying jobs. Finding that first full-time news position often takes patience and determination. Several industry trade journals, such as *Broadcasting,* RTNDA's *Communicator, Electronic Media,* and others, list news openings.

PREPARING THE NEWS STORY

Clean copy is imperative. News stories must be legible and intelligible and designed for effortless reading by the newscaster, or several different newscasters. Typos, mispunctuation, awkward phrasing, and incorrect spelling are anathema to the person at the microphone. Try reading the following news story aloud and imagine yourself in a studio broadcasting to thousands of perplexed listeners.

PRESIDENTREAGAN STATD TODXAY THAT

AMER. WILL NOLONGER TOLXRATE THE

BUILTUP OF SOVITE ARMS ALONXG THE

SYRIAN/LEBANON BORDERS,THE PRESIDENT

EGXPRESSED CONCENR ABOUT TH

GROWING TENSSIONS IN THE PART OF THE

WORLD, THIS MORNING AT A NEWS

CONFERANCE.

Going on the air with copy riddled with errors is inviting disaster. About the only things right about the preceding news copy are that it is typed in upper case and double-spaced. Here are a few suggestions to keep in mind when preparing a radio news story:

1. Type neatly. Avoid typos and x-outs. Eliminate a typing error completely. If it is left on the page, it could trip you up during a broadcast.
2. Use UPPER CASE throughout the story. It is easier to read. Don't forget, the story you are writing is going to be read on the air.
3. Double-space between lines for the same reason upper case is used—copy is easier to read. Space between lines of copy keeps them from merging together when read aloud.
4. Use one-inch margins. Don't run the copy off the page. Uniformity eliminates errors. At the same time, try not to break up words.
5. Avoid abbreviations, except for those meant to be read as such: YMCA, U.S.A., NAACP, AFL/CIO.
6. Write out numbers under ten, and use numerals for figures between 10 and 999. Spell out *thousand, million,* and so forth. For example, *21 million people,* instead of *21,000,000 people*. Numbers can be tricky, but a consistent approach prevents problems.
7. Use the phonetic spelling for words that may cause pronunciation difficulties, and underline the stressed syllable: Monsignor (Mon-<u>seen</u>-yor).
8. Punctuate properly. A comma out of place can change the meaning of a sentence.
9. When in doubt, consult a standard style guide. In addition, both AP and UPI publish handbooks on newswriting.

Notice how much easier it is to read a news story that is correctly prepared:

PRESIDENT REAGAN SAYS THAT AMERICA

NO LONGER WILL TOLERATE THE BUILD-UP

OF SOVIET ARMS ALONG THE SYRIAN AND

LEBANESE BORDERS. THE PRESIDENT

EXPRESSED GRAVE CONCERN ABOUT THE

GROWING TENSIONS IN THE MIDDLE EAST

AT A MORNING NEWS CONFERENCE.

Since radio news copy is written for the ear and not for the eye, its style must reflect that fact. In contrast to writing done for the printed page, radio writing is more conversational and informal. Necessity dictates this. Elaborately constructed sentences containing highly sophisticated language may effectively communicate to the reader but create serious problems for the listener, who must digest the text while it is being spoken. Whereas the reader has the luxury to move along at his or her own pace, the radio listener must keep pace with the newscaster or miss out on information. Radio writing must be accessible and immediately comprehensible. The most widely accepted and used words must be chosen so as to prevent confusion on the part of the listener, who usually does not have the time or opportunity to consult the dictionary. "Keep it simple and direct. No compound-complex sentences with dozens of esoteric phrases and terms. Try to picture the listener in your mind. He is probably driving a car or doing any number of things. Because of the nature of the medium, writing must be concise and conversational," contends KAPE's Judy Smith.

News stories must be well structured and organized. This adds to their level of understanding. The journalist's five W's—who, what, when, where, and why—should be incorporated into each story. If a story fails to provide adequate details, the listener may tune in elsewhere to get what radio commentator Paul Harvey calls "the rest of the story."

When quoting a source in a news story, proper attribution must be made. This increases credibility while placing the burden of responsibility for a statement on the shoulders of the person who actually made it:

THE DRIVER OF THE CAR THAT STRUCK THE

BUILDING APPEARED INTOXICATED,

ACCORDING TO LISA BARNES, WHO

VIEWED THE INCIDENT.

Uncorroborated statements can make a station vulnerable to legal actions. The reliability of news sources must be established. When there are doubts concerning the facts, the newsperson has a responsibility to seek verification.

ORGANIZING THE NEWSCAST

News on music-oriented radio stations commonly is presented in five-minute blocks and aired at the top or bottom of the hour. During drivetime periods stations often increase the length and/or frequency of newscasts. The five minutes allotted news generally is divided into segments to accommodate the presentation of specific information. A station may establish a format that allows for two minutes of local and regional stories, one minute for key national and international stories, one minute for sports, and fifteen seconds for weather information. A thirty- or sixty-second commercial break will be counted as part of the five-minute newscast.

The number of stories in a newscast may be preordained by program management or may vary depending on the significance and scope of the stories being reported. News policy may require that no stories, except in particular cases, exceed fifteen seconds. Here the idea is to deliver as many stories as possible in the limited time available, the underlying sentiment being that more is better. In five minutes, fifteen to twenty items may be covered. In contrast, other stations prefer that key stories be addressed in greater detail. As few as five to ten news items may be broadcast at stations taking this approach.

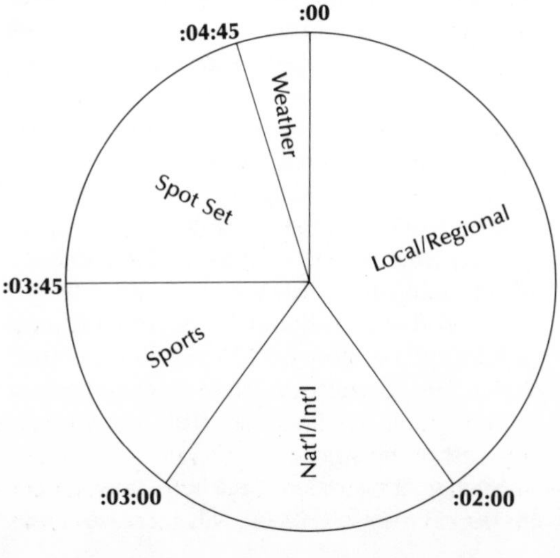

Stories are arranged according to their rank of importance, the most significant story of the hour topping the news. An informed newsperson will know what stories deserve the most attention. Wire services weigh each story and position them accordingly in news roundups. The local radio newsperson decides what wire stories will be aired and in what order.

Assembling a five-minute newscast takes skill, speed, and accuracy. Stories must be updated and rewritten to keep news broadcasts from sounding stale. This often requires that telephone calls be made for late-breaking information. Meanwhile, on-the-scene voicers (actualities) originating from audio news services (UPI, AP) or fed by local reporters must be taped and slotted in the newscast. "Preparing a fresh newscast each hour can put you in mind of what it must have been like to be a contestant on the old game show 'Beat the Clock.' A conscientious newsperson is a vision of perpetual motion," observes KGTM's Cecilia Mason.

Finally, most newspeople read their news copy before going on the air. "Reading stories cold is foolhardy and invites trouble. Even the most seasoned newscasters at metro market stations take the time to read over their copy before going on," comments WBCN's Whitman. Many newspeople read copy aloud in the news studio before airtime. This gives them a chance to get a feel for their copy. Proper preparation prevents unpleasant surprises from occurring while on the air.

FIGURE 5.7
Five-minute newscast format clock.

THE WIRE SERVICES

Without the aid of the major broadcast news wire services, radio stations would find it almost impossible to cover news on national and international levels. The wire services are a vital source of news information to nearly all of the nation's commercial radio stations. Both large and small stations rely on the news copy fed them by either Associated Press or United Press International, the two most prominent news wire services.

Broadcast wire services came into existence in the mid-1930s, when United Press (which became United Press International in 1958 after merging with International News Service) began providing broadcasters with news copy. Today UPI and AP serve over seventy-five hundred broadcast outlets.

Both news sources supply subscriber stations with around-the-clock coverage of national and world events. Over one hundred thousand stringers furnish stories from across the globe. AP and UPI also maintain regional bureaus for the dissemination of local news. Each wire service transmits over twenty complete news summaries daily. In addition, they provide weather, stock-market, and sports information, as well as a formidable list of features and data useful to the station's news and programming efforts. Rates for wire service vary depending on the size of the radio market, and audio service is available for an additional fee. Some eighteen hundred stations use UPI and AP audio news feeds.

Broadcasters are evenly divided over the question as to which is the best wire service. Each news service has about the same number of radio stations under contract.

Both major wire services have kept pace with the new technologies. In the mid-1980s, UPI alone purchased six thousand Z-15 desktop computers from Zenith Data Systems. Satellites also are utilized by the two news organizations for the transmission of teletype, teletext, and audio. The wire services have become as wireless as the wireless itself.

RADIO NETWORK NEWS

During the medium's first three decades, the terms "networks" and "news" were virtually synonymous. Most of the news broadcast over America's radio stations emanated from the networks. The public's dependence on network radio news reached its height during World War II. As television succeeded radio as the mainstay for entertainment programming in the 1950s and 1960s, the networks concentrated their efforts on supplying affiliates with news and information feeds. This approach helped the networks regain their footing in radio after a period of substantial decline. By the mid-1960s, the majority of the nation's stations utilized one of the four major networks for news programming.

FIGURE 5.8
AP and UPI teletype machines provide the bulk of a radio newsroom's broadcast copy.

In 1968, ABC decided to make available four distinct news formats designed for compatibility with the dominant sounds of the day. American Contemporary Radio Network, American FM Radio Network, American Entertainment Radio Network, and American Information Network each offered a unique style and method of news presentation. ABC's venture proved enormously successful. In the 1970s, over fifteen hundred stations subscribed to one of ABC's four news networks.

In response to a growing racial and ethnic awareness, the Mutual Broadcasting System launched two minority news networks in 1971. While the network's Black news service proved to be a fruitful venture, its Spanish news service ceased operation within two years of its inception. Mutual discovered that the ethnic group simply was too refracted and diverse to be effectively serviced by one network and that the Latin listeners they did attract did not constitute the numbers necessary to justify operation. In 1973 the network also went head-to-head with ABC by offering a network news service (Mutual Progressive Network) that catered to rock-oriented stations. Mutual's various efforts paid off by making it second only to ABC in number of affiliates.

The News and Information Service (NIS) was introduced by NBC in 1975 but ended in 1977. NIS offered client stations an All-News format. Fifty minutes of news was fed to stations each hour. The venture was abandoned after only moderate acceptance. CBS, which has offered its member stations World News Roundup since

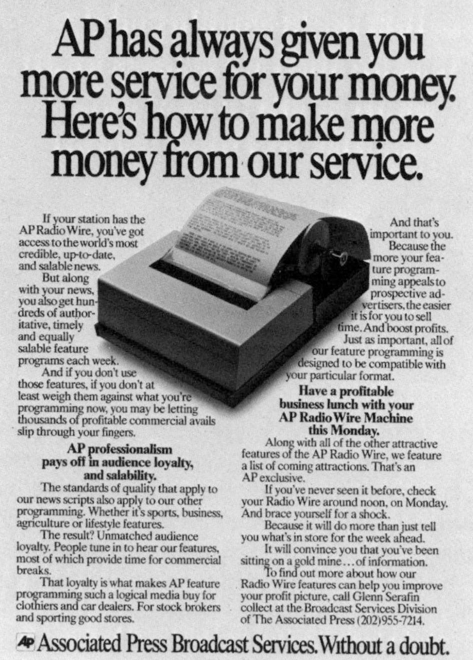

AP has always given you more service for your money. Here's how to make more money from our service.
If your station has the AP Radio Wire, you've got access to the world's most credible, up-to-date, and salable news.
But along with your news, you also get hundreds of authoritative, timely and equally salable feature programs each week.
And if you don't use those features, if you don't at least weigh them against what you're programming now, you may be letting thousands of profitable commercial avails slip through your fingers.
AP professionalism pays off in audience loyalty, and salability.
The standards of quality that apply to our news scripts also apply to our other programming. Whether it's sports, business, agriculture or lifestyle features.
The result? Unmatched audience loyalty. People tune in to hear our features, most of which provide time for commercial breaks.
That loyalty is what makes AP feature programming such a logical media buy for clothiers and car dealers. For stock brokers and sporting good stores.
And that's important to you.
Because the more your feature programming appeals to prospective advertisers, the easier it is for you to sell time. And boost profits.
Just as important, all of our feature programming is designed to be compatible with your particular format.
Have a profitable business lunch with your AP Radio Wire Machine this Monday.
Along with all of the other attractive features of the AP Radio Wire, we feature a list of coming attractions. That's an AP exclusive.
If you've never seen it before, check your Radio Wire around noon, on Monday. And brace yourself for a shock.
Because it will do more than just tell you what's in store for the week ahead.
It will convince you that you've been sitting on a gold mine... of information.
To find out more about how our Radio Wire features can help you improve your profit picture, call Glenn Serafin collect at the Broadcast Services Division of The Associated Press (202)955-7214.
AP Associated Press Broadcast Services. Without a doubt.

FIGURE 5.9
Associated Press promotional piece. AP claims more subscribers than any other wire service. Courtesy AP.

In August 1967, WCBS Radio became WCBS NEWSRADIO.
During those 15 years,
WCBS has become the station millions of people rely on for radio news.
And today, WCBS is listened to by more adults than any other radio station in the country.

For radio news,
the biggest...and....
the best in the business.

WCBS NEWSRADIO
New York

New York Arbitron, Spring '82 TSA TOTAL WEEK CUME ESTIMATES, ADULTS 18 +

FIGURE 5.10
Network radio is eager to attract young adult listeners. Courtesy WCBS.

1938, and NBC have under three hundred affiliates apiece.

Several state and regional news networks do well, but the big four, ABC, MBS, NBC, and CBS, continue to dominate. Meanwhile, independent satellite news and information networks have joined the field and more are planned.

The usual length of a network newscast is five minutes, during which affiliates are afforded an opportunity to insert local sponsor messages at designated times. The networks make their money by selling national advertisers spot availabilities in their widely broadcast news. Stations also pay the networks a fee for the programming they receive.

RADIO SPORTSCASTS

Sports is most commonly presented as an element within newscasts (see fig. 5.7). While many stations air sports as programming features unto themselves, most stations insert information, such as scores and schedules of upcoming games, at a designated point in a newscast and call it sports. Whether a station emphasizes sports largely depends on its audience. Stations gearing their format for youngsters or women often all but ignore sports. Adult-oriented stations, such as Middle-of-the-Road, will frequently offer a greater abundance of sports information, especially when the station is located in an area that has a major-league team.

Stations that hire individuals to do sports, and invariably these are larger outlets since few small stations can afford a full-time sportsperson, look for someone who is well versed in athletics. "To be good at radio sports, you have to have been involved as a participant somewhere along the line. That's for starters, in my opinion. This doesn't mean that you have to be a former major leaguer before doing radio sports, but to have a feel for what you're talking about it certainly helps to have been on the field or court yourself. A good sportscaster must have the ability to accurately analyze a sport through the eyes and body of the athlete," contends John Colletto, sports director, WPRO-AM, Providence, Rhode Island.

Unlike news that requires an impartial and somewhat austere presentation, sportscasts frequently are delivered in a casual and even opinionated manner. "Let's face it, there's a big difference between nuclear arms talks between the U.S. and the Soviets and last night's Red Sox/Yankees score. I don't think sports reports should be treated in a style that's too solemn. It's entertainment, and sportscasters should exercise their license to comment and analyze," says Colletto.

Although sports is presented in a less heavy-handed way than news, credibility is an important factor, contends Colletto. "There is a need for radio sportscasters to establish credibility just as there is for newspeople to do so. If you're not believable, you're not listened to. The best way to win the respect of your audience is by demonstrating a thorough knowledge of the game and by sounding like an insider, not just a guy reading the wire copy. Remember, sports fans can be as loyal to a sportscaster as they are to their favorite team. They want to hear the stories and scores from a person they feel comfortable with."

The style of a news story and a sports story may differ considerably. While news is written in a no-frills, straightforward way, sports stories often contain colorful colloquialisms and even popular slang. Here is an example by radio sportswriter Roger Crosley:

THE DEAN JUNIOR COLLEGE RED DEMON FOOTBALL TEAM RODE THE STRONG RUNNING OF FULLBACK BILL PALAZOLLO YESTERDAY TO AN 18–16 COME FROM BEHIND VICTORY OVER THE AMERICAN INTERNATIONAL COLLEGE JUNIOR VARSITY YELLOW JACKETS. PALAZOLLO CHURNED OUT A TEAM HIGH 93 YARDS ON TWENTY-FIVE CARRIES AND SCORED ALL THREE TOUCHDOWNS ON BLASTS OF 7, 2, AND 6 YARDS. THE DEMONS TRAILED THE HARD-HITTING CONTEST 16–6 ENTERING THE FINAL QUARTER. PALAZOLLO CAPPED A TWELVE PLAY 81 YARD DRIVE WITH HIS SECOND SIX-POINTER EARLY IN THE STANZA AND SCORED THE CLINCHER WITH 4:34 REMAINING. THE DEMONS WILL PUT THEIR 1 AND 0 RECORD ON THE LINE NEXT SUNDAY AT 1:30 AGAINST THE ALWAYS TOUGH HOLY CROSS JAYVEES IN WORCESTER.

Sportscasters are personalities, says WPRO's Colletto, and as such must be able to communicate on a different level than newscasters. "You're expected to have a sense of humor. Most successful sportscasters can make an audience smile or laugh. You have to be able to ad-lib, also."

The wire services and networks are the primary source for sports news at local stations.

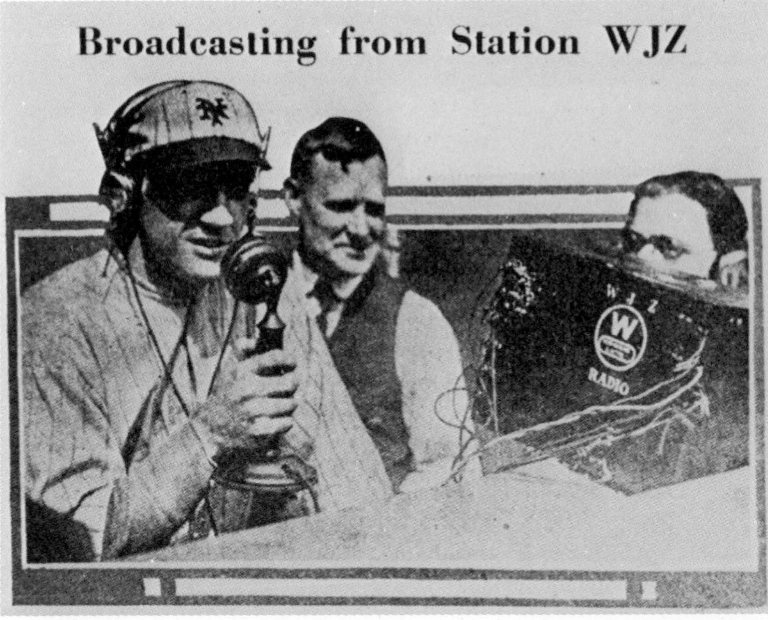

FIGURE 5.11
Sports on radio in 1923. Today sports programming generates a significant percentage of the medium's annual income. Courtesy Westinghouse Electric.

On the other hand, information about the outcome of local games, such as high school football and so forth, must be acquired firsthand. This usually entails a call to the team's coach or a direct report from a stringer or reporter.

RADIO NEWS AND THE FCC

The government takes a greater role in regulating broadcast journalism than it does print. Whereas it usually maintains a hands-off posi-

FIGURE 5.12
Promotional ad for radio sports features. Sports programming is more prevalent on the AM band. However, in the 1980s a growing number of FM stations began offering sports features. Courtesy WJR-AM.

tion when it comes to newspapers, the government keeps a watchful eye on radio to ensure that it meets certain operating criteria. Since the FCC perceives the airways as public domain, it expects broadcasters to operate in the public's interest.

In an effort to guarantee fair and equitable treatment of political and controversial issues, the FCC issued its Fairness Doctrine in 1949. Since then it has been the responsibility of broadcasters to report the news in a balanced and impartial manner. Partiality and bias in radio newscasts are a violation of the doctrine's tenets and can lead to punitive actions by the FCC. In the 1980s, an attempt was made, as part of the ongoing deregulation movement, to repeal the Fairness Doctrine. Thus far the effort has failed. Groups opposed to its abolishment fear that stations would be used to present one-sided views on important public issues. Meanwhile, proponents of repealing the doctrine argue that the vastness of the broadcast marketplace will guarantee a forum for all views and therefore mitigate the risk of exclusionism. Although the Fairness Doctrine remains on the books as of this writing, the issue continues to inspire debate.

The FCC also demands that radio reporters present news factually and in good faith. Stories that defame citizens through reckless or false statements may not only bring a libel suit from the injured party but action from the FCC, which views such behavior on the part of broadcasters as contrary to the public's interest. While broadcasters are protected under the First Amendment and therefore have certain rights, as public trustees they are charged with the additional responsibility of acting in a manner that benefits rather than harms members of society.

Broadcasters are free to express opinions and sentiments on issues through editorials. However, the FCC imposes certain conditions, as established in the Fairness Doctrine, to ensure that contrasting views are also provided a reasonable opportunity to be heard. A responsible party with an opposing viewpoint has the right to request airtime that reaches an audience comparable to that reached by the station's own editorial. In other words, it would be considered unfair for a station that broadcasts an editorial during prime time to schedule a contrasting opinion at midnight.

To avoid controversy, many radio stations choose not to editorialize even though the FCC encourages them to do so.

NEWS ETHICS

The highly competitive nature of radio places unusual pressure on newspeople. In a business where being first with the story is often equated with being the best, certain dangers exist. Being first at all costs can be costly, indeed, if information and facts are not adequately verified. As previously mentioned, it is the radio journalist's obligation to get the story straight and accurate before putting it on the air. Anything short of this is unprofessional.

The pressures of the clock can, if allowed, result in haphazard reporting. If a story cannot be sufficiently prepared in time for the upcoming news broadcast it should be withheld. Getting it on is not as important as getting it on right. Accuracy is the newsperson's first criterion. News accounts should never be fudged. It is tantamount to deceiving and misleading the public.

News reporters must exhibit discretion not only in the newsroom but also when on the scene of a story. It is commendable to assiduously pursue the facts and details of a story, but it is inconsiderate and insensitive to ignore the suffering and pain of those involved. For example, to press for comments from a grief-stricken parent whose child has just been seriously injured in an accident is callous and cruel and a disservice to all concerned, including the station the newsperson represents. Of course, a newsperson wants as much information as possible about an incident, but the public's right to privacy must be respected.

Objectivity is the cornerstone of good reporting. A newsperson who has lost his or her capacity to see the whole picture is handicapped. At the same time, the newsperson's job is to report the news and not create it. The mere presence of a member of the media can inspire a disturbance or agitate a volatile situation. Staging an event for the sake of increasing the newsiness of a story is not only unprofessional, but illegal. Groups have been known to await the arrival of reporters before initiating a disturbance for the sake of gaining publicity. It is the duty of reporters to remain as innocuous and uninvolved as possible when on an assignment.

Several industry associations, such as the National Association of Broadcasters, Radio and Television News Directors Association, and Society of Professional Journalists have established codes pertaining to the ethics and conduct of

Code of Broadcast News Ethics
Radio Television News Directors Association

The members of the Radio Television News Directors Association agree that their prime responsibility as journalists—and that of the broadcasting industry as the collective sponsor of news broadcasting—is to provide to the public they serve a news service as accurate, full and prompt as human integrity and devotion can devise. To that end, they declare their acceptance of the standards of practice here set forth, and their solemn intent to honor them to the limits of their ability.

Article One

The primary purpose of broadcast journalists—to inform the public of events of importance and appropriate interest in a manner that is accurate and comprehensive—shall override all other purposes.

Article Two

Broadcast news presentations shall be designed not only to offer timely and accurate information, but also to present it in the light of relevant circumstances that give it meaning and perspective.

This standard means that news reports, when clarity demands it, will be laid against pertinent factual background; that factors such as race, creed, nationality or prior status will be reported only when they are relevant; that comment or subjective content will be properly identified; and that errors in fact will be promptly acknowledged and corrected.

Article Three

Broadcast journalists shall seek to select material for newscast solely on their evaluation of its merits as news.

This standard means that news will be selected on the criteria of significance, community and regional relevance, appropriate human interest, service to defined audiences. It excludes sensationalism or misleading emphasis in any form; subservience to external or "interested" efforts to influence news selection and presentation, whether from within the broadcasting industry or from without. It requires that such terms as "bulletin" and "flash" be used only when the character of the news justifies them; that bombastic or misleading descriptions of newsroom facilities and personnel be rejected, along with undue use of sound and visual effects; and that promotional or publicity material be sharply scrutinized before use and identified by source or otherwise when broadcast.

Article Four

Broadcast journalists shall at all times display humane respect for the dignity, privacy and the well-being of persons with whom the news deals.

Article Five

Broadcast journalists shall govern their personal lives and such nonprofessional associations as may impinge on their professional activities in a manner that will protect them from conflict of interest, real or apparent.

Article Six

Broadcast journalists shall seek actively to present all news the knowledge of which will serve the public interest, no matter what selfish, uninformed or corrupt efforts attempt to color it, withhold it or prevent its presentation. They shall make constant effort to open doors closed to the reporting of public proceedings with tools appropriate to broadcasting (including cameras and recorders), consistent with the public interest. They acknowledge the journalist's ethic of protection of confidential information and sources, and urge unswerving observation of it except in instances in which it would clearly and unmistakably defy the public interest.

Article Seven

Broadcast journalists recognize the responsibility borne by broadcasting for informed analysis, comment and editorial opinion on public events and issues. They accept the obligation of broadcasters, for the presentation of such matters by individuals whose competence, experience and judgment qualify them for it.

Article Eight

In court, broadcast journalists shall conduct themselves with dignity, whether the court is in or out of session. They shall keep broadcast equipment as unobtrusive and silent as possible. Where court facilities are inadequate, pool broadcasts should be arranged.

Article Nine

In reporting matters that are or may be litigated, the journalist shall avoid practices which would tend to interfere with the right of an individual to a fair trial.

Article Ten

Broadcast journalists shall not misrepresent the source of any broadcast news material.

Article Eleven

Broadcast journalists shall actively censure and seek to prevent violations of these standards, and shall actively encourage their observance by all journalists, whether of the Radio Television News Directors Association or not.

FIGURE 5.13
RTNDA Code of Broadcast News Ethics.

broadcast reporters. One such set of criteria dealing with the responsibilities of radio newspeople is reproduced in full in figure 5.13.

TRAFFIC REPORTS

Traffic reports are an integral part of drivetime news programming at many metropolitan radio stations. Although providing listeners with traffic condition updates can be costly, especially air-to-ground reports which require the use of a helicopter or small plane, they can help strengthen a station's community service image and also generate substantial revenue. To avoid the cost involved in airborne observation, stations sometimes employ the services of local auto clubs or put their own mobile units out on the roads. A station in Providence, Rhode Island, broadcasts traffic conditions from atop a twenty-story hotel that overlooks the city's key arteries.

Traffic reports are scheduled several times an hour throughout the prime commuter periods on stations primarily catering to adults, and they range in length from thirty to ninety seconds. The actual reports may be done by a station employee who works in other areas of programming when not surveying the roads, or a member of the local police department or auto club may be hired for the job. Obviously, the prime criterion for such a position is a thorough knowledge of the streets and highways of the area being reported.

NEWS IN MUSIC RADIO

In the early 1980s, the FCC saw fit to eliminate the requirement that all radio stations devote a percentage of their broadcast day to news and public affairs programming. Opponents of the decision argued that such a move would mark the decline of news on radio. In contrast, proponents of the deregulation commended the FCC's actions that allow for the marketplace to determine the extent to which nonentertainment features are broadcast. Although a relatively short time has passed since the news requirement was dropped, the Radio and Television News Directors Association (RTNDA) contends that local news coverage has already declined. This they say has resulted in a decrease in the number of news positions around the country. Supporting their contention they point out that several major stations, such as KDKA, WOWO, and WIND, have cut back their news budgets.

Other members of the radio news industry contend that the FCC's decision to drop the news requirement has had little, if any, effect, and they see few real signs that suggest a decrease in news programming in the immediate future. "Recently I completed a study on the status of news at hundreds of music stations since the deregulation took effect and, surprisingly, the results show that only one percent had actually cut back or reduced their schedule of newscasts. At KAPE, the change has had no effect, nor will it," says Judy Smith.

WBCN's Sherman Whitman believes that the radio audience wants news even when a station's primary product is music. "The public has come to depend on the medium to keep it informed. It's a volatile world and certain events affect us all. Stations that aim to be full-service cannot do so without a solid news schedule."

Cecilia Mason of KGTM says that economics alone will help keep news a viable entity in radio. "While a lot of stations consider news departments expense centers, news is a money maker. This is especially true during drivetime periods when practically everyone tuned wants information, be it weather, sports, or news headlines. I don't see a growing movement to eliminate news. However, I do see a movement to soften things up, that is, to hire voices instead of radio journalists. In the long run, this means fewer news jobs, I suppose. Economics again. While stations, for the most part, are not appreciably reducing the amount of time devoted to airing news, I suspect that some may be thinning out their news departments. Hopefully, this is not a prelude to a measurable cutback. News is still big business, though."

Responsible broadcasters know that it is the inherent duty of the medium to keep the public apprised of what is going on, claims WTOD's Jewett. "While radio is primarily an entertainment medium, it is still one of the country's foremost sources of information. Responsible broadcasters, and most of us are, realize that we have a special obligation to fulfill. The tremendous reach and immediacy that is unique to radio forces the medium to be something more than just a jukebox."

While WARA's Frank Titus admits that he is

concerned about the long-term effects of deregulation on the quality of radio news presentation, he believes that stations will continue to broadcast news in the future. "There might be a tendency to invest less in news operations, especially at more music-oriented outlets, as the result of the regulation change, but news is as much a part of what radio is as are the deejays and songs. What it comes right down to is people want news broadcasts, so they're going to get them. That's the whole idea behind the commission's [FCC's] actions. There's no doubt in my mind that the marketplace will continue to dictate the programming of radio news."

CHAPTER HIGHLIGHTS

1. Although the first newscast occurred in 1910, broadcast journalism did not evolve until the early 1920s. The broadcast of the Harding–Cox election results in 1920 was a historical benchmark.

2. Because newspapers perceived radio as a competitive threat, United Press (UP), International News Service (INS), and Associated Press (AP) refused to sell to radio outlets from 1932 until 1934. Radio, however, proved it could provide its own news sources.

3. The advent of television led radio outlets to localize their news content, which meant less reliance on news networks and the creation of a station news department.

4. Surveys by the National Association of Broadcasters and the Radio Advertising Bureau found that more people tune in to radio news for their first daily source of information than turn to television or newspapers.

5. The size of a station's news staff depends upon the degree to which the station's format emphasizes news, the station's market size, and the emphasis of its competition. Small stations often have no newspeople and require deejays to rip 'n' read wireservice copy.

6. Large news staffs may consist of newscasters, writers, street reporters, and tech people, as well as stringers and interns.

7. Computers in radio newsrooms can be used as links to the various wire and news services, as display terminals for reading news copy on the air, and as word processors for writing and storing news.

8. The news director, who works with and for the program director, supervises news staff, develops and implements policy, handles the budget, assures the gathering of local news, is responsible for getting out breaking news stories rapidly and accurately, and plans public affairs programming.

9. News directors seek personnel with both college education and experience. However, finding a news slot at a small station is difficult, since their news staffs are small, so internships and experience at high school and college stations are important. Additionally, such personal qualities as enthusiasm, aggressiveness, energy, inquisitiveness, typing skills, a knowledge of the area where the station is located, announcing abilities, and a command of the English language are assets.

10. News stories must be legible, intelligible, and designed for effortless reading. They should sound conversational, informal, simple, direct, concise, and organized.

11. Actualities (on-the-scene voicers) are taped from news service feeds or recorded by station personnel at the scene.

12. Wire services are the primary, and often only, sources of national and international news for most stations. Associated Press and United Press International are the largest.

13. The FCC requires that broadcasters report the news in a balanced and impartial manner. Although protected under the First Amendment, broadcasters making reckless or false statements are subject to both civil and FCC charges.

14. Ethically, newspersons must maintain objectivity, discretion, and sensitivity.

15. Despite the FCC's deregulation of news and public affairs programming in the early 1980s, industry professionals believe that the demands of the listeners will ensure the continued importance of news on most commercial stations.

SUGGESTED FURTHER READING

Bartlett, Jonathan, ed. *The First Amendment in a Free Society*. New York: H.W. Wilson, 1979.

Bittner, John R., and Bittner, Denise A. *Radio Journalism*. Englewood Cliffs, N.J.: Prentice-Hall, 1977.

Bliss, Edward J., and Patterson, John M. *Writing News for Broadcast*, 2nd ed. New York: Columbia University Press, 1978.

Charnley, Mitchell. *News by Radio*. New York: Macmillan, 1948.

Culbert, David Holbrook. *News for Everyman: Radio and Foreign Affairs in Thirties America*. Westport, Conn.: Greenwood Press, 1976.

Fang, Irving. *Radio News/Television News*, 2nd ed. St. Paul, Minn.: Rada Press, 1985.

______. *Those Radio Commentators*. Ames: Iowa State University Press, 1977.

Friendly, Fred W. *The Good Guys, The Bad Guys, and the First Amendment: Free Speech vs. Fairness in Broadcasting*. New York: Random House, 1976.

Garvey, Daniel E. *Newswriting for the Electronic Media*. Belmont, Calif.: Wadsworth Publishing, 1982.

Gilbert, Bob. *Perry's Broadcast News Handbook*. Knoxville, Tenn.: Perry Publishing, 1982.

Hall, Mark W. *Broadcast Journalism: An Introduction to News Writing*. New York: Hastings House, 1978.

Hood, James R., and Kalbfeld, Brad, eds. *The Associated Press Handbook*. New York: Associated Press, 1982.

Hunter, Julius K. *Broadcast News*. St. Louis: C.V. Mosby Company, 1980.

Keirstead, Phillip A. *All-News Radio*. Blue Ridge Summit, Pa.: Tab Books, 1980.

Nelson, Harold L. *Laws of Mass Communication*. Mineola, N.Y.: The Foundation Press, 1982.

Simmons, Steven J. *The Fairness Doctrine and the Media*. Berkeley: University of California Press, 1978.

Stephens, Mitchell. *Broadcast News: Radio Journalism and an Introduction to Television*. New York: Holt, Rinehart and Winston, 1980.

UPI Stylebook: A Handbook for Writing and Preparing Broadcast News. New York: United Press International, 1979.

White, Ted. *Broadcast Newswriting*. New York: Macmillan, 1984.

Wulfemeyer, K. Tim. *Broadcast Newswriting*. Ames: Iowa State University Press, 1983.

6 Research

WHO IS LISTENING

As early as 1929, the question of listenership was of interest to broadcasters and advertisers alike. That year Cooperative Analysis of Broadcasting (CAB), headed by Archibald M. Crossley, undertook a study to determine how many people were tuned to certain network radio programs. Information was gathered by phoning a preselected sample of homes. One of the things the survey found was that the majority of listening occurred evenings between seven and eleven. This became known as radio's "prime time" until the 1950s.

On the local station level, various methods were employed to collect audience data, including telephone interviews and mail-out questionnaires. However, only a nominal amount of actual audience research was attempted during the late 1920s and early 1930s. For the most part, just who was listening remained somewhat of a mystery until the late 1930s.

In 1938 C.E. Hooper, Inc., began the most formidable attempt up to that time to provide radio broadcasters with audience information. Like Crossley's service, Hooper also utilized the telephone to accumulate listener data. While CAB relied on listener recall, Hooper required that interviewers make calls until they reached someone who was actually listening to the radio. This approach became known as the "coincidental" telephone method. Both survey services found their efforts limited by the fact that 40 percent of the radio listening homes in the 1930s were without a telephone.

As World War II approached, another major ratings service, known as The Pulse, began to measure radio audience size. Unlike its competitors, Pulse collected information by conducting face-to-face interviews.

Interest in audience research grew steadily throughout the 1930s and culminated in the establishment of the Office of Radio Research (ORR) in 1937. Funded by a Rockefeller Foundation grant, the ORR was headed by Paul F. Lazarsfeld, who was assisted by Hadley Cantril and Frank Stanton. The latter would go on to assume the presidency of CBS in 1946 and would serve in that capacity into the 1970s. Over a ten-year period, the ORR published several texts dealing with audience research findings and methodology. Among them were Lazarsfeld and Stanton's multivolume *Radio Research,* which covered the periods of 1941–1943 and 1948–1949. During the same decade, Lazarsfeld also published booklength reports on the public's attitude toward radio: *The People Look at Radio* (1946) and *Radio Listening in America* (1948). Both works cast radio in a favorable light by concluding that most listeners felt the medium did an exemplary job.

The Pulse and Hooper were the prevailing radio station rating services in the 1950s as the medium worked at regaining its footing following the meteoric rise of television. In 1965 Arbitron Ratings began measuring radio audience size through the use of a diary, which required respondents to document their listening habits over a seven-day period. By the 1970s Arbitron reigned as the leading radio measurement company, while Hooper and Pulse faded from the scene. To provide the radio networks and their affiliates and advertisers with much-needed ratings information, Statistical Research, Inc., of New Jersey, introduced Radio's All Dimension Audience Research (RADAR) in 1968. The company gathers its information through telephone interviews to over six thousand households. In the 1980s, Arbitron holds onto first place among services measuring local market radio audiences, but it is being challenged, especially by the Birch radio audience measurement report, which has gained considerable acceptance since its arrival on the scene in the late 1970s.

Ratings companies must be reliable. Credibility is crucial to success. Therefore measurement techniques must be tried and true. Information must be accurate, since millions of dollars are at stake. In 1963 the Broadcast Rating Council was established to monitor, audit, and accredit the various ratings companies. The council created performance standards to which rating services are expected to adhere. Those that fail to meet the council's operating criteria are not accredited. A nonaccredited ratings service will seldom succeed. In 1982 the Broadcast Rating Council was renamed the Electronic Media Planning Council to reflect a connection with the ratings services dealing with the cable television industry.

THE RATINGS SERVICES

The extreme fragmentation of today's listening audience created by the almost inestimable number of stations and formats makes the job of research a complex but necessary one. All stations, regardless of size, must put forth an effort to acquaint themselves with the characteristics of the audience, says Edward J. Noonan, co-director, Survey Research Associates. "A station cannot operate in a vacuum. It has to know who is listening and why before making any serious programming changes." Today this information is made available through several ratings services and research companies. More stations depend on Arbitron and Birch audience surveys than any other.

The front-runner in providing radio stations with ratings estimates for over two decades, Arbitron covers over 250 markets ranging in size from large to small. Arbitron claims over two thousand seven hundred radio clients and a staff of three thousand interviewers who collect listening information from two million households across the country. All markets are measured at least once a year during the spring; however, larger markets are measured up to four times a year. Until the early 1980s, metro markets traditionally were rated in the spring and fall. However, six months between surveys was considered too long in light of the volatile nature of the radio marketplace.

To determine a station's ranking, Arbitron follows an elaborate procedure. First the parameters of the area to be surveyed are established. Arbitron sees fit to measure listening both in the city or urban center, which it refers to as the Metro Survey Area (MSA), and in the surrounding communities or suburbs, which it classifies Total Survey Area (TSA). Arbitron classifies a station's primary listening locations as its Areas of Dominant Influence (ADI).

Once the areas to be measured have been ascertained, the next thing Arbitron does is select a sample base composed of individuals to be queried regarding their listening habits. Metromail provides Arbitron with computer tapes that contain telephone and mailing lists from which the rating company derives its randomly selected sample. Arbitron conducts its surveys over a three-to-four-week period, during which time new samples are selected weekly.

When the sample has been established, a letter is sent to each targeted household. The preplacement letter informs members of the sample that they have been selected to participate in a radio listening survey and asks their cooperation. Within a couple of days after the letter has been received, an Arbitron interviewer calls to describe the purpose of the survey as well as to determine how many individuals twelve or older reside in the household. Upon receiving the go-ahead, Arbitron mails its seven-day survey diary, which requires respondees to log their listening habits. An incentive stipend of fifty cents to a dollar accompanies the document. The diary is simple to deal with, and the information it requests is quite basic: time tuned to a station, station call letters or program name, whether AM or FM, where listening occurred—car, home, elsewhere. Although the diary asks for information pertaining to age, sex, and residence, the actual identity or name of those participating is not requested.

Prior to the start of the survey, a representative of Arbitron makes a presurvey follow-up call to those who have agreed to participate. This is done to make certain that the diary has been received and that everyone involved understands how to maintain it. Another follow-up call is made during the middle of the survey week to ascertain if the diary is being kept and to remind each participant to promptly return it upon completion of the survey. Outside the metro area, follow-ups take the form of a letter. The diaries are mailed to Beltsville, Maryland, for processing and computation.

Arbitron claims that sixty-five out of every one hundred diaries it receives are usable. Diaries that are inadequately or inaccurately filled out are not used. Upon arriving at Arbitron headquarters, diaries are examined by editors and rejected if they fail to meet criteria. Any diary received before the conclusion of the survey period is immediately voided, as are those that arrive more than twelve days after the end of the survey period. Diaries with blank or ambiguous entries also are rejected. Those diaries that survive the editors' scrutiny are then pro-

REPORT CONTENTS AND MAP PAGE

This is the first page of an Arbitron Radio Market Report. This page, in addition to showing a map of the Radio Market survey area and table of Report contents, contains a basic introduction to the Report, restrictions on the use of the Report, a copyright infringement warning, the survey time period, frequency of measurement and Arbitron's schedule of radio surveys for the year.

1 ► The map shows the survey area(s) included in a market definition and for which audience estimates are reported. Areas in grey are not included in a market definition (unless it is a part of an ADI to be reported for that particular market) and will have no audience estimates included in the Report.

2 ► The Metro Survey Area (MSA) of the reported market is shown by horizontal hatching.

3 ► The Total Survey Area (TSA) of the reported market is shown in white.

4 ► The Areas of Dominant Influence (ADI) of the reported market, if applicable, is shown by diagonal hatching.

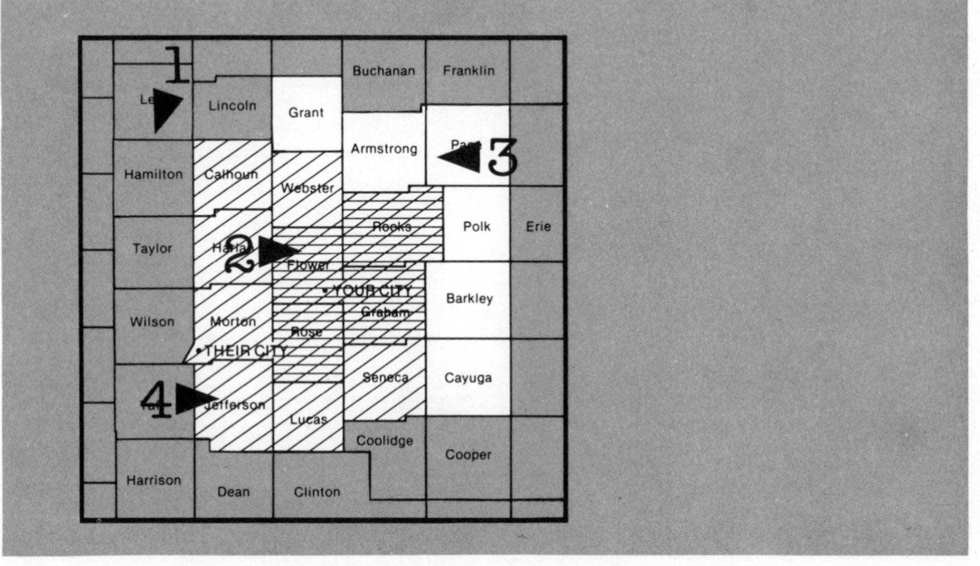

FIGURE 6.1
First page of an Arbitron Radio Market Report containing an explanation of what constitutes the survey area. Courtesy Arbitron.

FIGURE 6.2
Arbitron sample explanation. Courtesy Arbitron.

POPULATION ESTIMATES AND SAMPLE DISTRIBUTION BY AGE/SEX GROUP

1 ▶ The estimated populations of reported demographic categories within survey area(s) are reported for the Metro Survey Area, Area of Dominant Influence (where applicable), and the Total Survey Area.

2 ▶ The ratio (expressed as a percentage) of each reported demographic's estimated population to the estimated population of all persons age 12 or older within the relevant survey area.

3 ▶ The number of unweighted diaries in-tab, from persons in reported demographics, expressed as a percentage of the total number of unweighted in-tab diaries from all persons age 12 or older.

4 ▶ The distribution of in-tab diaries after weighting for reported demographics expressed as a percentage of the total number of tabulated diaries from all persons age 12 or older.

5 ▶ The total persons age 12 or older in the Metro and the percent of those persons living in military housing, college dormitories and other group quarters.

6 ▶ The number of estimated residences in the predesignated sample, number of homes contacted, number of homes in which diaries were placed, number of persons with whom diaries were placed and the number of persons who returned usable diaries (see The Sample, Section II, for further definition of these terms) are reported for Metro, ADI (if applicable), and TSA. Standard and ESF sample statistics are reported separately along with a total sample.

Total Survey Area

	1 Estimated Population	2 Estimated Population as Percent of Tot. Persons 12+	3 Percent of Unweighted In-Tab Sample	4 Percent of Weighted In-Tab Sample
Men 18-24	115,900	7.2	6.4	7.2
Men 25-34	139,900	8.7	8.5	8.7
Men 35-44	107,300	6.6	6.7	6.6
Men 45-49	49,400	3.1	2.5	3.1
Men 50-54	54,600	3.4	2.9	3.4
Men 55-64	96,300	6.0	7.0	6.0
Men 65 +	94,700	5.9	4.2	5.9

Area of Dominant Influence

Men 18-24	104,200	7.2	6.3	7.2
Men 25-34	124,900	8.6	8.1	8.6
Men 35-44	96,500	6.6	7.1	6.6
Men 45-49	44,800	3.1	2.3	3.1
Men 50-54	49,300	3.4	2.7	3.4
Men 55-64	86,300	5.9	7.0	5.9
Men 65 +	84,000	5.8	4.1	5.8

Metro Survey Area

Men 18-24	74,800	7.2	6.8	7.1
Men 25-34	90,900	8.6	8.5	8.6
Men 35-44	71,500	6.8	7.3	6.8
Men 45-49	33,500	3.2	2.4	3.2
Men 50-54	36,800	3.5	2.5	3.5
Men 55-64	62,300	5.9	6.7	5.9
Men 65 +	57,600	5.4	3.8	5.4

Metro Persons Living in Group Quarters

5 ▲

		% Military	% College	% Other Group Quarters
Total Persons 12 +	13,476,300	.1	.4	1.7

Diary Placement and Return Information

6 ▶

	Metro	ADI	TSA
Standard Residential Listings in Designated Sample	751	1,033	1,185
ESF Residential Listings in Designated Sample	359	380	383
Total Residential Listings in Designated Sample	1,110	1,413	1,568
Standard Contacts (homes in which telephone was answered)	740	1,023	1,568
ESF Contacts (homes in which telephone was answered)	353	359	374
Total Contacts (homes in which telephone was answered)	1,093	1,382	11,514
Standard Homes in Which Diaries Were Placed	657	920	1,037
ESF Homes in Which Diaries Were Placed	253	257	283
Total Homes in Which Diaries Were Placed	910	1,177	1,320
Standard Individuals Who Were Sent a Diary	1,562	2,210	2,487
ESF Individuals Who Were Sent a Diary	588	600	625
Total Individuals Who Were Sent a Diary	2,150	2,810	3,112
Standard Individuals Who Returned a Usable Diary (In-Tab)	1,031	1,421	1,610
ESF Individuals Who Returned a Usable Diary (In-Tab)	360	374	398
Total Individuals Who Returned a Usable Diary (In-Tab)	1,391	1,795	2,008

FIGURE 6.2
continued

POPULATION ESTIMATES AND TABULATED DIARIES BY SAMPLING UNIT

1 ▶ The specific designation of the Arbitron Sampling Unit and a coded indicator of its relationship to the market definition. The codes "M," "T" and, if applicable, "A" are used to identify Metro, TSA and ADI sampling units, respectively. Where a sampling unit qualifies for inclusion in more than one area, a code will appear for each area.

2 ▶ Estimated Population: The estimated population of persons 12 years or older in a sampling unit as determined by Market Statistics, Inc. (MSI), based on the 1980 U.S. Census estimates. These population estimates are updated annually as of the first of January, by MSI.

3 ▶ In-Tab: The number of usable diaries returned from diarykeepers in a sampling unit.

4 ▶ The county/sampling unit name and state included in the survey area.

5 ▶ Occasionally, no county estimated population and in-tab information will appear next to a county name if the county's in-tab has been combined (clustered). For Market Report production purposes, population and in-tab for the county have been included with the preceding population estimates and in-tab numbers, respectively. A footnote at the bottom of page 2 of the Radio Market Report pertaining to this "clustering" procedure has been added effective with the Winter 1983 radio survey. Specific county data breakouts are obtainable through Arbitron's Radio Policies and Procedures Department.

Population Estimates and Tabulated Diaries by Sampling Unit

1 Area	2 Estimated Population	3 In-Tab	4 Counties	State
MTA	76,100	67	GRAHAM	US
MTA	501,600	514	FLOWER	US
MTA	698,900	750	ROOKS	US
MTA	202,100	186	ROSE	US
TA	103,800	118	WEBSTER	US
			CALHOUN	US
TA	78,700 ◀ 5	▶ 51	HARLAN	US
			MORTON	US
			JEFFERSON	US
TA	142,500	132	LUCAS	US
TA	202,100	153	SENECA	US
T	41,900	20	GRANT	US
T	248,200	208	ARMSTRONG	US
T	61,300	30	PAGE	US
T	48,000	82	POLK	US
			BARKLEY	US
			CAYUGA	US

Area	Estimated Population	In-Tab	Counties	State

M — METRO SAMPLING UNIT T — TSA SAMPLING UNIT A — ADI SAMPLING UNIT

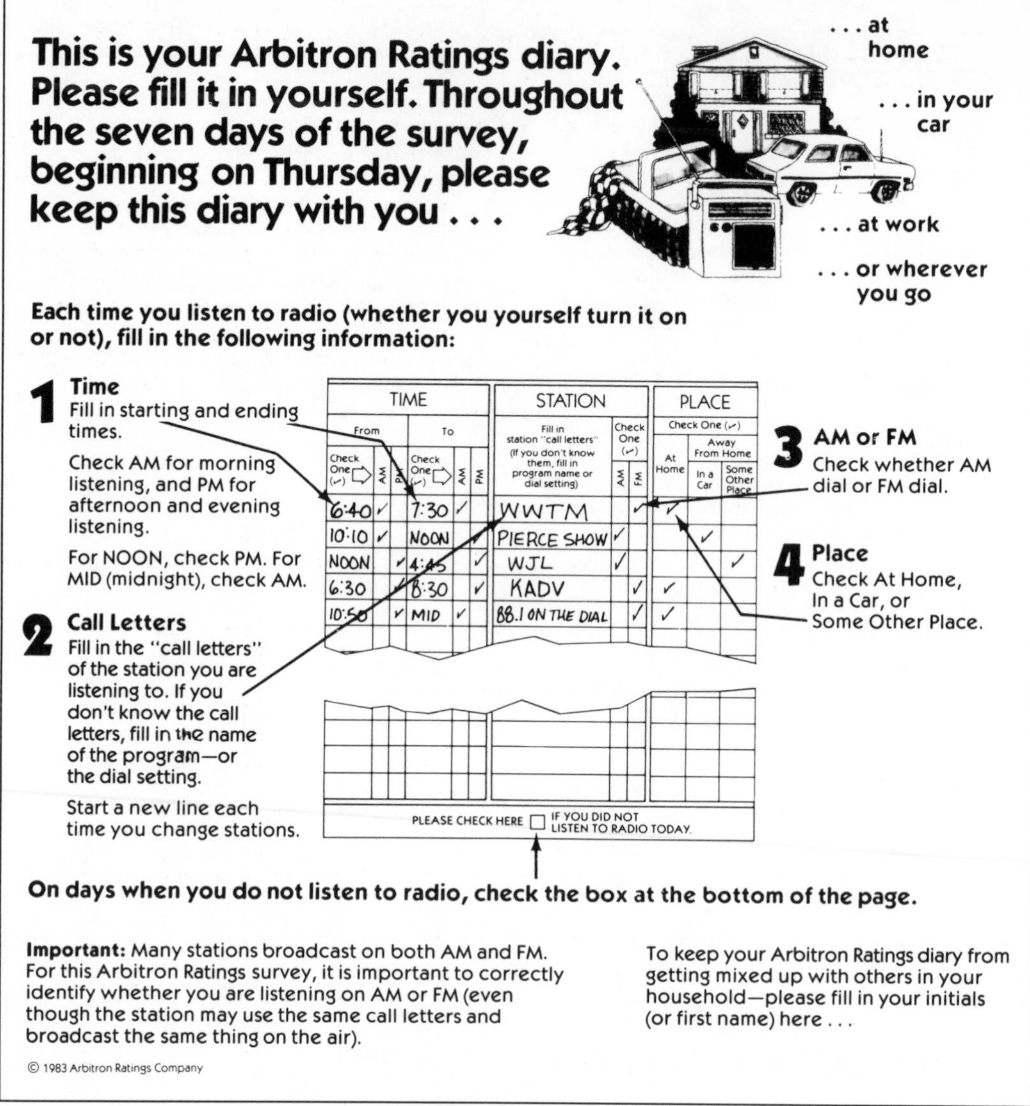

This is your Arbitron Ratings diary. Please fill it in yourself. Throughout the seven days of the survey, beginning on Thursday, please keep this diary with you . . .

. . . at home

. . . in your car

. . . at work

. . . or wherever you go

Each time you listen to radio (whether you yourself turn it on or not), fill in the following information:

1 Time
Fill in starting and ending times.

Check AM for morning listening, and PM for afternoon and evening listening.

For NOON, check PM. For MID (midnight), check AM.

2 Call Letters
Fill in the "call letters" of the station you are listening to. If you don't know the call letters, fill in the name of the program—or the dial setting.

Start a new line each time you change stations.

3 AM or FM
Check whether AM dial or FM dial.

4 Place
Check At Home, In a Car, or Some Other Place.

TIME						STATION			PLACE		
From			To			Fill in station "call letters" (If you don't know them, fill in program name or dial setting)	Check One (✓)		Check One (✓)		
Check One (✓)	AM	PM	Check One (✓)	AM	PM		AM	FM	At Home	Away From Home: In a Car	Away From Home: Some Other Place
6:40	✓		7:30	✓		WWTM		✓	✓		
10:10	✓		NOON		✓	PIERCE SHOW	✓			✓	
NOON		✓	4:45		✓	WJL	✓				✓
6:30		✓	8:30		✓	KADV		✓	✓		
10:50		✓	MID	✓		88.1 ON THE DIAL		✓	✓		

PLEASE CHECK HERE ☐ IF YOU DID NOT LISTEN TO RADIO TODAY.

On days when you do not listen to radio, check the box at the bottom of the page.

Important: Many stations broadcast on both AM and FM. For this Arbitron Ratings survey, it is important to correctly identify whether you are listening on AM or FM (even though the station may use the same call letters and broadcast the same thing on the air).

To keep your Arbitron Ratings diary from getting mixed up with others in your household—please fill in your initials (or first name) here . . .

FIGURE 6.3 Instructions for filling out diary. Accuracy is important. Courtesy Arbitron.

cessed through the computer, and their information is tabulated. Computer printouts showing audience estimates are sent to subscribers. Stations receive the "book" within a few weeks after the last day of the survey.

The most recent research tool made available by Arbitron to its subscribers is a telephone-delivered radio market service called Arbitrends, which utilizes the IBM-XT microcomputer to feed data to stations similarly equipped. Information regarding a station's past and current performances and those of competitors is available at the touch of a finger. Breakouts and tailor-made reports are provided on an ongoing basis by Arbitrends to assist stations in the planning of sales and marketing strategies. The survey company has over thirteen billion characters reserved on computer disk packs. Arbitron also makes available its Arbitrends Rolling Average Printed Reports to those stations without computers. To date, Arbitron has prepared over three hundred and seventy thousand radio market reports.

Arbitron's most formidable rival in recent years is Birch Radio, headquartered in New Jersey. As a radio audience measurement service, Birch provides clients with both quantitative and qualitative data on listening patterns, audience size, and demographics. Birch interviewers telephone a prebalanced sample of households

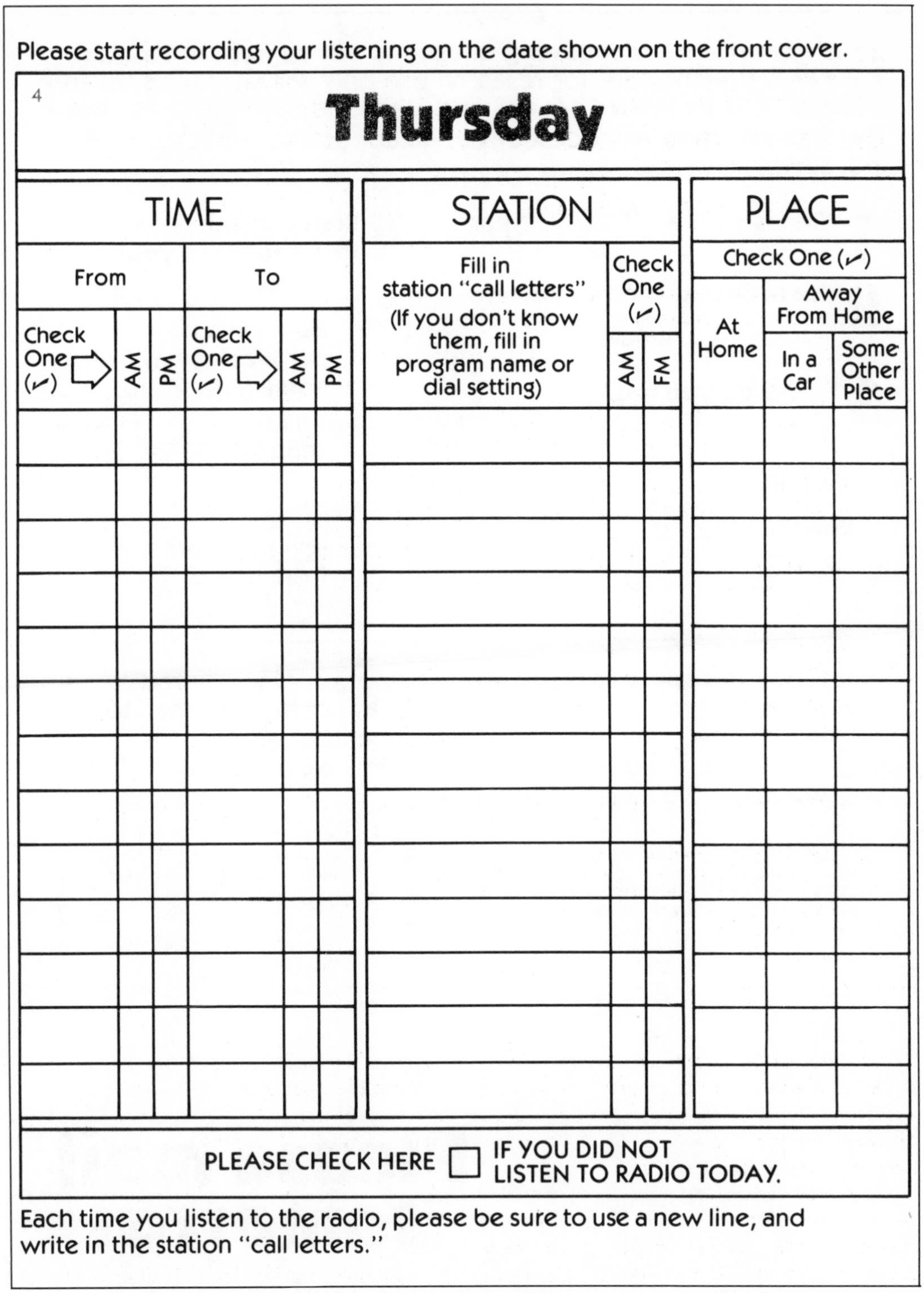

Please start recording your listening on the date shown on the front cover.

4

Thursday

TIME						STATION			PLACE		
From			To			Fill in station "call letters" (If you don't know them, fill in program name or dial setting)	Check One (✓)		Check One (✓)		
Check One (✓) ⇨	AM	PM	Check One (✓) ⇨	AM	PM		AM	FM	At Home	Away From Home: In a Car	Away From Home: Some Other Place

PLEASE CHECK HERE ☐ IF YOU DID NOT LISTEN TO RADIO TODAY.

Each time you listen to the radio, please be sure to use a new line, and write in the station "call letters."

FIGURE 6.4
Diary log. Courtesy Arbitron.

during the evening hours, Sunday through Friday, to acquire the information they need. Male and female respondents aged twelve or older are selected on alternating basis by interviewers to ensure gender equality. The sample sizes vary depending on the size of the market being surveyed. For example, Birch will contact approximately eleven hundred households in a medium

FIGURE 6.5
Diary holder demographic information page and further instructions. Courtesy Arbitron.

Your answers to the following questions will be kept in strictest confidence — they are for research use only. Please call us toll-free 800-638-7091 (in Alaska, Hawaii and Maryland call collect 301-441-3973) if you have any questions at all about the instructions or the survey itself.

1 **What is your age?**_______years

2 **Please check whether you are:**

☐ Male (1) ☐ Female (2)

3 **Where do you live?**

City_______________

County_______________

State_______________

Zip__________

4 **Please check (✓) the box that applies to you:**

☐ (1) I work away from home.

Hours I usually work per week away from home (check one):

Less than 20	20-29	30 or more
☐ 1	☐ 2	☐ 3

☐ (1) I do not work away from home.

Please use this space to tell us about any of your favorite radio stations or programs which you may or may not have listened to during this survey. Please make any comments or suggestions you might have about radio and tell us what you like or don't like about the stations.

Please mail tomorrow

LP-02-2

FIGURE 6.6
Arbitron estimates show where a station stands in its market. Courtesy Arbitron.

AVERAGE SHARE TRENDS

The Trends section provides an indication of individual station performance and the relative standing among stations for periods prior to the most recent survey. The duration reported in this section may include as many as five discrete Arbitron survey rating periods, always including the most recent. Trends may not always reflect actual changes over time due to changes in methodology, station operations, etc.

1 ▶ The estimates reported are average persons shares in the Metro Survey Area, by individual stations, on the basis of broad demographics in specific dayparts.
A share is the percent of all listeners in a demographic group that are listening to a specific station. This percent is calculated by dividing the Average Quarter-Hour Persons to a station by the Average Quarter-Hour Persons to all stations.

2 ▶ Total listening (Metro Totals) in the market, expressed as a rating, is also reported. This is the sum of all reported stations' Average Quarter-Hour estimates plus those for stations not meeting Minimum Reporting Standards and unidentified listening.

3 ▶ Reported dayparts are as follows:

Monday-Sunday 6AM-Midnight
Monday-Friday 6AM-10AM
Monday-Friday 10AM-3PM
Monday-Friday 3PM-7PM
Monday-Friday 7PM-Midnight

4 ▶ Reported demographics are as follows:

Total Persons 12+
Men 18+
Women 18+
Teens 12-17

5 ▶ A "+" indicates station changed call letters.

6 ▶ A "**" indicates station not reported for that survey.

YOUR CITY
SPRING 1984

Average Share Trends - Metro Survey Area

TOTAL PERSONS 12+

STATION CALL LETTERS	MON-SUN 6:00AM-MID					STATION CALL LETTERS	MON-FRI 6:00 AM-10:00 AM					STATION CALL LETTERS	MON-FRI 10:00 AM-3:00 PM					STATION CALL LETTERS
	SPRING 82	FALL 82	SPRING 83	FALL 83	SPRING 84		SPRING 82	FALL 82	SPRING 83	FALL 83	SPRING 84		SPRING 82	FALL 82	SPRING 83	FALL 83	SPRING 84	
WAAA	**	1.9	1.3	1.8	1.8	WAAA	**	1.0	1.2	1.8	1.5	WAAA	**	3.1	1.8	1.7	2.5	WAAA
WBBB	5.5	4.3	3.6	5.0	3.5	WBBB	3.6	3.4	2.8	4.3	2.5	WBBB	6.6	6.6	5.1	6.4	4.2	WBBB
+WCCC	13.9	13.3	14.3	12.2	14.3	+WCCC	21.5	21.6	20.3	19.7	19.2	+WCCC	11.9	10.3	11.3	10.5	10.8	WCCC
WDDD	7.6	8.3	[illegible]	9.1	7.8	WDDD	5.6	4.6	5.1	6.0	5.5	WDDD	6.7	9.6	7.4	8.0	6.8	WDDD
WEEE	3.2	3.8	[illegible]	5.4	5.9	WEEE	2.8	2.4	4.6	4.0	5.3	WEEE	2.7	3.5	6.0	5.2	4.4	WEEE
WFFF	1.0	.6	.8	**	1.0	WFFF	1.1	.8	.9	**	1.2	WFFF	.8	1.1	1.1	**	.8	WFFF
WGGG	1.0	1.5	2.3	1.4	1.6	WGGG	.5	1.0	.9	1.2	1.0	WGGG	1.2	1.4	[illegible]	1.4	.9	WGGG
WHHH	10.2	10.5	11.0	9.0	9.5	WHHH	12.8	12.8	12.4	9.5	11.8	WHHH	9.9	9.8	[illegible]	7.4	9.7	WHHH
WIII	5.1	5.4	3.6	6.7	6.0	WIII	3.6	3.5	2.4	3.8	4.4	WIII	4.8	4.5	2.7	5.9	5.1	WIII
WJJJ	.6	.9	.6	.6	.6	WJJJ	.4	1.0	1.0	1.0	.8	WJJJ	1.1	.9	.6	.8	.7	WJJJ
METRO TOTALS	15.6	15.9	16.0	15.9	16.3	METRO TOTALS	22.2	23.2	23.0	21.8	22.6	METRO TOTALS	17.6	18.8	19.2	18.9	18.9	METRO TOTALS

market and over two thousand in major metro markets.

A wide range of reports is available to clients, including the *Quarterly Summary Report:* estimates of listening by location, county by county, and other detailed audience information; *Standard Market Report:* audience analysis especially designed for small market broadcasters; *Capsule Market Report:* listening estimates in the nation's smallest radio markets; *Condensed Market Report:* designed specifically for radio outlets in markets not provided with regular syndicated measurements and where cost is a key consideration; *Monthly Trend Report:* an ongoing picture of the listening audience so that clients may benefit from current shifts in the marketplace; *Prizm:* lifestyle-oriented radio ratings book that defines radio audiences by lifestyle characteristics in more than eighty-five markets; and *BirchScan:* microcomputerized system (IBM-PC) ratings retrieval and analysis. Fees for Birch Radio services are based upon market size. Both Birch and Arbitron provide subscribers with general product consumption and media usage data.

Dozens of other research companies throughout the country provide broadcasters with a broad range of useful audience information. Many utilize approaches similar to Arbitron and Birch to collect data, while others use different

FIGURE 6.7
Sample page from an Arbitron Radio Report. Courtesy Arbitron.

Average Quarter-Hour Listening Estimates

PHILADELPHIA
SUMMER 1983

SUNDAY
3:00PM–7:00PM

AVERAGE PERSONS—TOTAL SURVEY AREA, IN HUNDREDS

TOT. PERS. 12+	MEN 18-24	MEN 25-34	MEN 35-44	MEN 45-54	MEN 55-64	WOMEN 18-24	WOMEN 25-34	WOMEN 35-44	WOMEN 45-54	WOMEN 55-64	TNS. 12-17	STATION CALL LETTERS
392		13	27	27	27	3	6	30	21	65	6	KYW
-1												WBCB
625	15	38	42	59	106	3	31	47	23	66	9	WCAU
562	59	16	41	5	26	65	78	53	8		178	WCAU FM
100		23				4	8	8	9			WDAS
437	107	38	7	25	13	48	107	11	13	15	53	WDAS FM
358		7	19	41	28	3	28	4	58	47		WEAZ
108	11	6	4	35		5	20	8	12		3	WFIL
12				3								WFLN
166			13	12	14	3	18	10	28	9		WFLN FM
178			13	15	14	3	18	10	28	9		TOTAL
109		10			8		5	14	33	26	13	WHAT
52		1				18					33	WIFI
269	14	99	31			28	74	1	6		16	WIOQ
172		4	12	4	20		15	24	16	40	27	WIP
96		2	5	19		13	10	30	17			WKSZ
427	18	94	11	34		29	100	33	38	14	30	WMGK
423	95	93	3	2	6	73	75	15			61	WMMR
551		15	13	70	65	16	28	39	73	41		WPEN
3			1								2	WSNI
146		28	21			23	49				25	WSNI FM
149		28	22			23	49				27	TOTAL
32	11					9	4				3	WSSJ
585	116	116	31		12	69	79	43	5	7	107	WUSL
19									19			WVCH
270	15			26		5	28	27	13	55	1	WWDB
97	10	4	8	11		2	15	14	20	4	2	WWSH
432	88	86		2	10	124	40	3	9	5	65	WYSP
40				17			17			4	2	WZZD
122				40	30			2		32		WJBR FM
113	12	20				54					27	WPST
52	12	11		2		16			8		3	WSTW
24			18		6							WTTM
55									17	13		WOR

AVERAGE PERSONS—METRO SURVEY AREA, IN HUNDREDS

STATION CALL LETTERS	TOT. PERS. 12+	MEN 18-24	MEN 25-34	MEN 35-44	MEN 45-54	MEN 55-64	WOMEN 18-24	WOMEN 25-34	WOMEN 35-44	WOMEN 45-54	WOMEN 55-64	TNS. 12-17
KYW	350		13	22	14	24	3	6	26	11	62	6
WBCB	-1											
WCAU	570	15	38	42	50	86	3	31	47	17	46	9
WCAU FM	420	59	16	27	5	26	59	51	28			126
WDAS	91		23				4	8	8			
WDAS FM	343	32	38	7	25	13	48	88	11	13	15	53
WEAZ	239		7	2	5	23	3		4	34	42	
WFIL	100	11	6	4	29		5	18	8	12		3
WFLN	12				3							
WFLN FM	144			13	12	14	3	5	10	28	9	
TOTAL	156			13	15	14	3	5	10	28	9	
WHAT	109		10			8		5	14	33	26	13
WIFI	23		1				14					8
WIOQ	263	14	94	31			28	74		6		16
WIP	144		4	12	4	20		13	24		30	27
WKSZ	76		2	5	10		9	10	30	10		
WMGK	374	18	83	11	27		29	70	28	38	14	30
WMMR	352	74	88	3	2	6	71	47	15			46
WPEN	533		15	13	70	47	16	28	39	73	41	
WSNI	3			1								2
WSNI FM	146		28	21			23	49				25
TOTAL	149		28	22			23	49				27
WSSJ	32	11					9	4				3
WUSL	466	75	87	31		12	51	79	38	5	7	81
WVCH	-1											
WWDB	237	15			26		5	28	27		55	1
WWSH	68	10	4	8	11		2	15	7	5	4	2
WYSP	289	63	52		2	10	75	8	3	9	5	62
WZZD	40				17			17			4	2
WJBR FM	18					13			2		3	
WPST	67	12	20				20					15
WSTW	18				2		16					
WTTM	24			18		6						
WOR	13										13	
TOTAL LISTENING IN METRO SURVEY AREA	6102	484	686	329	359	369	523	710	414	306	383	538

SHARES—METRO SURVEY AREA

STATION CALL LETTERS	TOT. PERS. 12+ %	MEN 18-24 %	MEN 25-34 %	MEN 35-44 %	MEN 45-54 %	MEN 55-64 %	WOMEN 18-24 %	WOMEN 25-34 %	WOMEN 35-44 %	WOMEN 45-54 %	WOMEN 55-64 %	TNS. 12-17 %
KYW	5.7		1.9	6.7	3.9	6.5	.6	.8	6.3	3.6	16.2	1.1
WBCB												
WCAU	9.3	3.1	5.5	12.8	13.9	23.3	.6	4.4	11.4	5.6	12.0	1.7
WCAU FM	6.9	12.2	2.3	8.2	1.4	7.0	11.3	7.2	6.8			23.4
WDAS	1.5		3.4				.8	1.1	1.9			
WDAS FM	5.6	6.6	5.5	2.1	7.0	3.5	9.2	12.4	2.7	4.2	3.9	9.9
WEAZ	3.9		1.0	.6	1.4	6.2	.6		1.0	11.1	11.0	
WFIL	1.6	2.3	.9	1.2	8.1		1.0	2.5	1.9	3.9		.6
WFLN	.2				.8							
WFLN FM	2.4			4.0	3.3	3.8	.6	.7	2.4	9.2	2.3	
TOTAL	2.6			4.0	4.1	3.8	.6	.7	2.4	9.2	2.3	
WHAT	1.8		1.5			2.2		.7	3.4	10.8	6.8	2.4
WIFI	.4		.1				2.7					1.5
WIOQ	4.3	2.9	13.7	9.4			5.4	10.4		2.0		3.0
WIP	2.4		.6	3.6	1.1	5.4		1.8	5.8		7.8	5.0
WKSZ	1.2		.3	1.5	2.8		1.7	1.4	7.2	3.3		
WMGK	6.1	3.7	12.1	3.3	7.5		5.5	9.9	6.8	12.4	3.7	5.6
WMMR	5.8	15.3	12.8	.9	.6	1.6	13.6	6.6	3.6			8.6
WPEN	8.7		2.2	4.0	19.5	12.7	3.1	3.9	9.4	23.9	10.7	
WSNI				.3								.4
WSNI FM	2.4		4.1	6.4			4.4	6.9				4.6
TOTAL	2.4		4.1	6.7			4.4	6.9				5.0
WSSJ	.5	2.3					1.7	.6				.6
WUSL	7.6	15.5	12.7	9.4		3.3	9.8	11.1	9.2	1.6	1.8	15.1
WVCH												
WWDB	3.9	3.1			7.2		1.0	3.9	6.5		14.4	.2
WWSH	1.1	2.1	.6	2.4	3.1		.4	2.1	1.7	1.6	1.0	.4
WYSP	4.7	13.0	7.6		.6	2.7	14.3	1.1	.7	2.9	1.3	11.5
WZZD	.7				4.7			2.4			1.0	.4
WJBR FM	.3					3.5			.5		.8	
WPST	1.1	2.5	2.9				3.8					2.8
WSTW	.3				.6		3.1					
WTTM	.4			5.5		1.6						
WOR	.2										3.4	

PAGE 84

Footnote Symbols: (*) means audience estimates adjusted for actual broadcast schedule.

ARBITRON RATINGS

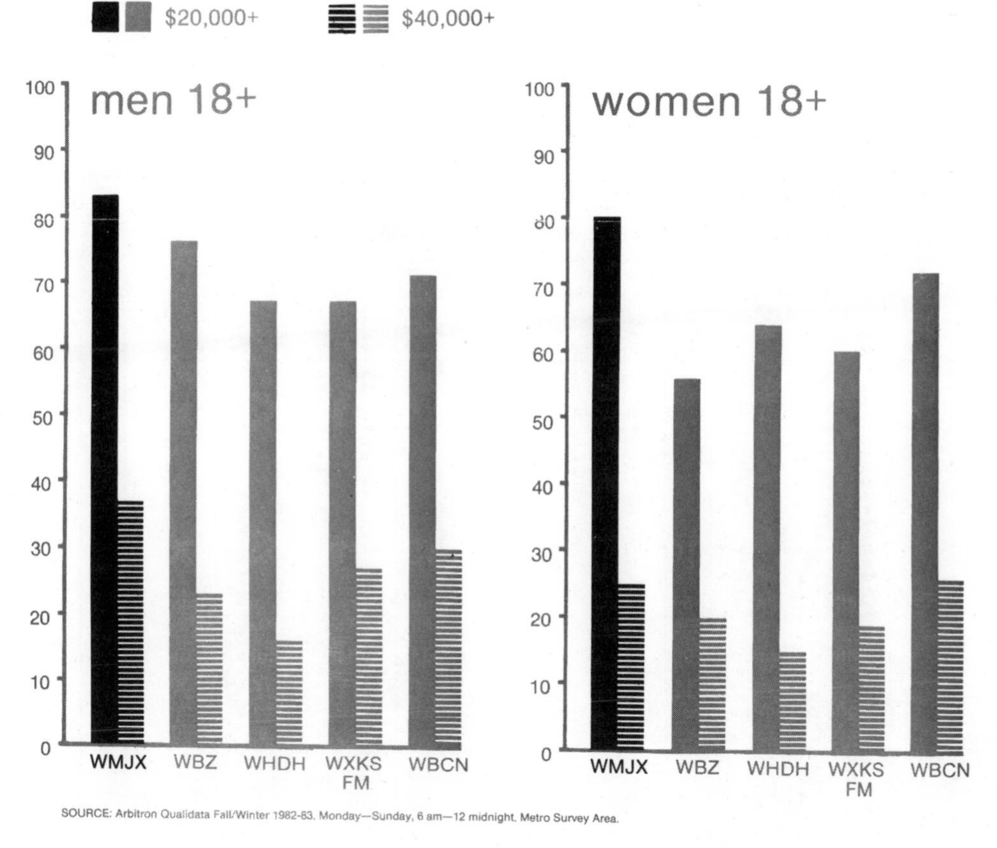

wmjx 106.7fm
Magic
WMJX leads in household income. . .
The WMJX men 18+ hold the lead in income earning among the major metro stations, with 83% earning over $20,000 and 37% earning $40,000 and above.
Not to be outdone, a full 80% of the WMJX women 18+ earn at least $20,000 per year, with 25% earning $40,000 or more.
These are impressive statistics when you consider the correlation between income earning and disposable income. . .
$20,000+
$40,000+
men 18+
100 90 80 70 60 50 40 30 20 10 0
WMJX WBZ WHDH WXKS FM WBCN
women 18+
100 90 80 70 60 50 40 30 20 10 0
WMJX WBZ WHDH WXKS FM WBCN
SOURCE: Arbitron Qualidata Fall/Winter 1982-83. Monday—Sunday, 6 am—12 midnight, Metro Survey Area.

FIGURE 6.8
Radio station promotional piece based upon survey statistics. Courtesy Arbitron.

methods. "Southeast Media Research offers four research methods: focus groups, telephone studies, mail intercepts, and music tests," explains Don Hagen, the company's president. Christopher Porter, associate director of Surrey Research, says that his company uses similar techniques.

Meanwhile, Dick Warner, president of S-A-M-S Research, claims that the telephone recall method is the most commonly used and effective approach to radio audience surveying. "The twenty-four-hour telephone recall interview, in my estimation, yields the most reliable information. Not only that, it is quick and current—important factors in a rapidly moving and hyperdynamic radio marketplace."

FIGURE 6.9
Station promotional piece based upon survey results. Courtesy WZPL-FM.

"Indy's Apple" dishes up huge slices of all the key demographics. Take your pick:

You can now buy the Indianapolis Market one station deep! WZPL gives you a huge, full spectrum audience from 12-54. We call it the "Apple Demo"...WZPL delivers a resounding 19.5 in 12-54.

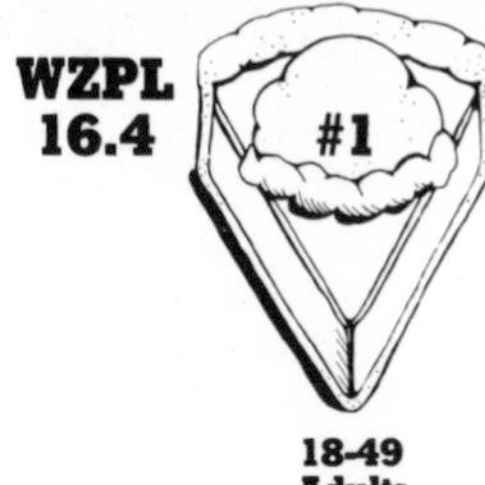

With the narrow formatting in radio today, there are few honest-to-goodness mass appeal stations left; stations whose popularity transcends every demographic and lifestyle group. Fortunately for the Indianapolis advertiser, WZPL is such a radio station.

And an incredible 47.6 in teens!

This information derived from copyrighted and proprietary estimates from: Fall '83 Arbitron. 6 AM to midnight, Monday through Sunday Metro.

FIGURE 6.10
Sample page from a Birch Radio report. Courtesy Birch Radio.

CHICAGO IL SCSA
APRIL - JUNE 1984

Target Demographics

WOMEN 25-64

AVERAGE QUARTER HOUR AND CUME ESTIMATES

	MON - SUN 6:00AM-12:00 MID				MON - FRI 6:00AM-10:00AM				MON - FRI 10:00AM-3:00PM				MON - FRI 3:00PM-7:00PM				MON - FRI 7:00PM-12:00 MID			
	AQH PRS (00)	AQH PRS RTG	AQH PRS SHR	CUME PRS (00)	AQH PRS (00)	AQH PRS RTG	AQH PRS SHR	CUME PRS (00)	AQH PRS (00)	AQH PRS RTG	AQH PRS SHR	CUME PRS (00)	AQH PRS (00)	AQH PRS RTG	AQH PRS SHR	CUME PRS (00)	AQH PRS (00)	AQH PRS RTG	AQH PRS SHR	CUME PRS (00)
WAGO-FM	11	.1	.3	90	8		.1	59	33	.2	.8	59	16	.1	.4	59				
WAIT	41	.2	1.1	629	55	.3	1.0	406	75	.4	1.8	184	37	.2	1.0	234	13	.1	.6	119
WAUR-FM	5		.1	97									8		.2	41				
WBBM	145	.7	4.0	2513	289	1.4	5.2	1678	106	.5	2.5	610	53	.3	1.4	412	64	.3	3.1	389
WBBM-FM	119	.6	3.3	1343	127	.6	2.3	646	175	.9	4.1	595	224	1.1	5.8	764	105	.5	5.0	582
WBMX-FM	88	.4	2.4	1351	115	.6	2.1	663	137	.7	3.2	594	29	.1	.7	340	92	.5	4.4	490
WCFL	28	.1	.8	527	64	.3	1.2	200	51	.3	1.2	159	8		.2	60	8		.4	51
WCLR-FM	164	.8	4.5	2153	201	1.0	3.6	1405	239	1.2	5.6	701	199	1.0	5.1	841	146	.7	7.0	751
WFMT-FM	54	.3	1.5	753	96	.5	1.7	430	54	.3	1.3	169	44	.2	1.1	239	59	.3	2.8	156
WFYR-FM	146	.7	4.0	1787	210	1.0	3.8	983	203	1.0	4.8	517	159	.8	4.1	871	68	.3	3.3	416
WGCI	116	.6	3.2	937	153	.8	2.8	491	136	.7	3.2	334	144	.7	3.7	470	105	.5	5.0	217
WGCI-FM	199	1.0	5.5	2482	244	1.2	4.4	1367	174	.9	4.1	758	165	.8	4.2	705	169	.8	8.1	914
WGN	381	1.9	10.6	3978	789	3.9	14.3	2917	387	1.9	9.1	1564	408	2.0	10.5	1526	112	.6	5.4	660
WIND	138	.7	3.8	1404	265	1.3	4.8	931	72	.4	1.7	363	95	.5	2.4	392	93	.5	4.5	410
WJEZ-FM	87	.4	2.4	1315	135	.7	2.5	674	137	.7	3.2	526	94	.5	2.4	698	28	.1	1.3	231
WJJD	102	.5	2.8	1214	92	.5	1.7	470	139	.7	3.3	535	115	.6	3.0	412	63	.3	3.0	245
WJOB	18	.1	.5	379	42	.2	.8	157					8		.2	38	21	.1	1.0	109
WJOL	9		.2	243	28	.1	.5	158	11	.1	.3	85	4		.1	42	11	.1	.5	43
WJPC	29	.1	.8	440	16	.1	.3	100	19	.1	.4	150	31	.2	.8	150	45	.2	2.2	100
WKQX-FM	175	.9	4.8	2064	317	1.6	5.8	1463	224	1.1	5.3	651	128	.6	3.3	627	69	.3	3.3	399
WLAK-FM	150	.7	4.2	2050	153	.8	2.8	839	222	1.1	5.2	750	75	.4	1.9	405	145	.7	7.0	737
WLOO-FM	131	.7	3.6	1384	130	.6	2.4	788	176	.9	4.1	543	194	1.0	5.0	920	54	.3	2.6	245
WLS	229	1.1	6.3	3491	453	2.3	8.2	2596	364	1.8	8.5	1269	226	1.1	5.8	1403	42	.2	2.0	350
WLS -FM	163	.8	4.5	2996	274	1.4	5.0	1763	165	.8	3.9	785	228	1.1	5.9	1284	109	.5	5.2	624
WLUP-FM	38	.2	1.1	751	44	.2	.8	358	51	.3	1.2	237	31	.2	.8	296	51	.3	2.5	178
WMAQ	130	.6	3.6	2641	197	1.0	3.6	1264	108	.5	2.5	854	201	1.0	5.2	1474	96	.5	4.6	485
WMET-FM	37	.2	1.0	811	37	.2	.7	236	70	.3	1.6	355	47	.2	1.2	292	17	.1	.8	127
WNIB-FM	18	.1	.5	345	8		.1	41	15	.1	.4	47	73	.4	1.9	265				
WOJO-FM	40	.2	1.1	298	33	.2	.6	99	56	.3	1.3	148	44	.2	1.1	106	13	.1	.6	42
WUSN-FM	118	.6	3.3	1138	151	.8	2.7	584	160	.8	3.8	445	152	.8	3.9	558	52	.3	2.5	329
WXRT-FM	111	.6	3.1	992	162	.8	2.9	801	74	.4	1.7	219	160	.8	4.1	506	74	.4	3.6	464
WYEN-FM	16	.1	.4	232	4		.1	59					20	.1	.5	119				
PERSONS USING RADIO	3610	18.0		19335	5508	27.4		16745	4260	21.2		12530	3885	19.4		13767	2080	10.4		9293

* ESTIMATES ADJUSTED FOR ACTUAL BROADCAST SCHEDULE

BIRCH RADIO

PAGE 84

QUALITATIVE AND QUANTITATIVE DATA

Since their inception in the 1930s, ratings services primarily have provided broadcasters with information pertaining to the number of listeners of a certain age and gender tuned to a station at a given time. It was on the basis of this quantitative data that stations chose a format and advertisers made a buy.

Due to the explosive growth of the electronic media in recent years, the audience is presented with many more options, and the radio broadcaster, especially in larger markets, must know more about his intended listeners in order to attract and retain them. Subsequently, the need for more detailed information arose. In the 1980s in-depth research is available to broadcasters from numerous sources, including both Arbitron and Birch. The latter survey service has concentrated its efforts in the area of qualitative research more than its prime competitor, Arbitron, which has begun to move more energetically in that direction as well. Birch Radio's *Prizm,* a semi-annual ratings report, examines the listening preferences of twelve lifestyle groups profiled in what it terms "Geo-Demographic Cluster Groups."

Birch's *Prizm* bases its Cluster Group profiles on statistics formulated by the United States Census Bureau. The study focuses on eighty-five specific markets. *Prizm* is particularly useful to media buying services that place advertising dollars on radio stations throughout the country. In this age of highly fragmented audiences, advertisers and agencies alike have become less comfortable with buying just numbers and look for audience qualities, notes Surrey's Christopher Porter. "The proliferation of stations has resulted in tremendous audience fragmentation. There are so many specialized formats out there, and many target the same piece of demographic pie. This predicament, if it can be called that, has made amply clear the need for qualitative, as well as quantitative, research. With so many stations doing approximately the same thing, differentiation is of paramount importance."

Today a station shooting for a top spot in the ratings surveys must be concerned with more than simply the age and sex of its target audience. Competitive programming strategies are built around an understanding and appreciation of the lifestyles, values, and behavior of those listeners sought by a station.

Why are more stations than ever choosing Surrey?

In the last year, Surrey Research has been privileged to experience its greatest clientele growth ever. Among reasons most commonly given for choosing Surrey are their superior quality, past performance and vast experience in radio research. Beyond this more general reasoning, some stations are pointing to more specific features. Here are just a few repeated most often.

Clearness of Research

As most programmers will agree, the clarity of research is vital to its usefulness. At Surrey our attempt is to design and display research in its clearest, most concise and easiest to put to use form. One example of this is offered in our *A & O Auditorium Music Testing.* Results are presented to you alphabetically by artist, in rank-order and also in a page-by-page cross-tabulation of each song's level of appeal to your key audience groups. To assist you in distinguishing the winners from the losers, Surrey applies *color-coding* to each ranked title. This exclusive Surrey feature puts an end to the once burdensome task of determining the cut-off point in Music Test results.

Stringent Recruiting Methodologies

One of the most important aspects of qualified research is the assembly of those individuals whose opinions most reflect that of your *core-listener.* This approach, as opposed to the often used targeted-demo concept, ensures you a representation of people more closely identified to your "perfect" listener. Not only does Surrey encourage this more defined sample of your desired audience, but additionally seeks those types of individuals most prone to participate in a rating study. We call them *diary-types.* Aside from your study providing you with listener data, this added recruiting technique lends itself well to the one thing you strive for most— ratings efficiency.

Affordability

Understandably, cost is a key in determining the amount of research you're able to acquire. Now for the first time, through Surrey's *investment planning* and *package discounts,* many stations are conducting research programs designed to accommodate their objectives rather than just their budgets.

Through *investment planning,* stations who contract for more than one study over a period of time are now afforded the luxury of monthly payments to accommodate the total investment. Aside from the benefits derived from investment planning, by contracting for more than one study over the course of a year, additional discounts are realized. These discounts are determined by the number of projects conducted.

These are just a few of the many reasons stations are choosing Surrey. Why not let us tell you more? To arrange for a personal visit from one of our representatives, call today **(303) 989-9980.** We'd like to become your choice.

SURREY

FIGURE 6.11
Research company promotional piece. Courtesy Surrey Research.

IN-HOUSE RESEARCH TECHNIQUES

Research data provided by the major survey companies can be costly. For this reason, and others, stations frequently conduct their own audience studies. Although stations seldom have the professional wherewithal and expertise of the research companies, they can derive useful information through do-it-yourself, in-house telephone, face-to-face, and mail surveys.

FIGURE 6.12
Birch Prizm's 12 Geo-Demographic Cluster Groups. Courtesy Birch Radio.

PROFILES ON RADIO
MONDAY-SUNDAY 6:00 AM - MIDNIGHT
PERSONS 12+

ATLANTA GA SMSA
JANUARY-JUNE 1984

WEEKLY CUME COMPOSITION INDEX PERCENTAGES

PRIZM CLUSTER GROUPS

	S1	S2	S3	U1	T1	S4	T2	U2	R1	T3	R2	U3	
WABE-FM	234	280	57	70	91		71	57		80	12	47	WABE-FM
WAOK		5	13	83	4		37	376		37	11	425	WAOK
WCLK-FM	108	59	21	132	23		40	346		60		257	WCLK-FM
WCNN	139	162	101	68	108		46	139		66	62	95	WCNN
WGST	199	208	68	89	70		49	106		48	11	126	WGST
WIGO		15	17	98			10	304		40	7	448	WIGO
WJYF-FM	134	158	117	104	97		191	76		24	52	58	WJYF-FM
WKHX-FM	70	62	146	90	119		134	34		145	159	27	WKHX-FM
WKLS-FM	146	131	131	93	103		108	43		94	100	45	WKLS-FM
WPCH-FM	183	167	104	98	110		77	65		62	49	71	WPCH-FM
WPLO	103	75	127	126	138		119	48		115	98	52	WPLO
WQXI	131	84	127	114	74		43	58		74	78	107	WQXI
WQXI-FM	136	126	122	98	107		109	67		88	75	60	WQXI-FM
WRAS-FM	141	247	30	153	144		112	45		72	18	61	WRAS-FM
WRMM-FM	130	104	137	101	106		90	84		71	111	39	WRMM-FM
WSB	71	112	94	92	154		89	76		112	136	76	WSB
WSB -FM	116	120	96	100	108		65	102		86	75	116	WSB -FM
WVEE-FM	26	43	38	116	29		51	293		91	36	277	WVEE-FM
WWLT-FM	310	74	158	38	87		50	62		116	87	44	WWLT-FM
WZGC-FM	105	84	134	83	104		108	75		110	101	76	WZGC-FM
OTHER	40	77	70	78	55		99	144		195	168	102	OTHER

BIRCH RADIO

Telephone surveying is the most commonly used method of deriving audience data on the station level. It generally is less costly than the other forms of in-house research, and sample selection is less complicated and not as prone to bias. It also is the most expedient method. There are, however, a few things that must be kept in mind when conducting call-out surveys. To begin with, not everyone has a phone and many numbers are unlisted. People also are wary

FIGURE 6.12
continued

HOW TO USE PROFILES ON RADIO

This report redefines radio listening audiences according to the 12 PRIZM Geo-Demographic Cluster Groups. The Cluster groups used in this report are:

GROUP	DEFINITION
S1	Educated, Affluent, Executives and Professionals in Elite Metro Suburbs.
S2	Pre and Post-Child Families and Singles in Upscale White-Collar Suburbs.
S3	Upper-Middle, Child-Raising Families in Outlying, Owner-Occupied Suburbs.
U1	Educated, White-Collar Singles and Ethnics in Upscale, Urban Areas.
T1	Educated, Young, Mobile Families in Exurban Suburbs and Boom Towns.
S4	Middle-Class, Post-Child Families in Aging Suburbs and Retirement Areas.
T2	Middle-Class, Child-Raising, Blue-Collar Families in Remote Suburbs and Towns.
U2	Middle-Class Immigrants and Minorities in Dense Urban Row and High-Rise Areas.
R1	Rural Towns and Villages Amidst Farms and Ranches Across Agrarian Mid-America.
T3	Mixed Gentry and Blue-Collar Labor in Lo-Mid Rustic Mill and Factory Towns.
R2	Mixed Whites, Blacks, Hispanics and Indians in Poor Rural Towns and Farms.
U3	Mixed Blacks, Hispanics and Immigrants in Aging Urban Row and Hi-Rise Areas.

of phone interviews for fear that the ultimate objective of the caller is to sell something. The public is inundated by phone solicitors. Finally, extensive interviews are difficult to obtain over the phone. Five to ten minutes usually is the extent to which an interviewee will submit to questioning. Call-out interview seminars and instructional materials are available from a variety of sources, including the telephone company itself.

The face-to-face or personal interview also is a popular research approach at stations, although the cost can be higher than call-out, especially if a vast number of individuals are being surveyed in an auditorium setting. The primary advantages of the in-person interview are that questions can be more substantive and greater time can be spent with the respondees. Of course, more detailed interviews are time consuming and usually require refined interviewing skills, both of which can be cost factors.

Mail surveys can be useful for a host of reasons. To begin with, they eliminate the need to hire and train interviewers. This alone can mean a great deal in terms of money and time. Since there are no interviewers involved, one source of potential bias also is eliminated. Perhaps most important is that individuals questioned through the mail are somewhat more inclined toward candor since they enjoy greater anonymity. The major problem with the mail survey approach stems from the usual low rate of response. One in every five questionnaires mailed may actually find their way back to the station. The length of the questionnaire must be kept relatively short, and the questions succinct and direct. Complex questions create resistance and may result in the survey being ignored or discarded.

Large and major market outlets usually employ someone to direct research and survey efforts. This person works closely with upper management and department heads, especially the program director and sales manager. These

FIGURE 6.13
Call-out music research helps a station determine what and what not to program. In this example, callers were played bits of songs over the phone and asked to rank them. Courtesy Balon Associates.

'HATE IT' - TOP 50

	RESPONSES
MICKEY - TONI BASIL	24
GUITAR MAN - ELVIS PRESLEY	19
I'LL TUMBLE FOR YA - CULTURE CLUB	18
UNDERCOVER OF THE NIGHT - ROLLING STONES	14
RIO - DURAN DURAN	13
LITTLE RED CORVETTE - PRINCE	13
PLAY THAT FUNKY MUSIC - WILD CHERRY	12
PLEASE COME TO BOSTON - DAVE LOGGINS	12
AFTER THE LOVING - ENGELBERT HUMPERDINCK	12
COME ON EILEEN - DEXY'S MIDNIGHT RUNNERS	12
A WOMAN NEEDS LOVE - RAY PARKER	12
THE TIDE IS HIGH - BLONDIE	11
I WRITE THE SONGS - BARRY MANILOW	11
CARS - GARY NUMAN	11
CRUMBLIN' DOWN - JOHN COUGAR	11
ANY DAY NOW - RONNIE MILSAP	11
SUDDENLY LAST SUMMER - MOTELS	11
NIGHT FEVER - BEE GEES	10
JACK AND DIANE - JOHN COUGAR	10
SQUEEZE BOX - THE WHO	10
HELLO MY LOVER - ERNIE KADOE	10
MORNING TRAIN - SHEENA EASTON	10
MINUTE BY MINUTE - DOOBIE BROTHERS	10
YOU MAKE MY DREAMS COME TRUE - HALL AND OATES	10
THE ONE THAT YOU LOVE - AIR SUPPLY	10
FUNKY TOWN - LIPPS, INC.	10
ANOTHER ONE BITES THE DUST - QUEEN	10
EMOTIONAL RESCUE - ROLLING STONES	10
DR. HECKYL AND MR. JIVE - MEN AT WORK	10
POISON ARROW - ABC	10
WE TOO - LITTLE RIVER BAND	10
ONE THING LEADS TO ANOTHER - FIXX	9
KEY LARGO - BERTIE HIGGINS	9
LOVIN' YOU - MINNIE RIPERTON	9
WHAT ARE WE DOING IN LOVE - DOTTIE WEST AND KENNY ROGERS	9
TRY AGAIN - CHAMPAGNE	9
PHYSICAL - OLIVIA NEWTON-JOHN	9
TELL IT LIKE IT IS - AARON NEVILLE	8
YOU DON'T MESS AROUND WITH JIM - JIM CROCE	8
I CAN'T GO FOR THAT - HALL AND OATES	8
TOO MUCH HEAVEN - BEE GEES	8
'65 LOVE AFFAIR - PAUL DAVIS	8
SOMEWHERE DOWN THE ROAD - BARRY MANILOW	8
PRIVATE EYES - HALL AND OATES	8
IT'S RAINING - IRMA THOMAS	8
SLOW HAND - POINTER SISTERS	8
I LOVE THE NIGHT LIFE - ALICIA BRIDGES	8
SOLITARY MAN - NEIL DIAMOND	8
YOU NEEDED ME - ANNE MURRAY	8
LOVE TAKES TIME - ORLEANS	8

FIGURE 6.13
continued

'ONE OF MY FAVORITES' - TOP 50

	RESPONSES
STILL - COMMODORES	40
YESTERDAY - BEATLES	35
ALWAYS AND FOREVER - HEATWAVE	35
JUST ONCE - JAMES INGRAM	34
LADY - KENNY ROGERS	33
DON'T STOP TILL YOU GET ENOUGH - MICHAEL JACKSON	33
THE WAY WE WERE - BARBRA STREISAND	32
SWEET HOME ALABAMA - LEONARD SKYNARD	31
TOTAL ECLIPSE OF THE HEART - BONNIE TYLER	30
TIME IN A BOTTLE - JIM CROCE	29
MANIAC - MICHAEL SEMBELLO	28
FLASHDANCE (WHAT A FEELING) - IRENE CARA	27
MAGGIE MAY - ROD STEWART	25
PYT - MICHAEL JACKSON	25
ENDLESS LOVE - DIANA ROSS AND LIONEL RICHIE	24
ROCK WITH YOU - MICHAEL JACKSON	23
HOTEL CALIFORNIA - EAGLES	23
THAT'S THE WAY OF THE WORLD - EARTH, WIND AND FIRE	23
NEW ORLEANS LADIES - LOUISIANA LEROUX	23
(THEY LONG TO BE) CLOSE TO YOU - CARPENTERS	22
I JUST WANNA STOP - GINO VANNELLI	22
OPEN ARMS - JOURNEY	22
IN THE MIDNIGHT HOUR - WILSON PICKETT	22
CELEBRATE - KOOL AND THE GANG	22
YOU NEEDED ME - ANNE MURRAY	22
ONE IN A MILLION YOU - LARRY GRAHAM	22
YOU DON'T BRING ME FLOWERS - STREISAND AND DIAMOND	21
ALWAYS ON MY MIND - WILLIE NELSON	21
BABY COME BACK - PLAYER	21
IT'S RAINING - IRMA THOMAS	21
IF YOU COULD READ MY MIND - GORDON LIGHTFOOT	21
SO VERY HARD TO GO - TOWER OF POWER	21
SHE WORKS HARD FOR THE MONEY - DONNA SUMMER	21
YOU'VE GOT A FRIEND - JAMES TAYLOR	20
SOMETIMES WHEN WE TOUCH - DAN HILL	20
HEATWAVE - DANDELLAS	20
PIANO MAN - BILLY JOEL	20
PHYSICAL - OLIVIA NEWTON-JOHN	20
SARA SMILE - HALL AND OATES	20
100 WAYS - JAMES INGRAM	20
COME SAIL AWAY - STYX	19
TELL IT LIKE IT IS - AARON NEVILLE	19
NIGHT FEVER - BEE GEES	19
I GO CRAZY - PAUL DAVIS	19
EVERGREEN - BARBRA STREISAND	19
DUST IN THE WIND - KANSAS	19
AIN'T NO STOPPIN' US NOW - MCFADDEN AND WHITEHEAD	19
LOVIN' YOU - MINNIE RIPERTON	19
MAKING LOVE OUT OF NOTHING AT ALL - AIR SUPPLY	19
LOVE ME DO - BEATLES	19

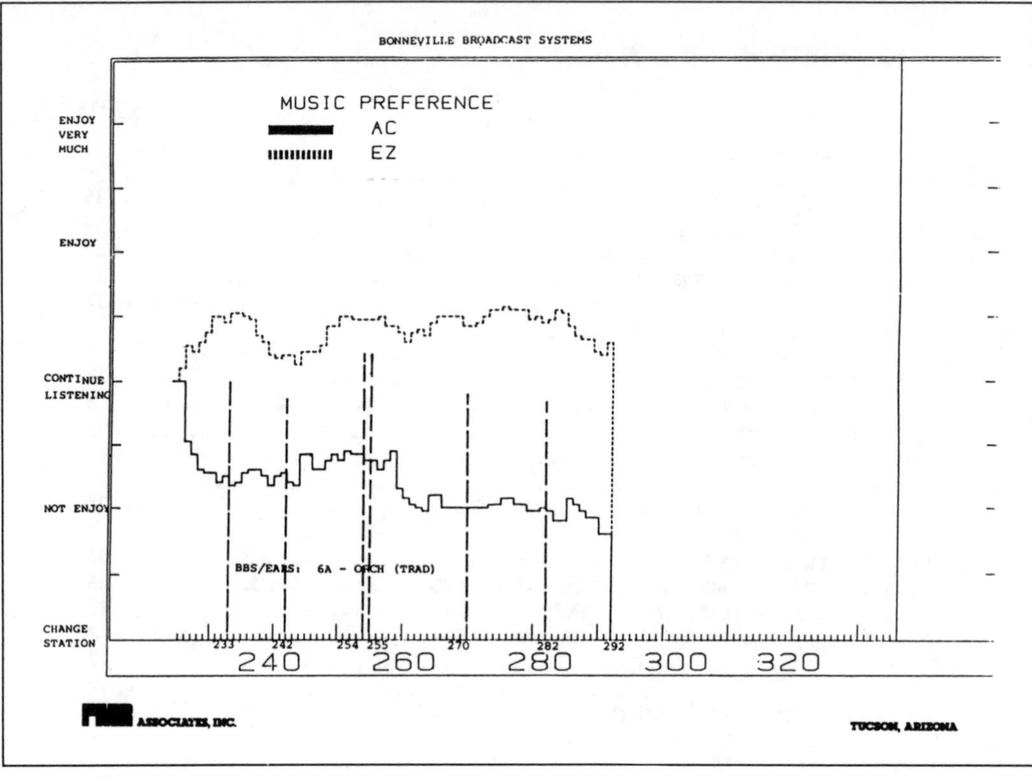

FIGURE 6.14
The electronic music testing graph shows where listening drops off. In this test the preferences of Adult Contemporary and Easy Listening listeners were gauged as certain tunes were played. Courtesy FMR Associates.

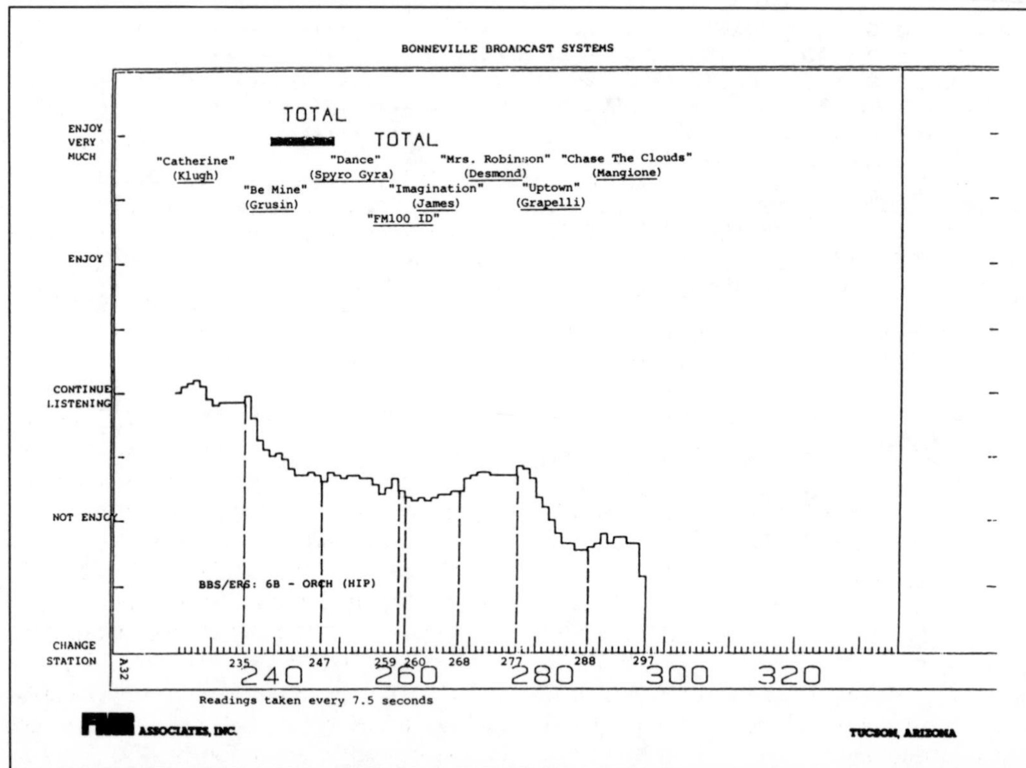

FIGURE 6.15
The electronic music testing graph shows audience reaction to specific songs. Even the ID is tested. Courtesy FMR Associates.

two areas require data on which to base programming and marketing decisions. At smaller outlets, area directors generally are responsible for conducting surveys relevant to their department's needs. A case in point would be the PD who plans a phone survey during a special broadcast to help ascertain whether it should become a permanent program offering. To accomplish this task, the programmer enlists the aid of a secretary and two interns from a local college. Calls are made, and data are collected and analyzed.

The objective of a survey must be clear from the start, and the methodology used to acquire data should be as uncomplicated as possible. Do-it-yourself surveys are limited in nature, and overly ambitious goals and expectations are seldom realized. However, in-house research can produce valuable information that can give a station a competitive edge. Today, no radio station can operate in a detached way and expect to prosper.

Every station has numerous sources of information available to it. Directories containing all manner of data, such as population statistics and demographics, manufacturing and retailing trends, and so on, are available at the public library, city hall, chamber of commerce, and various business associations. The American Marketing Association and American Research Foundation also possess information designed to guide stations with their in-house survey efforts.

RESEARCH DEFICITS

Although broadcasters refer deferentially to the ratings surveys as the "Book" or "Bible," the stats they contain are audience-listening "estimates," no more, and, hopefully, no less. Since their inception, research companies have been criticized for the methods they employ in collecting audience listening figures. The most prevalent complaint has had to do with the selection of samples. Critics have charged that they invariably are limited and exclusionary. Questions have persisted as to whether those surveyed are truly representative of an area's total listenership. Can one percent of the radio universe accurately reflect general listening habits? The research companies defend their tactics and have established a strong case for their methodology.

In the 1970s ratings companies were criticized for neglecting minorities in their surveys. In efforts to rectify this deficiency, both Arbitron and Birch have established special sampling procedures. The incidence of nontelephone households among Blacks and Hispanics tends to be higher. The survey companies also had to deal with the problem of measuring Spanish-speaking people. Arbitron found that using the personal-retrieval technique significantly increased the response rate in the Spanish community, especially when bilingual interviewers were used. The personal-retrieval technique did not work as well with Blacks, since it was difficult to recruit interviewers to work in many of the sample areas. Thus, Arbitron used a telephone retrieval procedure that involved call-backs to selected households over a seven-day period to document listening habits. In essence, the interviewer filled out the diaries for those being surveyed. In 1982, Arbitron implemented Differential Survey Treatment (DST), a technique designed to increase the response rate among Blacks. The survey company provides incentives over the customary fifty cents to a dollar to certain Black households. Up to five dollars is paid to some respondents. DST employs follow-up calls to retrieve diaries.

Birch Radio uses special sampling procedures and bilingual interviewers to collect data from the Hispanic population. According to the company, its samples have yielded a high response rate among Blacks. Thus Birch has not utilized other special sampling controls. Ethnic listening reports containing average quarter-hour and cume estimates for Hispanics, Blacks, and others are available from the company in a format similar to that of its Capsule Market Report.

Both survey companies have employed additional procedures to survey other nontelephone households, especially in markets that have a large student or transient population. In the late 1970s, a Boston station targeting young people complained that Arbitron failed to acknowledge the existence of over two hundred thousand college students who did not have personal phone listings. The station, which was rated among the top five in the market at the time, contended that a comprehensive survey of the city's listening audience would bear out the fact that they were, in fact, number one.

Similar complaints of skewed or inconclusive surveys persist today, but the procedures and methods used by the major radio audience research companies, although far from perfect, are more effective than ever. Christopher Porter of Surrey Research says the greatest misconception about research data is that they are absolutes etched in granite. "The greatest fallacy is that research findings are gospel. This goes not only for the quantitative studies but for focus groups as well. Regardless of the methodology, any findings should be used as a 'gut adjuster,' rather than a 'gut replacer.' Sampling error is often ignored in a quantitative study, even in an Arbitron report. When we report that 25 percent of a sample feels some way about something, or when a station with a 4.1 beats one with a 3.8 in a book, most station managers and PDs take all these statistics at face value."

Rip Ridgeway, vice president of Arbitron Ratings, believes that stations place too much emphasis on survey results. "I think that station hierarchy puts too much credence on the ratings estimates. They're an indicator, a sort of report card on a station's performance. They're not the absolute end all. To jump at the next numbers and make sweeping changes based on them generally is a big mistake."

WROR's Lorna Ozman concurs with both Porter and Ridgeway and warns that research should help direct rather than dictate what a station does. "I use research, rather than letting it use me. The thing to remember is that no methodology is without a significant margin of error. To treat the results of a survey as gospel is dangerous. I rely on research to provide me with the black and white answers and depend on myself to make determinations on the gray areas. Research never provided a radio station with the glitter to make it sparkle."

Richard Bremkamp of WRCH/WRCQ also expresses concern over what he perceives as an almost obsessive emphasis on survey statistics. "The concern for numbers gets out of hand. There are some really good-sounding stations out there that don't do good book, but the money is in the numbers. Kurt Vonnegut talks about the Universal Will to Become in his books. In radio that can be expanded to the Universal Will to Become Number One. This is good if it means the best, but that's not always what it means today."

The proliferation of survey services has drawn criticism from broadcasters who feel that they are being oversurveyed and overresearched. When Arbitron introduced its computerized monthly ratings service (Arbitrends), the chairman of its own radio advisory council opposed the venture on the grounds that it would cause more confusion and create more work for broadcasters. He further contended that the monthly service would encourage short-term buying by advertisers. Similar criticism has been lodged against Birch Radio's own computerized service, Birchscan. However, both services have experienced steady growth since their inception.

HOW AGENCIES BUY RADIO

The primacy of numbers perhaps is best illustrated through a discussion of how advertising agencies place money on radio stations. It is the media buyer's job to effectively and efficiently invest the advertiser's money, in other words, to reach the most listeners with the budget allotted for radio use.

According to Lynne Price, media buyer for Kenyon and Eckhardt, Boston, the most commonly employed method determines the cost per point (CPP) on a given station. Lynne explains the procedure: "A media buyer is given a budget and a gross rating point (GRP) goal. Our job is to buy to our GRP goal, without going over budget, against a predetermined target audience, i.e., adults twenty-five to fifty-four, teens, men eighteen to thirty-four, etc. Our CPP is derived by taking the total budget and dividing by the GRP goal, or total number of rating points we would like to amass against our target audience. Now, using the CPP as a guideline, we take the cost per spot on a given station, and divide by the rating it has to see how close to the total CPP the station is. This is where the negotiation comes in. If the station is way off, you can threaten not to place advertising until they come closer to what you want to spend."

The other method used to justify station buys is cost per thousand (CPM). Using this technique, the buyer determines the cost of reaching a thousand people at a given station. The CPM of one station is then compared with that of another's to ascertain efficiency. In order to find out a station's CPM, the buyer must know the station's average quarter-hour audience (AQH Persons) estimate in the daypart targeted and the cost of a commercial during that time frame. The computation below will provide the station's CPM:

$$\frac{\$30 \text{ for } 60 \text{ seconds}}{25\ (000)\ \text{AQH}} = \$1.20 \text{ CPM}$$

By dividing the number of people reached into the cost of the commercial, the cost per thousand is deduced. Thus, the lower the CPM, the more efficient the buy. Of course, this assumes that the station selected delivers the target audience sought. Again, this is the responsibility of the individual buying media for an agency. It should be apparent by now that many things are taken into consideration before airtime is purchased.

CAREERS IN RESEARCH

The number of media research companies has grown rapidly since the late 1960s. Today over 150 research houses nationwide offer audience measurement and survey data to the electronic media and allied fields. Job opportunities in research have increased proportionately. Persons wanting to work in the research area need sound educational backgrounds, says Surrey's Porter. "College is essential. An individual attempting to enter the field today without formal training is at a serious disadvantage. In fact, a master's degree is a good idea." Dr. Rob Balon, president of Balon and Associates, agrees with Porter. "Entering the research field today requires substantial preparation. College research courses are where to start."

Ed Noonan of Survey Research is of the same opinion. "It is a very competitive and demanding profession. Formal training is very important. I'd advise anyone planning a career in broadcast research to get a degree in communications or some related field and heavy-up on courses in research methodology and analysis, statistics, marketing, and computers. Certain business courses are very useful, too."

To Don Hagen of Southeast Media Research a strong knowledge of media is a key criterion when hiring. "One of the things that I look for in a job candidate is a college background in electronic media. That's the starting point. You have to know more today than ever before. Audience research has become a complex science."

As might be expected, research directors also place considerable value on experience. "The job prospect who offers some experience in the research area, as well as a diploma, is particularly attractive," notes Dick Warner of S-A-M-S Research. Ed Noonan concurs. "Actual experience in the field, even if it is gained in a summer or part-time job, is a big plus." Christopher Porter advises aspiring researchers to work in radio to get a firsthand feel for the medium. "A hands-on knowledge of the broadcast industry is invaluable, if not vital, in this profession."

In the category of personal attributes, Dick Warner puts inquisitiveness at the top of the list. "An inquiring mind is essential. The job of the researcher is to find and collect facts and information. Curiosity is basic to the researcher's personality." Don Hagen adds objectivity and perceptiveness to the list, while Dwight Douglas of Burkhart, Abrams, Michaels, Douglas and Associates emphasizes interactive skills. "People skills are essential, since selling and servicing research clients are as important as the research itself."

While not everyone is suited for a career in audience research, those who are find the work intellectually stimulating and financially rewarding.

THE FUTURE OF RESEARCH IN RADIO

Most experts agree that the role of research in radio will continue to grow. They base their predictions on the ever-increasing fragmentation of the listening audience, which makes the job of targeting and positioning a more complex one. "The field of broadcast research has grown considerably in the last decade, and there is every reason to suspect that the growth will continue. As demographic targets and formats splinter, there will be an increasingly greater need to know. Much of the gut feel that has propelled radio programming will give way to objective research that is based on a plan," contends Dwight Douglas.

Christopher Porter sees the proliferation of radio frequencies as creating a greater demand for research. "With the inevitability of many more frequencies crowding already overcrowded markets, the need to stay abreast of market developments is critical. Yes, the role of research will grow proportionately."

While the role of research in the programming of large market stations is significant, the expense involved will continue to limit many smaller stations to in-house methods, claims Ed Noonan. "Professional research services can be very costly. This will keep research to a minimum in lesser markets, although there will be more movement there than in the past. Call-out research will continue to be a mainstay for the small station."

Computer consultant Vicki Cliff believes that microcomputers will become feasible research tools at small stations as they become more affordable. "Cost-effective ways to perform and utilize sophisticated psychographic data will make the computer standard equipment at most stations, small and large alike. Research is becoming a way of life everywhere, and computers are an integral part of the information age. Computers will encourage more do-it-yourself research at stations, as well."

S-A-M-S's Dick Warner contends that technological advances also will improve the nature and quality of research. "We'll see more improvement in methodology and a greater diversity of applicable data as the result of high-tech innovations. I think the field of research will take a quantum leap in the 1990s. It has in the 1970s and 1980s, but the size of the leap will be greater in the last decade of this century."

Today it is common for stations to budget 5 to 10 percent of their annual income to research, and it is probable that this figure will increase, says Christopher Porter. "As it evolves, it is likely that the marketplace will demand that more funds be allocated for research purposes. Research may not guarantee success, but it's not getting any easier to be successful without it."

Research has been a part of radio broadcasting since its modest beginnings in the 1920s, and it appears that it will play an even greater role in the operations of stations as the next century unfolds.

CHAPTER HIGHLIGHTS

1. Beginning in the late 1920s surveys were conducted to determine the most popular stations and programs with various audience groupings. Early surveys (and their methods) included C.E. Hooper, Inc. (telephone), Cooperative Analysis of Broadcasting (telephone), and The Pulse (in-person). In 1968 Radio's All Dimension Audience Research (telephone to six thousand households) began to provide information for networks. The current leader among local market audience surveys is Arbitron (week-long diary), although Birch Radio (telephone) is making inroads.

2. In 1963 the Broadcast Rating Council was established to monitor, audit, and accredit ratings companies. In 1982 it was renamed the Electronic Media Planning Council to reflect its involvement with cable television ratings.

3. Arbitron measures listenership in the Metro Survey Area (MSA), the city or urban center, and the Total Survey Area (TSA), the surrounding communities.

4. A station's primary listening locations are called Areas of Dominant Influence (ADI).

5. The Arbitron daily diary logs time tuned to a station; station call letters or program name; whether AM or FM; where listening occurred; and the listener's age, sex, and area of residence.

6. Birch Radio gathers data by calling equal numbers of male and female listeners aged twelve and over. Clients are offered seven different report formats, including a microcomputerized data retrieval system.

7. With today's highly fragmented audiences, advertisers and agencies are less comfortable buying just ratings numbers and look for audience qualities. Programmers must consider, not only the age and sex of the target audience, but also their lifestyles, values, and behavior.

8. Station in-house surveys utilize telephone, face-to-face, and mail methods.

9. In response to complaints about "missed" audiences, the major survey companies adjusted their survey techniques to assure inclusion of minorities and nontelephone households. Today's survey results are more accurate.

10. Ratings data should only direct, not dictate, what a station does.

11. Media buyers for agencies use station ratings to determine the most cost-effective buy for their clients. Two methods they use are the cost per rating point (CPP) and the cost per thousand (CPM).

12. Although the significant increase in the number of broadcast research companies (over 150 nationwide) has created a growing job market, a college education is necessary. Courses in communications, research methods, statistics, marketing, computers, and business are useful. Beneficial personal traits include inquisitiveness, objectivity, perceptiveness, and interpersonal skills.

APPENDIX 6A: RAB'S GLOSSARY OF RESEARCH TERMS

RADIO RESEARCH GLOSSARY

RAB Guide To Key Terms And Formulas

A wealth of terminology has come to describe the many dimensions of Radio audience research. As new buying and planning techniques are developed, new terms are added to the vocabulary and even established terms take on new meanings. To help you understand and use today's language of Radio audience research, RAB has prepared this brief guide.

APPENDIX 6A: continued

AUDIENCE RESEARCH TERMS

AVERAGE QUARTER-HOUR AUDIENCE (AQH PERSONS)

An average of the number of people listening for at least five minutes in each quarter-hour over a specified period of time. In modern Radio average quarter-hour measurement should be considered a measure of Total Time Spent Listening (see below).

AVERAGE QUARTER-HOUR RATING (AQH RATING)

Average Quarter-Hour Audience expressed as a percentage of the population being measured. *

AQH PERSONS ÷ POPULATION = AQH RATING

SHARE OF AUDIENCE (SHARE)

The percentage of those listening to Radio in the AQH that are listening to a particular station.

AQH PERSONS / ONE STATION ÷ AQH PERSONS / RADIO = SHARE

Because the AQH actually reflects Total Time Spent Listening, Share Of Audience is the share of Total Time Spent Listening to Radio.

CUMULATIVE AUDIENCE (CUME PERSONS)

Also called Unduplicated Audience, it is the number of different people listening for at least five minutes during a specified period of time. Cume Audience is the potential group that can be exposed to advertising on a radio station, just as readership is the potential exposure group for a magazine or newspaper.

CUMULATIVE RATING (CUME RATING)

Cumulative Audience expressed as a percentage of the population being measured. *

CUME PERSONS ÷ POPULATION = CUME RATING

TOTAL TIME SPENT LISTENING (TTSL)

The number of quarter-hours of listening to Radio or to a station by the population group being measured. *

AQH PERSONS × QUARTER-HOURS IN TIME PERIOD = TTSL (IN QUARTER HOURS)

AVERAGE TIME SPENT LISTENING (TSL)

The time spent listening by the average person who listens to Radio or to a station.

TTSL ÷ CUME PERSONS = TSL

Average Time Spent Listening is an indicator of audience availability to advertising messages. The more time spent listening, the greater opportunity for exposure and ability to develop frequency.

AUDIENCE TURNOVER (T/O)

The number of times the Average Quarter-Hour Audience is replaced by new listeners in a specified period of time. Audience Turnover is also the number of announcements required to reach approximately 50% of the station's Cumulative Audience in the time period.

CUME PERSONS ÷ AQH PERSONS = T/O

* The population being measured can be all people or any demographic group.

APPENDIX 6A: continued

SCHEDULE MEASUREMENT TERMS

REACH

The number of different people who are exposed to a schedule of announcements, i.e., those listening during a quarter-hour when announcements are aired.

Reach can also be expressed as a Rating (percentage of the population being measured):

PERSONS REACHED ÷ POPULATION = REACH RATING

GROSS IMPRESSIONS

The total number of exposures to a schedule of announcements. Not a measure of the number of different people exposed to a commercial.

AQH PERSONS × NUMBER OF ANNOUNCEMENTS = GROSS IMPRESSIONS

FREQUENCY

The average number of times the audience reached by an advertising schedule (those listening during a quarter-hour when an announcement is aired) is exposed to a commercial.

GROSS IMPRESSIONS ÷ REACH = FREQUENCY

GROSS RATING POINTS (GRPs)

Gross Impressions expressed as a percentage of the population being measured.* One Rating Point equals one percent of the population.

GROSS IMPRESSIONS ÷ POPULATION = GRPs

It can also be derived by combining AQH Ratings:

AQH RATING × NUMBER OF ANNOUNCEMENTS = GRPs

COST PER THOUSAND (CPM)

The basic term to express Radio's unit cost. It establishes 1000 as the basic unit for comparing Radio values. Most frequently used to compare the cost of 1000 Gross Impressions on different stations, it can also be used to compare the cost of reaching 1000 people.

$$\frac{\text{SCHEDULE COST}}{\text{GROSS IMPRESSIONS (IN THOUSANDS)}} = \text{CPM}$$

COST PER RATING POINT (CPP)

An expression of Radio's unit cost using a Rating Point, which is one percent of the population being measured*. Cost Per Rating Point is often used for planning Radio in conjunction with GRPs.

$$\frac{\text{SCHEDULE COST}}{\text{GROSS RATING POINTS}} = \text{CPP}$$

* The population being measured can be all people or any demographic group.

APPENDIX 6A: continued

REACH / FREQUENCY EVALUATIONS

REACH / FREQUENCY FORMULAS

There are three factors in any Reach / Frequency formula: **1) Reach, 2) Frequency and 3) GRPs**. Their relationship is expressed in these formulas, with any two factors predicting the third:

REACH × FREQUENCY = GRPs
GRPs ÷ FREQUENCY = REACH
GRPs ÷ REACH = FREQUENCY

For example, if 100 GRPs are purchased and the advertiser has determined a 4 Frequency is necessary, the Reach will be 25. These formulas make Radio planning extremely flexible.

EFFECTIVE FREQUENCY

The minimum level of frequency—number of exposures—determined to be effective in achieving the goals of an advertising campaign (e.g., awareness, recall, sales, etc.). This level will vary with individual products or services and the marketing objectives of the campaign.

FREQUENCY DISTRIBUTION

A tabulation separating those reached by a schedule, according to their minimum levels of exposure: 2 or more times, 3 or more times, 4 or more times, etc.

EFFECTIVE REACH

The number of people reached by a schedule at the pre-determined level of Effective Frequency.

EFFECTIVE RATING POINTS (ERPs)

Effective Reach expressed as a percentage of the population being measured.*

EFFECTIVE REACH ÷ POPULATION = ERPs

* The population being measured can be all people or any demographic group.

For further information please contact RAB Research.

Radio Advertising Bureau, Inc.
485 Lexington Avenue, New York, N.Y. 10017 • (212) 599-6666

APPENDIX 6B: ARBITRON'S GLOSSARY OF RESEARCH TERMS

XII. GLOSSARY OF TERMS

ADVANCE RATINGS/An Arbitron Ratings Radio Special Service's service that provides a client, via telephone, with selected estimates which will appear in his market report (RMR) as soon as the report has been approved for printing.

AGE/SEX POPULATIONS/Estimates of population, broken out by various age/sex groups within a county.

AM-FM TOTAL/A figure shown in market reports for AM-FM affiliates in time periods when they are simulcast.

ARBITRENDS/An Arbitron Ratings service, introduced in 1984 and available through Radio Special Services. Delivers averages of tabulated Radio audience listening estimates directly to clients' microcomputers in two types of reports: (1) Rolling Average Report, containing averages of listening estimates from three consecutive Arbitron Radio survey months; and (2) Quarterly Report, containing estimates from a three-month Arbitron survey.

ARBITRON INFORMATION ON DEMAND (AID)/An Arbitron Ratings Radio Special Services information service for direct access clients (via terminals) and indirect access clients (via AID division of Radio Special Services Department). Provides audience estimates and Reach and Frequency information, based on the same diaries that are used in the processing of the Radio Market Reports (RMRs).

ARBITRON SURVEY WEEK NUMBER/All fifty-two weeks in a year are assigned a number from 01-53 consecutively, beginning with the week in which January 1 falls. Appears on the diary label and serves as a quality check to ensure that a diary is placed in the correct week of a survey.

AREA OF DOMINANT INFLUENCE (ADI)/An exclusive geographic area, defined by Arbitron Television, consisting of sampling units in which the home-market television stations receive a preponderance of viewing. Every county in the United States (excluding Alaska and Hawaii) is allocated exclusively to one ADI.

A-SALE TAPE/An Arbitron Ratings Radio Special Services data tape of ADI estimates for one or more of the top fifty ADI markets available to agency and station clients that subscribe to the RMRs. Also known as "ASL" or "ADI" tape.

ASCRIPTION/A statistical technique that allocates radio listening proportionate to each conflicting station's diaries as calculated on a county basis using up to four surveys' TALO from the previous year, excluding the most recently completed survey for those markets with back-to-back surveys. Diary credit is randomly assigned automatically to a station based on its share of total diaries in the county.

AUDIENCE/A group of households, or a group of individuals, that are counted in a radio audience according to any one of several alternative criteria.

AVERAGE QUARTER-HOUR PERSONS/The estimated number of persons who listened at home and away to a station for a minimum of five minutes within a given quarter-hour. The estimate is based on the average of the reported listening in the total number of quarter-hours the station was on the air during a reported time period. This estimate is shown for the MSA, TSA and, where applicable, the ADI.

AVERAGE QUARTER-HOUR RATING/The Average Quarter-Hour Persons estimate expressed as a percentage of the universe. This estimate is shown in the MSA and, where applicable, the ADI.

AVERAGE QUARTER-HOUR SHARE/The Average Quarter-Hour estimate for a given station expressed as a percentage of the Average Quarter-Hour Persons estimate for the total listening in the MSA within a given time period. This estimate is shown only in the MSA.

AWAY FROM HOME LISTENING/Estimates of listening for which the diarykeeper indicated listening was done away from home, either "in a car" or "some other place."

CLIENT TAPE/A magnetic tape containing the same data as the Arbitron Ratings reports, sent to clients who subscribe to the printed report.

CONDENSED RADIO MARKET/Generally a small to middle-sized radio market; most are surveyed only once, in the Spring. The Metro and TSA sample objectives are considerably less than those for Standard Radio Markets and an abbreviated version of the Standard Radio Market Report is produced.

CONFLICT/Two or more stations using the same or similar slogan/program/personality/sports identification in the same county and qualifying under Arbitron's "One Percent TALO" criteria.

CONSOLIDATED METROPOLITAN STATISTICAL AREA (CMSA)/As defined by the U. S. Government's Office of Management and Budget; a grouping of closely related Primary Metropolitan Statistical Areas.

COUNTY SLOGAN EDIT FILE/A county-by-county listing of stations whose signals penetrate a county. Denotes all One Percent TALO qualifying stations. Includes each station's call letters, slogan ID, city and county of license, exact frequency and network affiliation(s). An internal document used to process diary entries.

APPENDIX 6B: continued

CUME PERSONS/The estimated number of different persons who listened at home and away to a station for a minimum of five minutes within a given daypart. (Cume estimates may also be referred to as "cumulative," "unduplicated," or "reach" estimates.) This estimate is shown in the MSA, TSA and ADI.

CUME RATING/The estimated number of Cume Persons expressed as a percentage of the universe. This estimate is shown for the MSA only.

DAYPART/The days of a week and the portion of those days for which listening estimates are calculated (e.g., Monday-Sunday 6AM-Midnight, Monday-Friday 6AM-10AM).

DEMOGRAPHICS/Statistical identification of human populations according to sex, age, race, income, etc.

DISCRETE DEMOGRAPHICS/Uncombined or non-overlapping sex/age groupings for listening estimates (e.g., men and/or women 18-24, 25-34, 35-44, etc.) as opposed to "target" group demographics (e.g., men and/or women 18+, 18-34, 18-49, 25-49).

DUAL CITY OF IDENTIFICATION/A multi-city identification, with the city of license required to be named first in all multi-city identification announcements, according to FCC guidelines (see section 73.1201 (B) (2), amended October 19, 1983). Also known as Home Market Guidelines, at Arbitron.

EFFECTIVE SAMPLE BASE (ESB)/An estimate of the size of simple random samples (in which all diaries have equal values) that would be required to provide the same degree of statistical reliability as the sample actually used to produce the estimates in a report.

ELECTRONIC MEDIA RATING COUNCIL (EMRC)/An organization that accredits broadcast ratings services; performs annual audits of the compliance of a service with certain minimum standards.

ETHNIC CONTROLS/Arbitron Ratings placement and weighting techniques used in certain sampling units to establish proper representation of the black and/or Hispanic populations in the Metros of qualifying Arbitron Radio Markets.

EXCLUSIVE CUME LISTENING/The estimated number of Cume Persons who listened to one and only one station within a given daypart.

EXPANDED SAMPLE FRAME (ESF)/A universe that consists of unlisted telephone households — households which do not appear in the current or available telephone directories because either (a) they have requested their telephone number not to be listed or (b) they are households where the assigned telephone number is not listed in the directory because of the date of installation and the date of telephone directory publication.

FACILITY FORM/(See Station Information Form.)

FLIP/A computerized edit procedure that assigns aberrated call letters to legal call letters, or the AM designation of a set of call letters may be changed to an FM designation, e.g., WODC-AM flips to WOBC-AM and WOBC-FM flips to WOBC-AM.

GROUP QUARTERS/Residences of all persons not living in nuclear households. The population in group quarters includes, for example, persons living in college dormitories, homes for the aged, military barracks, rooming houses, hospitals and institutions.

HOME MARKET GUIDELINES/The criteria by which a radio station with multi-city identification can be reported Home to an Arbitron Radio Metro Area. Also known as Dual City of Identification.

HOME NUMBER/A unique four-digit number assigned to each household within a county being sampled.

IN-TAB/The number of usable diaries actually tabulated in producing an Arbitron Ratings report.

LISTED SAMPLE/Names, addresses, and telephone numbers of selected potential diary-keepers derived from telephone directories.

LOCAL MARKET REPORT (LMR)/A syndicated report for a designated market; also known as SRMR (Standard Radio Market Report) and RMR (Radio Market Report).

MARKET TOTALS/The estimated number of persons in the market who listened to reported stations, as well as to stations that did not meet the Minimum Reporting Standards, and/or to unidentified stations.

MENTION/The number of different diaries in which a station is mentioned once with at least five minutes of listening, in a quarter-hour, (does not indicate all the entries to a station in one diary); appears in county and station TALO.

METROPOLITAN STATISTICAL AREA (MSA)/ As defined by the U.S. Government's Office of Management and Budget; a free-standing metropolitan area, surrounded by nonmetropolitan counties and not closely associated with other metropolitan areas.

METRO TOTALS AND ADI TOTALS (Total Listening in Metro Survey Area or Total Listening in the ADI)/The Metro and ADI total estimates include estimates of listening to reported stations as well as to stations that did not meet the Minimum Reporting Standards plus estimates of listening to unidentified stations.

NETWORK AFFILIATE/A broadcasting station, usually independently owned, in contractual agreement with a network in which the station grants the network an option on specific time periods for the broadcast of network-originated programs.

APPENDIX 6B: continued

NEW ENGLAND COUNTY METROPOLITAN AREA (NECMA)/As defined by the U.S. Government's Office of Management and Budget; New England MSA or PMSA definition adjusted to a whole county definition.

ONE PERCENT (1%) TALO RULE/An Arbitron radio procedure that establishes a cutoff point for resolving conflicts. The cutoff is one percent of the *previous year's* TALO by county, by station. All "potential" conflicting stations are analyzed to determine whether they qualify for conflict resolution. If only one of the two or more stations "potentially" in conflict receives one percent or more of the mentions in that county, then that station will receive credit for the contested entries in that county. However, if two or more of the stations "potentially" in conflict receive one percent or more of the total mentions in that county, each is considered in conflict. Ascription procedures are then instituted to determine proper listening credit.

PREMIUM/A token cash payment most often mailed with the diaries; serves as an inducement for a diarykeeper to participate in the survey and to return the diary to Arbitron. A premium is sent for each person twelve years of age and older in a household. The amount of the premium may vary.

PRIMARY METROPOLITAN STATISTICAL AREA (PMSA)/As defined by the U. S. Government's Office of Management and Budget; a metropolitan area that is closely related to another.

RATING/(See Average Quarter-Hour Rating and Cume Rating.)

REACH (Station)/Each county in which it has been determined by Arbitron Ratings that the signal for a specific radio station may be received.

R-SALE TAPE/An Arbitron Radio Special Services data tape of Metro and TSA estimates from one or more of the RMRs; available to clients who subscribe to the RMRs; also known as "RSL" or "Market Report data tape."

SHARE/The percentage of individuals listening to radio, who are listening to a specific station at a particular time.

SLOGAN/An on-air identifier used in place of or in conjunction with a station's call letters or exact frequency.

STATION INFORMATION FORM/A computer-generated form that lists essential station information including: power (day and night), frequency, sign-on/sign-off times, simulcasting (if any), slogan ID, network affiliates and national representative (if applicable). The Station Information Form is forwarded for verification to the applicable station prior to the survey period.

STATION INFORMATION PACKET/A set of forms mailed by Arbitron Ratings to a radio station approximately fifty days prior to each survey; allows station to change its slogan ID, sign-on/sign-off times, and make routine programming changes; included are forms for: Station Information, Programming Schedule Information, and Sports Programming.

STATION REACH FILE/A county-by-county file of stations that can be received in a county. This file is based on previous diary history and is updated with recent diary information as well as changes in power/antenna height; replaces the subjective review of all diary mentions within a county each survey to determine whether or not the mentions are "logical." Also, a station-by-station file of total counties reached.

TARGET DEMOGRAPHICS/Audience groupings containing multiple discrete demographics (e.g., men and/or women 18+, 18-34, 18-49, 25-49, etc.) as opposed to discrete demographics (e.g., men and/or women 18-24, 25-34, 35-44, etc.).

TECHNICAL DIFFICULTIES (TD)/Time period(s) of five or more consecutive minutes, in a quarter-hour, during the survey period in which a station listed in an Arbitron Ratings market report notified Arbitron Ratings in writing of technical difficulties including, but not limited to, times it was off the air or operating at reduced power.

TIME SPENT LISTENING (TSL)/An estimate of the amount of time the average person spends listening to a radio or to a station during a specified time period.

$$\frac{\text{Quarter-Hours in Time Period} \times \text{Average Quarter-Hours Persons Audience}}{\text{Cume Audience}} = \text{TSL}$$

TOTAL AUDIENCE LISTENING OUTPUT (TALO)/ The number of diaries in which a station is "mentioned" in (a) a market, (b) a county, or (c) another designated geographic area; a county-by-county printout showing the stations that are mentioned in the in-tab diaries and the number of mentions for each station; can be used to rank stations, to calculate weekly cumes and raw bases.

UNCOMBINED LISTENING ESTIMATES/ (See Discrete Demographics.)

UNIVERSE/The estimated total number of persons in the sex/age group and geographic area being reported.

UNLISTED SAMPLE/(See Expanded Sample Frame.)

APPENDIX 6B: continued

UNUSABLE DIARIES/Returned diaries determined to be unusable according to established Arbitron Ratings Radio Edit procedures.

UUUU/Unidentified; listening that could not be interpreted as belonging to a specific station.

FREQUENTLY USED ABBREVIATIONS

ADI	Area of Dominant Influence
AID	Arbitron Information on Demand
AQH	Average Quarter-Hour
CMSA	Consolidated Metropolitan Statistical Area
CRMR	Condensed Radio Market Report
CSB	Client Service Bulletin
DST	Differential Survey Treatment
EMRC	Electronic Media Rating Council
ESB	Effective Sample Base
ESF	Expanded Sample Frame
HDBA	High Density Black Area
HDHA	High Density Hispanic Area
LMR	Local Market Report
MMAC	MetroMail Advertising Company
MRS	Minimum Reporting Standards
MSA	Metro Survey Area Metropolitan Statistical Area
MSI	Market Statistics, Inc.
NECMA	New England County Metropolitan Area
PMSA	Primary Metropolitan Statistical Area
PPDV	Persons Per Diary Value
PPH	Persons Per Household
PSA	Primary Signal Area
PUR	Persons Using Radio
QM	Quarterly Measurement
RMR	Radio Market Report
SRDS	Standard Rate and Data Service, Inc.
SRMR	Standard Radio Market Report
SU	Sampling Unit
TALO	Total Audience Listening Output
TAR	Trading Area Report
TD	Technical Difficulty
TSA	Total Survey Area
TSL	Time Spent Listening

SUGGESTED FURTHER READING

Arbitron Company. *Research Guidelines for Programming Decision Makers*. Beltsville, Md.: Arbitron Company, 1977.

Aspen Handbook on the Media—1977–79 Edition: A Selective Guide to Research, Organizations, and Publications in Communications. New York: Praeger, 1977.

Bartos, Rena. *The Moving Target, What Every Marketer Should Know About Women*. New York: The Free Press, 1982.

Broadcast Advertising Reports. New York: Broadcast Advertising Research, periodically.

Broadcasting Yearbook. Washington, D.C.: Broadcasting Publishing, 1935 to date, annually.

Browne, Bortz, and Coddington (consultants). *Radio Today—and Tomorrow*. Washington, D.C.: NAB, 1982.

Chappell, Matthew N., and Hooper, C.E. *Radio Audience Measurement*. New York: Stephen Day Press, 1944.

Compaine, Benjamin et al. *Who Owns the Media? Confrontation of Ownership in the Mass Communication Industry*, 2nd ed. White Plains, N.Y.: Knowledge Industry Publications, 1982.

Duncan, James. *American Radio*. Kalamazoo, Mich.: the author, twice yearly.

———. *Radio in the United States: 1976–82. A Statistical History*. Kalamazoo, Mich.: the author, 1983.

Eastman, Susan Tyler et al., eds. *Broadcast Programming: Strategies for Winning Television and Radio Audiences*. Belmont, Calif.: Wadsworth Publishing, 1981.

Electronic Industries Association. *Electronic MarketData Book*. Washington, D.C.: EIA, annually.

Fletcher, James E., ed. *Handbook of Radio and Television Broadcasting: Research Procedures in Audience, Programming, and Revenues*. New York: Van Nostrand Reinhold, 1981.

Jamieson, Kathleen Hall, and Campbell, Karlyn Kohrs. *The Interplay of Influence: Mass Media and Their Publics in News, Advertising, and Politics*. Belmont, Calif.: Wadsworth Publishing, 1983.

Lazarsfeld, Paul F., and Kendall, Patrick. *Radio Listening in America*. Englewood Cliffs, N.J.: Prentice-Hall, 1948.

Lichty, Lawrence W., and Topping, Malachi C., eds. *American Broadcasting: A Sourcebook on the History of Radio and Television*. New York: Hastings House, 1975.

National Association of Broadcasters. *Radio Financial Report*. Washington, D.C.: NAB, 1955 to date, annually.

Radio Facts. New York: Radio Advertising Bureau, 1985.

Sterling, Christopher H. *Electronic Media: A Guide to Trends in Broadcasting and Newer Technologies, 1902–1983*. New York: Praeger, 1984.

7 Promotion

PAST AND PURPOSE

FIGURE 7.1

In 1959 author Norman Mailer published a book of essays entitled *Advertisements for Myself*. The title could be used to describe a text on radio promotion. Of course, Mailer's book is not about radio promotion, which is a form of self-advertisement. "WXXX—The station without equal," "For the best in music and news tune WXXX," or "You're tuned to the music giant—WXXX" certainly illustrate this point. Why must stations practice self-glorification? The answer is simple: to keep the listener interested and tuned. The highly fragmented radio marketplace has made promotion a basic component of station operations. Five times as many stations vie for the listening audience today as did when television arrived on the scene. It is competition that makes promotion necessary.

Small stations as well as large promote themselves. In the single station market, stations promote to counter the effects of other media, especially the local newspaper, which often is the archrival of small town outlets. Since there is only one station to tune to, promotion serves to maintain listener interest in the medium. In large markets, where three stations may be offering the same format, promotion helps a station differentiate itself from the rest. In this case, the station with the best promotion often wins the ratings war.

Radio recognized the value of promotion early. In the 1920s and 1930s, stations used newspapers and other print media to inform the public of their existence. Remote broadcasts from hotels, theaters, and stores also attracted attention for stations and were a popular form of promotion during the medium's golden age. Promotion-conscious broadcasters of the pre-television era were just as determined to get the audience to take notice as they are today. From the start, stations used whatever was at hand to capture the public's attention. Call letters were configured in such a way as to convey a particular sentiment or meaning: WEAF/New York = Water, Earth, Air, Fire; WOW/Omaha = Woodmen Of the World; WIOD/Miami = Wonderful Isle Of Dreams. Placards were affixed to vehicles, buildings, and even blimps as a means of heightening the public's awareness.

As ratings assumed greater prominence in the age of specialization, stations became even more cognizant of the need to promote. The relationship between good ratings and effective promotion became more apparent. In the 1950s and 1960s, programming innovators such as Todd Storz and Gordon McLendon used promotions and contests with daring and skill, if not a bit of lunacy, to win the attention of listeners.

The growth of radio promotion has paralleled the proliferation of frequencies. "The more stations you have out there, the greater the necessity to promote. Let's face it, a lot of stations are doing about the same thing. A good promotion sets you apart. It gives you greater identity, which means everything when a survey company asks a listener what station he or she tunes. Radio is an advertising medium in and of itself. Promotion makes a station salable. You sell yourself so that you have something to sell advertisers," observes Charlie Morriss, promotion director, KOMP, Las Vegas.

Stations that once confined the bulk of their promotional effort to spring and fall to coincide with rating periods now find it necessary to engage in promotional campaigns on an ongoing basis throughout the year, notes Morriss. "More competition and monthly audience surveys mean that stations have to keep the promotion fire burning continually, the analogy being that if the flame goes out you're likely to go cool in the ratings. So you really have to hype your outlet every opportunity you get. The significance of the role of promotion in contemporary radio cannot be overstated."

The vital role that promotion plays in radio is not likely to diminish as hundreds of more stations enter the airways by the turn of the century. "Promotion has become as much a part of radio as the records and the deejays who spin them, and that's not about to change," contends John Grube, promotion director, WGNG-AM, Providence, Rhode Island.

PROMOTIONS PRACTICAL AND BIZARRE

The idea behind any promotion is to win listeners. Over the years, stations have used a variety of methods, ranging from the conventional to the outlandish, to accomplish this goal. "If a promotion achieves top-of-the-mind awareness in the listener, it's a winner. Granted, some strange things have been done to accomplish this," admits Bob Lima, WVMI/WQLD, Biloxi, Mississippi.

Promotions designed to captivate the interest of the radio audience have inspired some pretty bizarre schemes. In the 1950s Dallas station KLIF placed overturned cars on freeways with a sign on their undersides announcing the arrival of a new deejay, Johnny Rabbitt. It would be hard to calculate the number of deejays who have lived atop flagpoles or in elevators for the sake of a rating point.

In the 1980s the shenanigans continue. To gain the listening public's attention, a California deejay set a world record by sitting in every seat of a major league ballpark that held sixty-five thousand spectators. In the process of the stunt, the publicity-hungry deejay injured his leg. However, he went on to accomplish his goal by garnering national attention for himself and his station. Another station offered to give away a mobile home to contestants who camped out the longest on a platform at the base of a billboard. The challenge turned into a battle of wills as three contestants spent months trying to outlast each other. In the end, one of the three was disqualified and the station, in an effort to cease what had become more of an embarrassment than anything else, awarded the two holdouts recreational vehicles.

One of the most infamous examples of a promotion gone bad occurred when a station decided to air-drop dozens of turkeys to a waiting crowd of listeners in a neighborhood shopping center parking lot. Unfortunately, the station discovered too late that turkeys are not adept at flying at heights above thirty feet. Consequently, several cars were damaged and witnesses traumatized as turkeys plunged to the ground. This promotion-turned-nightmare was depicted in an episode of the television sitcom "WKRP in Cincinnati."

The list of glitches is seemingly endless. In the late 1960s a station in central Massachusetts asked listeners to predict how long its air personality could ride a carousel at a local fair. The hardy airman's effort was cut short on day three

As advertised in
Advertising Age/Adweek/Radio & Records/Radio Only
WHTZ...#1 IN NEW YORK...IN 74 DAYS!
In Birch Radio's latest published ratings, WHTZ-FM...The New Z 100.. placed first with a Total Persons 12+ share of 8.9.
WHTZ signed on the air August 2, 1983. Z 100's personality-oriented contemporary hit format came from nowhere to #1 in only 74 days.
Represented Nationally by
EASTMAN RADIO
Z-100 FM
WHTZ - FM 100.3
TOP TEN
Birch Radio Monthly Trend Report*
New York Metropolitan Area
1-15 Sept/1-14 Oct 1983
Mon-Sun 6am-12mid
Total Persons 12+ AQH Shares
#1 WHTZ 8.9
#2 WOR 6.2
#3 WKTU 4.8
#4 WRKS 4.8
#5 WAPP 4.6
#6 WINS 4.6
#7 WBLS 4.4
#8 WPLJ 3.5
#9 WYNY 3.4
#10 WABC 3.3
A MALRITE COMMUNICATIONS GROUP STATION

FIGURE 7.2 WHTZ promotional piece.

FIGURE 7.3
Station promotion in the early 1920s with an airborne antenna. Courtesy Westinghouse Electric.

when motion sickness got the best of him and he vomited on a crowd of spectators and newspaper photographers. A station in California came close to disaster when a promotion that challenged listeners to find a buried treasure resulted in half the community being dug up by overzealous contestants.

These promotions did indeed capture the attention of the public, but in each case the station's image was somewhat tarnished. The axiom that any publicity, good or bad, is better than none at all can get a station into hot water, contends Chuck Davis, promotion director, WSUB/WQGN, New London, Connecticut. "It's great to get lots of exposure for the station, but if it makes the station look foolish, it can work against you."

The vast majority of radio contests and promotions are of a more practical nature and run without too many complications. Big prizes, rather than stunts, tend to draw the most interest and thus are offered by stations able to afford them. In the mid-1980s WASH-AM, Washington, D.C., and KSSK-AM, Honolulu, both gave a lucky listener a million dollars. Cash prizes always have attracted tremendous response. Valuable prizes other than cash also can boost ratings. For example, Los Angeles station KHTZ-AM experienced a sizable jump in its ratings when it offered listeners a chance to win a one hundred and twenty-two thousand dollar house. Increased ratings also resulted when KHJ-AM, Los Angeles, gave away a car every day during the month of May.

Promotions that involve prizes, both large and small, spark audience interest, says Rick Peters, vice president of programming, TK Communications. "People love to win something or, at least, feel that they have a shot at winning a prize. That's basic to human nature, I believe. You really don't have to give away two city blocks, either. A listener usually is thrilled and delighted to win a pair of concert tickets."

Although there are numerous examples to support the view that big prizes get big audiences, there is an ample amount to support the contention that low-budget giveaways, involving T-shirts, albums, tickets, posters, dinners, and so forth are very useful in building and maintaining audience interest. In fact, some surveys have revealed that smaller, more personalized prizes may work better for a station than the high-priced items. Record albums and dinners-for-two rank among the most popular contest prizes according to surveys. Cheaper items usually also mean more numerous or frequent giveaways.

THE PROMOTION DIRECTOR'S JOB

Not all stations employ a full-time promotion director. But most stations designate someone to handle promotional responsibilities. At small outlets, promotional chores are assumed by the program director or even the general manager. Larger stations with bigger operating budgets typically hire an individual or individuals to work exclusively in the area of promotion. "At major market stations, you'll find a promotion department that includes a director and possibly assistants. In middle-sized markets, such as ours, the promotion responsibility is often designated to someone already involved in programming," says WVMI's Bob Lima.

The promotion director's responsibilities are manifold. Essential to the position are a knowledge and understanding of the station's audience. A background in research is important, contends WGNG's Grube. "Before you can initiate any kind of promotion you must know something about who you're trying to reach. This requires an ability to interpret various research data that you gather through in-house survey efforts or from outside audience research companies, like Birch and Arbitron. You don't give away beach balls to fifty-year-old men. Ideas must be confined to the cell group you're trying to attract."

Writing and conceptual skills are vital to the job of promotion director, says KOMP's Morriss. "You prepare an awful lot of copy of all types. One moment you're composing press releases about programming changes, and the next you're writing a thirty-second promo about the station's expanded news coverage or upcoming remote broadcast from a local mall. A knowledge of English grammar is a must. Bad writing reflects negatively on the station. The job also demands imagination and creativity. You have to be able to come up with an idea and bring it to fruition."

WSUB's Chuck Davis agrees with Morriss and adds that while the promotion person should be able to originate concepts, a certain number of ideas come from the trades and other stations. "When this is the case, and it often is, you have to know how to adapt an idea to suit your own station. Of course, the promotion must reflect your location. Lifestyles vary almost by region. A promotion that's successful at a station with a similar format in Louisiana may bear no relevance to a station in Michigan. On the other hand, with some adjustments, it may work as effectively there. The creativity in this example exists in the adaptation."

Promotion directors must be versatile. A familiarity with graphic art generally is necessary, since the promotion director will be involved in developing station logos and image IDs for advertising in the print media and billboards. The promotion department also participates in the design and preparation of visuals for the sales area.

The acquisition of prize materials through direct purchase and trades is another duty of the promotion person, who also may be called upon to help coordinate sales co-op arrangements. "You work closely with the sales manager to arrange tie-ins with sponsors and station promotions," contends KOMP's Morriss.

Like other radio station department heads, it is the promotion director's responsibility to ensure that the rules and regulations established by the FCC, relevant to the promotions area, are observed. This will be discussed further in a subsequent section of this chapter.

FIGURE 7.4
Promoting success in the ratings strengthens a station's image as a winner. Courtesy WRCH-FM.

WHOM PROMOTION DIRECTORS HIRE

In each section devoted to hiring in preceding and subsequent chapters, college training is listed as a desirable, if not necessary, attribute. This is no less true in the area of radio station promotion. "My advice to an individual interested in becoming a promotion person would be to get as much formal training as possible in

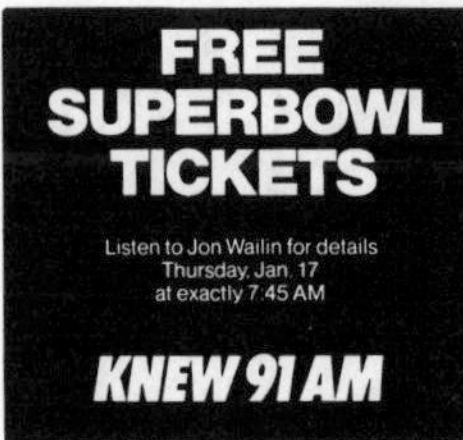

FIGURE 7.5
Giving away "hot" items generates audience interest. Courtesy KNEW-AM.

marketing, research, graphics, writing, public relations and, of course, broadcasting. The duties of the promotion director, especially at a large station, are diverse," notes TK Communications' Rick Peters.

KOMP's Morriss agrees with Peters and adds, "A manager reviewing the credentials of candidates for a promotion position will expect to find a statement about formal training, that is, college. Of course, nothing is a substitute for a solid track record. Experience is golden. This is a very hands-on field. My advice today is to get a good education and along the way pick up a little experience, too."

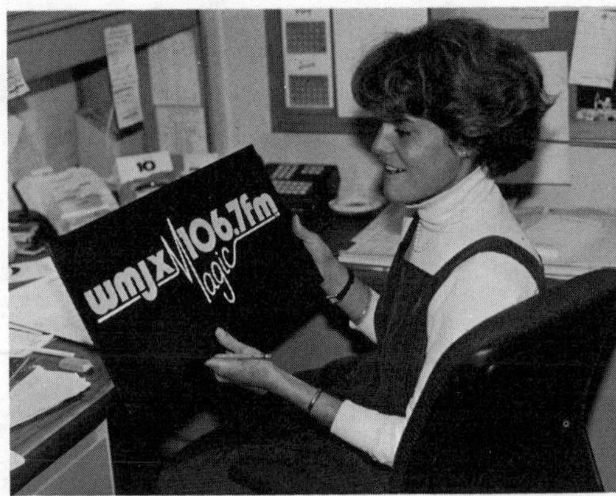

FIGURE 7.6
WMJX Promotion Director Shelby Rogerson with station logo graphic. Courtesy WMJX-FM.

FIGURE 7.7
Station sales packet folders are designed to correspond with the station's on-air image. Courtesy WMJX.

Familiarity with programming is important, contends WVMI's Lima, who suggests that prospective promotion people spend some time on the air. "Part-timing it on mike at a station, be it a small commercial outlet or a college facility, gives a person special insight into the nature of the medium that he or she is promoting. Working in sales also is valuable. In the specific skills department, I'd say the promotion job candidate should have an eye for detail, be well organized, and possess exemplary writing skills. It goes without saying that a positive attitude and genuine appreciation of radio are important as well."

Both John Grube and Chuck Davis cite wit and imagination as criteria for the job of promotion. "It helps to be a little wacky and crazy. By that I mean able to conceive of entertaining, fun concepts," says Grube. Davis concurs, "This is a convivial medium. The idea behind any promotion or contest is to attract and amuse the listener. A zany, off-the-wall idea is good, as long as it is based in sound reasoning. Calculated craziness requires common sense and creativity, and both are qualities you need in order to succeed in promotion."

The increasing competition in the radio marketplace has bolstered job opportunities in promotion. Thus the future appears bright for individuals planning careers in this facet of the medium.

TYPES OF PROMOTION

There are two primary categories of station promotion—"on-air" and "off-air." The former will be examined first since it is the most prevalent form of radio promotion. Broadcasters already possess the best possible vehicle to reach listeners, so it should come as no surprise that on-air promotion is the most common means of getting word out on a station. The challenge confronting the promotion director is how to most effectively market the station so as to expand and retain listenership. To this end, a number of promotional devices are employed, beginning with the most obvious—station call letters. "The value of a good set of call letters is inestimable," says WRCH's Richard Bremkamp, whose own call letters have long been associated with the term "rich" and all that it implies: "Hartford's Rich Music Station—WRCH."

Call letters convey the personality of a station. For instance, try connecting these call letters with a format: WHOG, WNWS, WEZI,, WOLF, WJZZ, WIND, WHTT. If you guessed Country, News, Easy Listening, Oldies, Jazz, Talk, and Hits you were correct. The preceding call letters not only identify their radio stations, but they literally convey the nature or content of the programming offered.

When stations do not possess call letters that create instant recognition, they often couple their frequency with a call letter or two, such as JB-105 (WPJB-FM 105) or KISS-108 (WXKS-FM 108). This also improves the retention factor. Slogans frequently are a part of the on-air ID. "Music Country—WSOC-FM, Charlotte"; "A Touch of Class—WTEB-FM, New Bern"; "Texas Best Rock—KTXQ-FM, Fort Worth" are some examples. Slogans exemplify a station's image. When effective, they capture the mood and flavor of the station and leave a strong impression in the listener's mind. It is standard programming policy at many stations to announce the station's call letters and even slogan each time a deejay opens the microphone. This is especially true during ratings sweeps when listeners are asked by survey companies to identify the stations they tune in to. "If your calls stick in the mind of your audience, you've hit a home run. If they don't, you'll go scoreless in the book. You've got to carve them into the listener's gray matter and you start by making IDs and signatures that are as memorable as possible," observes TK Communications' Rick Peters.

It is a common practice for stations to "bookend"—place call letters before and after all breaks between music. For example, "WHJJ. Stay tuned for a complete look at local and national news at the top of the hour on HJJ." Deejays also are told to graft the station call letters onto all bits of information: "WHJJ Time," "WHJJ Temperature," "WHJJ Weather," and so on. There is a rule in radio that call letters can never be overannounced. The logic behind this is clear. The more a station tells its audience whom it is tuned to, the more apt it is to remember, especially during rating periods.

On-air contests are another way to capture and hold the listener's attention. Contests must be easy to understand (are the rules and requirements of the contest easily understood by the listener?) and possess entertainment value (will nonparticipants be amused even though they are not actually involved?). A contest should engage the interest of all listeners, players and nonplayers alike.

FIGURE 7.8
Call letters tell the story. Courtesy WHTT-FM.

A contest must be designed to enhance a station's overall sound or format. It must fit in, be compatible. Obviously, a mystery sound contest requiring the broadcast of loud or shrill noises would disrupt the tranquility and continuity of an Easy Listening station and result in tune-out.

Successful contests are timely and relevant to the lifestyle of the station's target audience, says WVMI's Bob Lima. "A contest should offer prizes that truly connect with the listener. An awareness of the needs, desires, and fantasies of the listener will help guide a station. For example, giving away a refrigerator on a hot hit station would not really captivate the sixteen-year-old tuned. This is obvious, of course. But the point I'm making is that the prizes that are up for grabs should be something the listener really wants to win, or you have apathy."

The importance of creativity already has been stated. Contests that attract the most attention often are the ones that challenge the listener's imagination, contends KOMP's Morriss. "A contest should have style, should attempt to be different. You can give away what is perfectly suitable for your audience, but you can do it in a way that creates excitement and adds zest to the programming. The goal of any promotion is to set you apart from the other guy. Be daring within reason, but be daring."

On-air promotion is used to inform the audience of what a station has to offer: station personalities, programs, special features and events. Rarely does a quarter hour pass on any station that does not include a promo that highlights some aspect of programming:

"Tune-in WXXX's News at Noon each weekday for a full hour of . . ."

FIGURE 7.9 Newspapers are popular radio promotion vehicles. Courtesy WAAF, WBCN, WHK, and WFNX.

"Irv McKenna keeps 'Nightalk' in the air midnight to six on the voice of the valley—WXXX. Yes, there's never a dull moment . . ."

"Every Saturday night WXXX turns the clock back to the fifties and sixties to bring you the best of the golden oldies . . ."

"Hear the complete weather forecast on the hour and half hour throughout the day and night on your total service station—WXXX . . ."

On-air promotion is a cost-efficient and effective means of building an audience when done correctly, says WGNG's John Grube. "There are good on-air promotions and weak or ineffective on-air promotions. The latter can inflict a deep wound, but the former can put a station on the map. As broadcasters, the airtime is there at our disposal, but we sometimes forget just how potent an advertising tool we have."

Radio stations employ "off-air" promotional

techniques to reach people not tuned. Billboards are a popular form of outside promotion. To be effective they must be both eye-catching and simple. Only so much can be stated on a billboard, since people generally are in a moving vehicle and only have a limited amount of time to absorb a message. Placement of the billboard also is a key factor. In order to be effective, billboards must be located where they will reach a station's intended audience. Whereas an All-News station would avoid the use of a billboard facing a high school, a rock music outlet may prefer the location.

Bus cards are a good way to reach the public. Cities often have hundreds of buses on the streets each day. Benches and transit shelters also are used by billboard companies to get their client's message across to the population. Outside advertising is an effective and fairly cost-efficient way to promote a radio station, although certain billboards at heavy traffic locations can be extremely expensive to lease.

Newspapers are the most frequent means of off-air promotion. Stations like the reach and targeting that newspapers can provide. In large metro areas, alternative newspapers, such as the *Boston Phoenix,* are very effective in delivering certain listening cells. The *Phoenix* enjoys one of the largest readerships of any independent press in the country. Its huge college-age and young professional audience makes it an ideal promotional medium for stations after those particular demographics. While the readership of the more conventional newspapers traditionally is low among young people, it is high in older adults, making the mainstream publications useful to stations targeting the over-forty crowd.

While newspapers with large circulations provide a great way to reach the population at large, they also can be very costly, although some stations are able to trade airtime for print space. Newspaper ads must be large enough to stand out and overcome the sea of advertisements that often share the same page. Despite some drawbacks, newspapers usually are the first place radio broadcasters consider when planning an off-air promotion.

Television is a costly but effective promotional tool for radio. A primary advantage that television offers is the chance to target the audience that the station is after. An enormous amount of information is available pertaining to television viewership. Thus a station that wants to reach the eighteen- to twenty-four-year-olds is able to ascertain the programs and features that best draw that particular cell.

The costs of producing or acquiring ready-made promos for television can run high, but most radio broadcasters value the opportunity to actually show the public what they can hear when they tune to their station. WBZ-AM in Boston has used local television extensively to promote its morning personality, Dave Maynard. Ratings for the Westinghouse-owned station have been consistently high, and management points to their television promotion as a contributing factor.

Bumper stickers are manufactured by the millions for distribution by practically every commercial radio station in the country. The primary purpose of stickers is to increase call letter awareness. Over the years, bumper stickers have developed into a unique pop-art form, and hundreds of people actually collect station decals as a hobby. Some station bumper stickers are particularly prized for the lifestyle or image they portray. Youths, in particular, are fond of displaying their favorite station's call letters. Stations appealing to older demographics find that their audiences are somewhat less enthusiastic about bumper stickers.

Stations motivate listeners to display bumper stickers by tying them in with on-air promotions:

"WXXX WANTS TO GIVE YOU A THOUSAND DOLLARS. ALL YOU HAVE TO DO IS PUT AN X-100 BUMPER STICKER ON YOUR CAR TO BE ELIGIBLE. IT'S THAT SIMPLE. WHEN YOUR CAR IS SPOTTED BY THE X-100 ROVING EYE, YOUR LICENSE NUMBER WILL BE ANNOUNCED OVER THE AIR. YOU WILL THEN HAVE THIRTY MINUTES TO CALL THE STATION TO CLAIM YOUR ONE THOUSAND DOLLARS. . . ."

Hundreds of ways have been invented to entice people to display station call letters. The idea is to get the station's name out to the public, and ten thousand cars exhibiting a station's bumper sticker is an effective way to do that.

Thousands of items displaying station call letters and logos are given away annually by stations. Among the most common promotional items handed out by stations are posters, T-shirts, calendars, key-chains, coffee mugs, music hit-lists, book covers, pens, and car litter bags. The list is vast.

Plastic card promotions have done well for many stations. Holders are entitled to a variety of benefits, including discounts at various stores and valuable prizes. The bearer is told to listen to the station for information as to where to use the card. In addition, holders are eligible for special on-air drawings.

Another particularly effective way to increase a station's visibility is to sponsor special activities, such as fairs, sporting events, theme dances, and concerts. Hartford's Big Band station, WRCQ-AM, has received significant at-

FIGURE 7.10
Bumper stickers visually convey a station's sound.

tention by presenting an annual music festival that has attracted over twenty-five thousand spectators each year, plus the notice of other media, including television and newspapers.

Personal appearances by station personalities are one of the oldest forms of off-air promotion but are still very effective. Remote broadcasts from malls, beaches, and the like also aid in getting the word of the station out to the public.

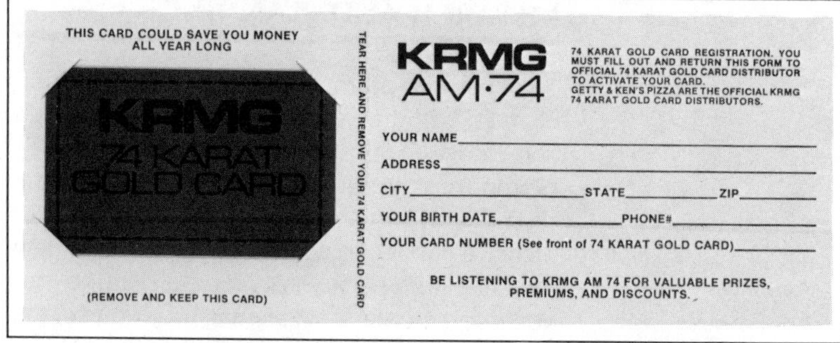

THIS CARD COULD SAVE YOU MONEY ALL YEAR LONG

KRMG 74 KARAT GOLD CARD

(REMOVE AND KEEP THIS CARD)

TEAR HERE AND REMOVE YOUR 74 KARAT GOLD CARD

KRMG AM·74

74 KARAT GOLD CARD REGISTRATION. YOU MUST FILL OUT AND RETURN THIS FORM TO OFFICIAL 74 KARAT GOLD CARD DISTRIBUTOR TO ACTIVATE YOUR CARD.
GETTY & KEN'S PIZZA ARE THE OFFICIAL KRMG 74 KARAT GOLD CARD DISTRIBUTORS.

YOUR NAME__________

ADDRESS__________

CITY__________ STATE__________ ZIP__________

YOUR BIRTH DATE__________ PHONE#__________

YOUR CARD NUMBER (See front of 74 KARAT GOLD CARD)__________

BE LISTENING TO KRMG AM 74 FOR VALUABLE PRIZES, PREMIUMS, AND DISCOUNTS.

FIGURE 7.11
Plastic cards draw listeners and generate business for participating sponsors. Courtesy KRMG-AM.

SALES PROMOTION

Promoting a station can be very costly, as much as half a million dollars annually in some metro markets. To help defray the cost of station promotion, advertisers are often recruited. This way both the station and the sponsor stand to benefit. The station gains the financial wherewithal to execute certain promotions that it could not do on its own, and the participating advertiser gains valuable exposure by tying in with special station events. Stations actually can make money and promote themselves simultaneously if a substantial spot schedule is purchased by a client as part of a promotional package.

There are abundant ways to involve advertisers in station promotion efforts. They run the gamut from placing advertisements and coupons on the back of bumper stickers to joining the circus for the day: for example, "WXXX brings the 'Greatest Show on Earth' to town this Friday night, and you go for half price just by mentioning the name of your favorite radio station—WXXX." The ultimate objective of a station/sponsor collaborative is to generate attention in a cost-efficient manner. If a few dollars are made for the station along the way, all the better.

As stated previously, the promotion director also works closely with the station's sales department in the preparation and design of sales promotion materials, which include items such as posters, coverage maps, ratings breakouts, flyers, station profiles, rate cards, and much more.

FIGURE 7.12 (left)
Oldies station WHK maintains a high profile by promoting events consistent with its image.

FIGURE 7.13 (below)
WLS brings rock group the "Go-Go's" to Chicago. Courtesy WLS.

RESEARCH AND PLANNING

To effectively promote a station, the individual charged with the task must possess a thorough knowledge of the station and its audience. This person must then ascertain the objective of the promotion. Is it the aim of the promotion to increase call letter awareness, introduce a new format/feature/personality, or bolster the station's community service image? Of course, the ultimate goal of any promotion is to enhance listenership.

Effective promotions take into account both internal and external factors. An understanding of the product, consumer, and competition is essential to any marketing effort, including radio. Each of these three areas presents the promotion director with questions that must be addressed before launching a campaign. As stated earlier in this chapter, it is imperative that the promotion or contest fit the station's sound. In other words, be compatible with the format. This accomplished, the next consideration is the relevancy of the promotion to the station's audience. For example, does it fit the listener's lifestyle? Thirdly, is the idea fresh enough in the market to attract and sustain interest?

Concerning the basic mechanics of the contest, the general rule is that if it takes too long to explain, it is not appropriate for radio. "Contests that require too much explanation don't work well in our medium. That is not to say that they have to be thin and one-dimensional. On the contrary, radio contests can be imaginative and captivating without being complicated or complex," notes WGNG's John Grube.

The planning and implementation of certain promotions may require the involvement of consultants who possess the expertise to ensure smooth sailing. Contests can turn into bad dreams if potential problems are not anticipated. Rick Sklar, who served as program director for WABC in New York for nearly twenty years, was responsible for some of the most successful radio promotions ever devised, but not all went without a hitch. In his autobiographical book, *Rocking America* (St. Martin's Press, 1984), Sklar tells of the time that he was forced to hire, at great expense, sixty office temporaries for a period of one month to count the more than 170 million ballots received in response to the station's "Principal of the Year" contest. The previous year the station had received a paltry six million ballots.

On another occasion, Sklar had over four million WABC buttons manufactured as part of a promotion that awarded up to twenty-five thousand dollars in cash to listeners spotted wearing one. What Sklar did not anticipate was the huge cost involved in shipping several million metal buttons from various points around the country. The station had to come up with thousands of unbudgeted dollars to cover air freight. Of course, both miscalculations were mitigated by the tremendous success of the promotions, which significantly boosted WABC's ratings.

Careful planning during the developmental phase of a promotion generally will prevent any unpleasant surprises, says WVMI's Bob Lima. "Practical and hypothetical projections should be made. Radio can fool you by its pulling power. If a promotion catches on, it can exceed all expectations. You've got to be prepared for all contingencies. These are nice problems to have, but you can get egg on your face. Take a good look at the long and short of things before you bolt from the starting gate. Don't be too hasty or quick to execute. Consider all the variables, then proceed with care."

Lima tells of a successful promotion at WVMI that required considerable organization and planning. "We called it 'The Great Easter Egg Hunt.' What was basically an Easter egg hunt at a local park here in Biloxi attracted over eleven thousand people. A local bottler co-sponsored the event and provided many of the thousands of dollars in prizes. The station hyped the event for several weeks over the air, and a little off-air promotion was done. The reason the promotion worked so well is that there actually was a need for a large, well-organized Easter event. We did our homework in selecting and executing this promotion, which turned out to be a big winner."

KOMP's Charlie Morriss shares an account of a successful promotion at his radio station. "The KOMP jocks recently flew around the Las Vegas skies in several World War II fighter planes owned by Miller Beer while our call letters were written by a skywriter. The effect was stunning. This promotion worked because skywriting is so rare these days and not many people have seen a squadron of vintage warplanes. It also worked because it didn't cost us a penny. It was a trade agreement with Miller. We gave them the exposure, and they gave us the airshow. Of course, a lot of details had to be worked out in advance."

The most effective promotion in recent years at New London's WSUB involves awarding contestants an elaborate night out on the town. Chuck Davis relates: "Our 'Night Out' promotion has been popular for some time. The station gets premium concert tickets through a close alliance with a New York concert promoter. It then finds a sponsor to participate in the giveaway and provides him with counter signs and an entry box so that people may register in his store. The sponsor then becomes part of the promotion and in return purchases an air schedule. Contestants are told to go to the store to register for the 'Night Out,' thus increasing store traffic even more. In addition, the sponsor agrees to absorb the expense of a limousine to transport the winners, who also are treated to a preconcert dinner at a local restaurant that provides the meals in exchange for promotional consideration—a mention on the air in our 'Night Out' promos. In the end, the station's cost amounts to a couple of phone calls, a cardboard sign, and a box. Reaction has always been great from all parties. The sponsor likes the tie-in with the promotion. The restaurant is very satisfied with the attention it receives for providing a few dinners, and the concert promoter gets a lot of exposure for the acts that he books simply by giving the station some tickets. It works like a charm. We please our audience and also put a few greenbacks in the till."

The point already has been made that a station can ill-afford not to promote itself in today's highly competitive marketplace. Promotions are an integral part of contemporary station operations, and research and planning are what make a promotion a winner.

BUDGETING PROMOTIONS

Included in the planning of a promotion are cost projections. The promotion director's budget may be substantial or all but nonexistent. Stations in small markets often have minuscule budgets compared to their giant metro market counterparts. But then again, the need to promote in a one- or two-station market generally is not as great as it is in multistation markets. To a degree, the promotion a station does is commensurate with the level of competition.

A typical promotion at an average-size station may involve the use of newspapers, plus additional hand-out materials, such as stickers, posters, buttons, and an assortment of other items depending on the nature of the promotion. Television and billboards may also be utilized. Each of these items will require an expenditure unless some other provision has been made, such as a trade agreement in which airtime is swapped for goods or ad space.

The cost involved in promoting a contest often constitutes the primary expense. When WASH-FM in Washington, D.C., gave away a million dollars, it spent two hundred thousand dollars to purchase an annuity designed to pay the prize recipient twenty thousand dollars a year for fifty years. The station spent nearly an equal amount to promote the big giveaway. Most of the promotional cost resulted from a heavy use of local television.

KHTZ-FM in Los Angeles spent over three hundred thousand dollars on billboards and television to advertise its dream-house giveaway. The total cost of the promotion approached a half million dollars. The price tag of the house was $122,000. Both of these high-priced contests accomplished their goals—increased ratings. In a metro market, one rating point can mean a million dollars in ad revenue. "A promotion that contributes to a two- or three-point jump in the ratings is well worth the money spent on it," observes TK Communications' Rick Peters.

The promotion director works with the station manager in establishing the promotion budget. From there, it is the promotion director's job to allocate funds for the various contests and promotions run throughout the station's fiscal period. Just as in every other area of a station, computers are becoming a prominent fixture in the promotion department. "If you have a large budget, a microcomputer can make life a lot easier. The idea is to control the budget and not let it control you. Computers can help in that effort," states Bonneville Broadcasting's Marlin Taylor, who also contends that large sums of money need not be poured into promotions if a station is on target with its programming. "In 1983, the Malrite organization came to New York and launched Z-100, a contemporary hit-formated outlet, moving it from 'worst to first' in a matter of months. They did a little advertising and gave away some money. I estimate that their giveaways totaled less dollars than some of their competitors spent on straight advertising. But the station's success was built on three key factors: product, service, and employee incentives. Indeed, they do have a quality product. Secondly, they are providing a service to their customers or listeners, and, thirdly, the care and feeding of the air staff and support team are obvious at all times. You don't necessarily have to spend a fortune on promotion."

Since promotion directors frequently are expected to arrange trade agreements with merchants as a way to defray costs, a familiarity with and understanding of the station's rate structure is necessary. Trading airtime for use in promotions is less popular at highly rated stations that can demand top dollars for spots. Most stations, however, prefer to exchange available airtime for goods and services needed in a promotion, rather than pay cash.

PROMOTIONS AND THE FCC

Promotions and contests must be conducted with propriety. The FCC has dealt some stiff penalties to stations that have misled or duped the public. Section 73.1216 of the FCC's rules and regulations (as printed in the *Code of Federal Regulations*) outlines the do's and don't's of contest presentations.

Stations are prohibited from running a contest in which contestants are required to pay in order to play. The FCC regards as lottery any contest in which the elements of prize, chance, and consideration exist. In other words, contestants must not have to risk something in order to win.

Contests must not place participants in any danger or jeopardize property. Awarding prizes to the first five people who successfully scale a treacherous mountain or swim a channel filled with alligators certainly would be construed by the FCC as endangering the lives of those involved. Contestants have been injured and stations held liable more than once. In the case of the station in California that ran a treasure hunt resulting in considerable property damage, it incurred the wrath of the public, town officials, and the FCC.

Stations are required by the FCC to disclose the material terms of all contests and promotions conducted. These include:

- Entering procedures
- Eligibility requirements
- Deadlines
- When or if prizes can be won
- Value of prizes
- Procedure for awarding prizes
- Tie-breaking procedures

The public must not be misled concerning the nature of prizes. Specifics must be stated. Implying that a large boat is to be awarded when, in fact, a canoe is the actual prize would constitute misrepresentation, as would suggesting that an evening in the Kontiki Room of the local Holiday Inn is a great escape weekend to the exotic South Seas.

The FCC also stipulates that any changes in contest rules must be promptly conveyed to the public. It makes clear, too, that any rigging of contests, such as determining winners in ad-

FIGURE 7.14 Stations are obliged to make contest rules clear to the public. Courtesy WHTT-FM.

CONTEST RULES and REGULATIONS

1) YOU HAVE TWO WAYS TO ENTER TO WIN CASH AND/OR PRIZES FROM WHTT. DISPLAY THE LARGE PART OF THIS BUMPER STICKER ON YOUR VEHICLE *AND* FILL OUT THE OFFICIAL WHTT ENTRY FORM AND MAIL IT TO: WHTT-HITRADIO; BOSTON, MA 02199. ENTRIES MUST BE MAILED SEPARATELY AND RECEIVED BY DECEMBER 24, 1984.
2) IF YOUR VEHICLE IS SEEN DISPLAYING THE WHTT BUMPER STICKER IN THE WHTT GREATER BOSTON LISTENING AREA BY OUR SPOTTERS, THEY WILL WRITE DOWN YOUR LICENSE NUMBER AND THE MAKE AND MODEL OF YOUR VEHICLE AND ENTER THAT INFORMATION INTO THE PRIZE DRAWING.
3) SELECTED DAYS NOW THROUGH DECEMBER 31, 1984, WHTT WILL CHOOSE LICENSE NUMBERS BY RANDOM DRAWING FROM ALL THE OFFICIAL ENTRIES RECEIVED, AND, FROM THE SPOTTER'S INFORMATION COLLECTED, AND ANNOUNCE THE SELECTED LICENSE NUMBERS ON THE AIR. WHEN YOU HEAR YOUR LICENSE NUMBER, YOU HAVE 103 MINUTES TO CALL US AND IDENTIFY YOURSELF AND YOUR VEHICLE. YOU THEN MUST BRING YOUR VEHICLE REGISTRATION TO THE WHTT OFFICE FOR VERIFICATION TO CLAIM YOUR PRIZE. PRIZES NOT CLAIMED WITHIN 30 DAYS WILL NOT BE AWARDED.
4) ONLY ONE PRIZE PER LICENSE NUMBER, PER MONTH WILL BE AWARDED. CBS EMPLOYEES AND THEIR FAMILIES ARE NOT ELIGIBLE TO ENTER. ONLY REGISTERED OWNERS OF PRIVATE, NON-COMMERCIAL VEHICLES ARE ELIGIBLE. THE DECISION OF THE JUDGES IS FINAL. VOID WHERE PROHIBITED.
5) THIS IS A FREE BUMPER STICKER. NO PURCHASE REQUIRED TO ENTER. FREE BUMPER STICKER AND ENTRY FORM AVAILABLE WHILE SUPPLIES LAST AT THE WHTT OFFICES, OR BY SENDING A SELF-ADDRESSED, STAMPED, BUSINESS SIZED ENVELOPE TO: WHTT BUMPER STICKER; 4418 PRUDENTIAL TOWER; BOSTON, MA 02199. WRITTEN REQUESTS MUST BE RECEIVED BY DECEMBER 1, 1984.

OFFICIAL W-H-T-T ENTRY FORM

(Please print)

NAME ________ AGE ____

ADDRESS ________

CITY ________ STATE ____ ZIP ____

HOME PHONE ________ WORK PHONE ________

CAR MAKE & MODEL ________

LICENSE NUMBER ________ STATE ____

HOURS OF DAY YOU LISTEN TO WHTT ________

mail this entry to:

103 FM WHTT HITRADIO

BOSTON, MA 02199

vance, is a direct violation of the law and can result in a substantial penalty, or even license revocation.

Although the FCC does not require that a station keep a contest file, most do. Maintaining all pertinent contest information, including signed prize receipts and releases by winners, can prevent problems should questions or a conflict arise later.

Stations that award prizes valued at six hundred dollars or more are expected by law to file a 1099 MISC form with the IRS. This is done strictly for reporting purposes and stations incur no tax liability. However, failure to do so puts a station in conflict with the law.

BROADCASTERS PROMOTION ASSOCIATION

The Broadcasters Promotion Association (BPA) was founded in 1956 as a nonprofit organization expressly designed to provide information and services to station promotion directors around the world. The objectives of the BPA are as follows:

- Increase the effectiveness of broadcast promotion personnel.
- Improve broadcast promotion methods, and research principles and techniques.
- Enhance the image and professional status of its members, and members of the broadcast promotion profession.
- Facilitate liaison with allied organizations in broadcasting, promotion, and government.
- Increase awareness and understanding of broadcast promotion at stations, in the community, and at colleges and universities.

The BPA conducts national seminars and workshops on promotion-related subjects. Further information about the organization may be obtained by contacting: Administrative Secretary, BPA Headquarters, 248 W. Orange Street, Lancaster, PA 17603.

CHAPTER HIGHLIGHTS

1. To keep listeners interested and tuned, stations actively promote their image and call letters. Small market stations promote to compete for audience with other forms of media. Major market stations promote to differentiate themselves from competing stations.

2. Radio recognized the value of promotion early and used print media, remote broadcasts, and billboards to inform the public. Later, ratings surveys proved the importance of effective promotions.

3. Greater competition because of the increasing number of stations and monthly audience surveys means today's stations must promote continually.

4. The most successful (attracting listenership loyalty) promotions involve large cash or merchandise prizes.

5. A successful promotion director possesses knowledge and understanding of the station's audience, a background in research, writing, and conceptual skills, the ability to adapt existing concepts to a particular station, and a familiarity with graphic art. The promotion director is responsible for acquiring prizes through trade or purchase and for compliance with FCC regulations covering promotions.

6. On-air promotions are the most common method used to retain and expand listenership. Such devices as slogans linked to the call letters and contests are common.

7. To "bookend" call letters means to place them at the beginning and conclusion of each break. To "graft" call letters means to include them with all informational announcements.

8. Contests must have clear rules and must provide entertainment for players and nonplayers alike. Successful contests are compatible with the station's sound, offer prizes attractive to the target audience, and challenge the listener's imagination.

9. Off-air promotions are intended to attract new listeners. Popular approaches include billboards, bus cards, newspapers, television, bumper stickers, discount cards, giveaway items embossed with call letters or logo, deejay personal appearances, special activity sponsorship, and remote broadcasts.

10. To offset the sometimes substantial cost of an off-air promotion, stations often collaborate with sponsors to share both the expenses and the attention gained.

11. FCC regulations governing promotions are contained in Section 73.1216. Basically, stations may not operate lotteries, endanger contestants, rig contests, or mislead listeners as to the nature of the prize.

12. The nonprofit Broadcasters Promotion Association was founded in 1956 to provide information and services for station promotion directors worldwide.

SUGGESTED FURTHER READING

Aaker, David A., and Myers, John G. *Advertising Management,* 2nd ed. Englewood Cliffs, N.J.: Prentice-Hall, 1982.

Bergendorff, Fred L. *Broadcast Advertising and Promotion: A Handbook for Students and Professionals.* New York: Hastings House, 1983.

Eastman, Susan Tyler, and Klein, Robert A., eds. *Strategies in Broadcast and Cable Promotion.* Belmont, Calif.: Wadsworth Publishing, 1982.

Gompertz, Rolf. *Promotion and Publicity Handbook for Broadcasters.* Blue Ridge Summit, Pa.: Tab Books, 1977.

Macdonald, Jack. *The Handbook of Radio Publicity and Promotion.* Blue Ridge Summit, Pa.: Tab Books, 1970.

Nickels, William. *Marketing Communications and Promotion,* 3rd ed. New York: John Wiley and Sons, 1984.

Peck, William A. *Radio Promotion Handbook.* Blue Ridge Summit, Pa.: Tab Books, 1968.

Savage, Bob. *Perry's Broadcast Promotion Sourcebook.* Oak Ridge, Tenn.: Perry Publications, 1982.

Stanley, Richard E. *Promotions,* 2nd ed. Englewood Cliffs, N.J.: Prentice-Hall, 1982.

8 Traffic and Billing

THE AIR SUPPLY

A station sells airtime. That is its inventory. The volume or size of a given station's inventory chiefly depends on the amount of time it allocates for commercial matter. For example, some stations with Easy Listening and Adult Contemporary formats deliberately restrict or limit commercial loads as a method of enhancing overall sound and fostering a "more music–less talk" image. Other outlets simply abide by commercial load stipulations as outlined in their license renewal applications. The NAB has long suggested that stations not exceed eighteen minutes of commercial time hourly, and most stations regard this recommendation as an industry standard, although the FCC does not dictate or impose a ceiling on the amount of time stations may devote to commercial messages.

A full-time station has over ten thousand minutes to fill each week. This computes to approximately three thousand minutes for commercials, based on NAB's suggested limit. In the eyes of the sales manager, this means anywhere from three to six thousand availabilities or slots—assuming that a station sells sixty- and thirty-second spot units—in which to insert commercial announcements.

From the discussion up until now, it should be apparent that inventory control and accountability at a radio station are no small job. They are, in fact, the primary duty of the person called the traffic manager.

THE TRAFFIC MANAGER

A daily log is prepared by the traffic manager (also referred to as the traffic director). This document is at once a schedule of programming elements (commercials, features, public service announcements) to be aired and a record of what was actually aired. It serves to inform the on-air operator of what to broadcast and at what time, and it provides a record for, among other things, billing purposes.

Let us examine the process involved in logging a commercial for broadcast beginning at the point at which the salesperson writes an order for a spot schedule.

The salesperson writes an order and returns it to the station.
Order is then checked and approved by the sales manager.
Order is typed by the sales secretary.
Copies of formalized order are distributed to: traffic manager, sales manager, billing, salesperson, and client.
Order is placed in the traffic scheduling book or entered into the computer for posting to the log by traffic manager.
Order is logged commencing on the start date according to the stipulations of the buy.

Although the preceding is both a simplification and generalization of the actual process, it does convey the basic idea. Keep in mind that not all stations operate in exactly the same manner. The actual method for preparing a log will differ, too, from station to station depending on whether it is done manually or by computer. Those outlets using the manual system often simplify the process by preparing a master or semipermanent log containing fixed program elements and even long-term advertisers. Short-term sponsors and other changes will be entered on an ongoing basis. This method significantly reduces typing. The master may be imprinted on plastic or mylar, and entries can be made and erased according to need. Once the log is prepared, it is then copied and distributed.

It is the traffic manager's responsibility to see that an order is logged as specified and that each client is treated fairly and equitably. A sponsor who purchases two spots, five days a week during morning drive, can expect to receive good rotation for maximum reach. It is up to the traffic

FIGURE 8.1
Unless the traffic manager keeps a close eye on the air supply, it can be exhausted, placing the station in an oversold situation. Courtesy Marketron.

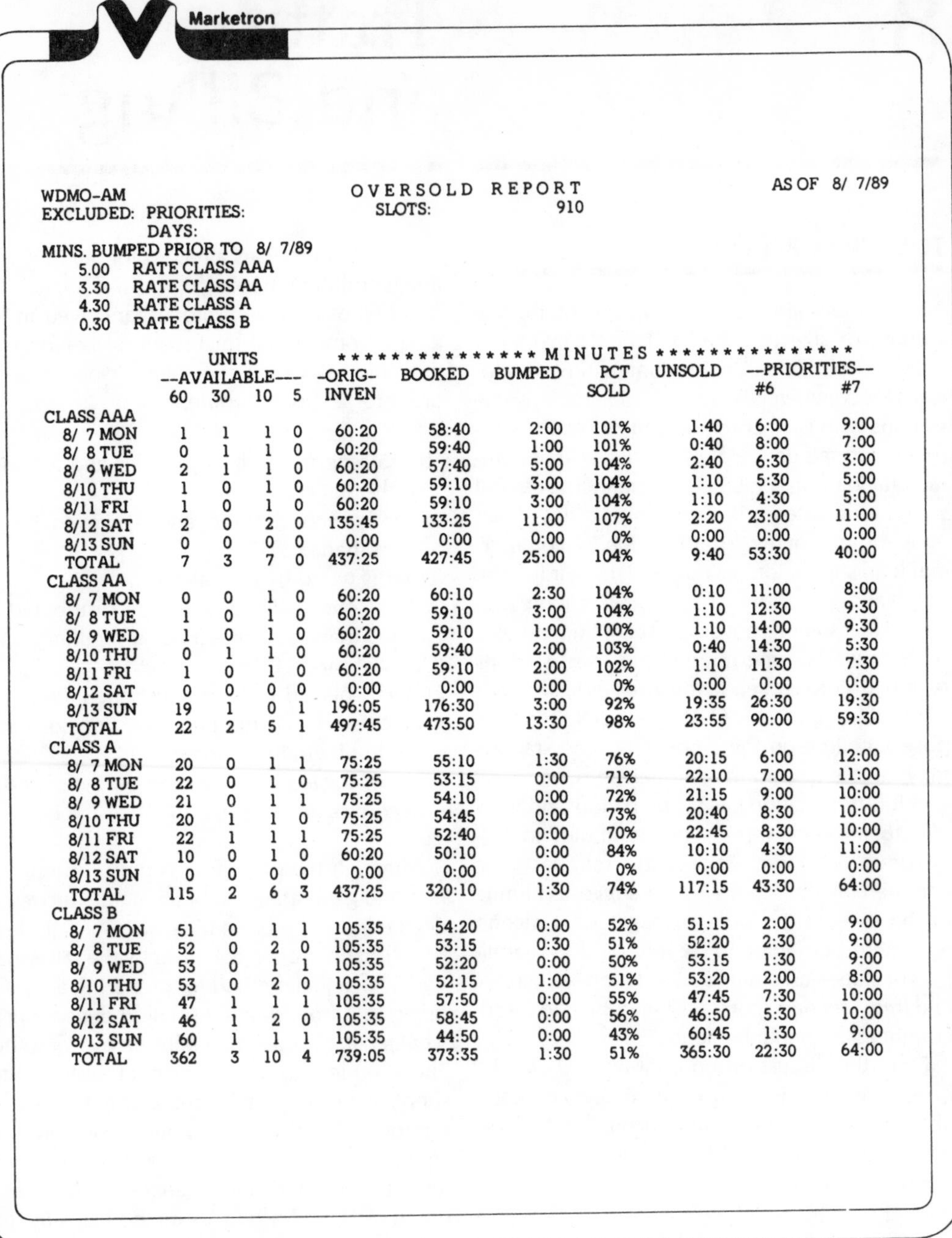

WDMO-AM OVERSOLD REPORT AS OF 8/ 7/89
EXCLUDED: PRIORITIES: SLOTS: 910
DAYS:
MINS. BUMPED PRIOR TO 8/ 7/89
5.00 RATE CLASS AAA
3.30 RATE CLASS AA
4.30 RATE CLASS A
0.30 RATE CLASS B

	UNITS --AVAILABLE---				*************** MINUTES ***************					--PRIORITIES--	
	60	30	10	5	-ORIG- INVEN	BOOKED	BUMPED	PCT SOLD	UNSOLD	#6	#7
CLASS AAA											
8/ 7 MON	1	1	1	0	60:20	58:40	2:00	101%	1:40	6:00	9:00
8/ 8 TUE	0	1	1	0	60:20	59:40	1:00	101%	0:40	8:00	7:00
8/ 9 WED	2	1	1	0	60:20	57:40	5:00	104%	2:40	6:30	3:00
8/10 THU	1	0	1	0	60:20	59:10	3:00	104%	1:10	5:30	5:00
8/11 FRI	1	0	1	0	60:20	59:10	3:00	104%	1:10	4:30	5:00
8/12 SAT	2	0	2	0	135:45	133:25	11:00	107%	2:20	23:00	11:00
8/13 SUN	0	0	0	0	0:00	0:00	0:00	0%	0:00	0:00	0:00
TOTAL	7	3	7	0	437:25	427:45	25:00	104%	9:40	53:30	40:00
CLASS AA											
8/ 7 MON	0	0	1	0	60:20	60:10	2:30	104%	0:10	11:00	8:00
8/ 8 TUE	1	0	1	0	60:20	59:10	3:00	104%	1:10	12:30	9:30
8/ 9 WED	1	0	1	0	60:20	59:10	1:00	100%	1:10	14:00	9:30
8/10 THU	0	1	1	0	60:20	59:40	2:00	103%	0:40	14:30	5:30
8/11 FRI	1	0	1	0	60:20	59:10	2:00	102%	1:10	11:30	7:30
8/12 SAT	0	0	0	0	0:00	0:00	0:00	0%	0:00	0:00	0:00
8/13 SUN	19	1	0	1	196:05	176:30	3:00	92%	19:35	26:30	19:30
TOTAL	22	2	5	1	497:45	473:50	13:30	98%	23:55	90:00	59:30
CLASS A											
8/ 7 MON	20	0	1	1	75:25	55:10	1:30	76%	20:15	6:00	12:00
8/ 8 TUE	22	0	1	0	75:25	53:15	0:00	71%	22:10	7:00	11:00
8/ 9 WED	21	0	1	1	75:25	54:10	0:00	72%	21:15	9:00	10:00
8/10 THU	20	1	1	0	75:25	54:45	0:00	73%	20:40	8:30	10:00
8/11 FRI	22	1	1	1	75:25	52:40	0:00	70%	22:45	8:30	10:00
8/12 SAT	10	0	1	0	60:20	50:10	0:00	84%	10:10	4:30	11:00
8/13 SUN	0	0	0	0	0:00	0:00	0:00	0%	0:00	0:00	0:00
TOTAL	115	2	6	3	437:25	320:10	1:30	74%	117:15	43:30	64:00
CLASS B											
8/ 7 MON	51	0	1	1	105:35	54:20	0:00	52%	51:15	2:00	9:00
8/ 8 TUE	52	0	2	0	105:35	53:15	0:30	51%	52:20	2:30	9:00
8/ 9 WED	53	0	1	1	105:35	52:20	0:00	50%	53:15	1:30	9:00
8/10 THU	53	0	2	0	105:35	52:15	1:00	51%	53:20	2:00	8:00
8/11 FRI	47	1	1	1	105:35	57:50	0:00	55%	47:45	7:30	10:00
8/12 SAT	46	1	2	0	105:35	58:45	0:00	56%	46:50	5:30	10:00
8/13 SUN	60	1	1	1	105:35	44:50	0:00	43%	60:45	1:30	9:00
TOTAL	362	3	10	4	739:05	373:35	1:30	51%	365:30	22:30	64:00

manager to schedule the client's commercials in as many quarter-hour segments of the daypart as possible. The effectiveness of a spot schedule is reduced if the spots are logged in the same quarter-hour each day. If a spot is logged at 6:45 daily, it is only reaching those people tuned at that hour each day. However, if on one day it is logged at 7:15 and then at 8:45 on another, and so on, it is reaching a different audience each day. It also would be unfair to the advertiser who purchased drivetime to only have spots logged prior to 7:00 A.M., the beginning of the prime audience period.

The traffic manager maintains a record of when a client's spots are aired to help ensure effective rotation. Another concern of the traffic

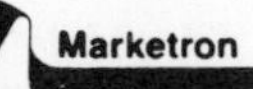

WDMO-AM PROJECTED AVAILS SUMMARY 5:30A–10:00 A 8/ 7/89–10/15/89 AS OF 8/ 7/89
EXCLUDED: PRIORITIES- SLOTS- 8 9 10
: DAYS- SA SU

WEEK ENDING	8/13	8/20	8/27	9/ 3	9/10	9/17	9/24	10/ 1	10/ 8	10/15
# PTY 1.0	2	2	0	0	0	0	0	0	0	0
# PTY 2.0	26	10	10	12	0	0	3	0	0	0
# PTY 3.0	14	14	5	5	5	0	0	0	0	0
# PTY 4.0	7	4	0	0	0	0	0	0	0	0
# PTY 5.0	212	185	143	115	100	67	72	75	65	70
# PTY 6.0	9	0	0	0	0	0	0	0	0	0
# PTY 7.0	38	37	15	10	10	10	6	5	0	0
UNITS SOLD	307	252	173	141	115	77	81	80	65	70
TIME SOLD	4:34	3:48	2:38	2:11	1:45	1:07	1:11	1:10	0:55	1:00
INVENTORY	5:40	5:40	5:40	5:40	5:40	5:40	5:40	5:40	5:40	5:40
AVAILABLE	1:06	1:52	3:03	3:29	3:56	4:34	4:29	4:30	4:45	4:40
% SOLD/WEEK (ALL SPOTS)	81%	68%	47%	39%	31%	20%	21%	21%	17%	18%
% SOLD/WEEK (DAILIES ONLY)	44%	39%	30%	24%	21%	11%	9%	11%	9%	11%
% SOLD/DAY (ALL SPOTS)										
MONDAY	80%	67%	47%	39%	31%	20%	21%	21%	17%	18%
TUESDAY	80%	68%	48%	40%	32%	21%	22%	22%	17%	19%
WEDNESDAY	80%	66%	47%	39%	31%	20%	21%	21%	17%	18%
THURSDAY	81%	70%	47%	39%	31%	20%	21%	21%	17%	18%
FRIDAY	84%	67%	47%	39%	31%	20%	21%	21%	17%	18%
SATURDAY	0%	0%	0%	0%	0%	0%	0%	0%	0%	0%
SUNDAY	0%	0%	0%	0%	0%	0%	0%	0%	0%	0%

FIGURE 8.2
Stations often project availabilities to avoid being oversold. Computers simplify this task. Courtesy Marketron.

manager is to keep adequate space between accounts of a competitive nature. Running two restaurants back-to-back or within the same spot set likely would result in having to reschedule both at different times at no cost to the client.

It also falls within the traffic manager's purview to make sure copy and production tapes are in on time. Most stations have a policy, often stated in their rate card, requiring that commercial material be on hand at least forty-eight hours before it is scheduled for broadcast. Carol Bates, traffic manager of WGNG-AM, Providence, says that getting copy before the air date can be a problem. "It is not unusual to get a tape or copy a half-hour before it is due to air. We ask that copy be in well in advance, but

CONTRACT NO. 1 SALES ORDER DATE 7/20

X NEW ☐ CHANGE ☐ ADDITION ☐ CANCEL ☐ PRICE CHANGE X AM ☐ FM ☐ COMBO

X CASH ☐ TRADE ☐ NON-CMML START DATE 7/31/89

X RETAIL ☐ POLITICAL ☐ SPONSORSHIPS

☐ NON-RETAIL ☐ SPORTS ☐ PROGRAMS ☐ NETWORK END DATE 9/10/89

ADVERTISER SEARS ADV NO. 8400 ☐ REMIT X CASH IN ADVANCE

PRODUCT SUMMER SALE PCC NO. ___ SPC NO. ___ BILLING CYCLE X STANDARD ☐ CALENDAR ☐ WEEKLY ☐ SPECIAL

AGENCY NW AYER & CO. AGY NO. 300 SPOT PRICES PRINTED X Yes ☐ No AGENCY COM X Yes ☐ No

BILLING ADDRESS 111 EAST WACKER DR. ☐ ALT AGENCY ☐ MBS (Address Below)

CITY CHICAGO STATE ILL ZIP 60601 PLAN TYPE NONE RATE CARD 9

PHONE 854-5301 BUYER PENNY SCHOB X VARIOUS PRODUCTS SEPARATE INVOICES ☐ Yes X No CONTRACT PRIORITY (VARIES)

SALESPERSON CRONIN NO. 1 COMM. ___%

FEATURE/SPONSORSHIP LOCAL NEWS

OPTIONS FOR MAKE GOODS
☐ LIMITED
X ADVANCED APPROVAL
☐ UNRESTRICTED

DATES		TIMES		SPOTS										
WEEKS TO RUN	# OF WEEKS	CLASS	START/END (PREFER)	LEN	PER WEEK	M	T	W	TH	F	SAT	SUN	RATE	PTY
7/31-9/10	6	AAA	5:30-10A (6-10A)	60	5	1	1	1	1	1			60	3
7/31-9/10	6	AA	3-7P	60	4	←———				———→			48.	3
7/31, 8/7, 9/4	3	A	10-3P (LOCAL NEWS)	60	6	←———				———→		←→	45.	1

SALES MGR. Cummings

ENTERED BY

BILLING NAME & ADDRESS

NOTES TO TRAFFIC & ACCOUNTING COPY IN HOUSE

STANDARD REMARK # 1

SPECIAL REMARKS

# SPOTS/WEEK	# SPOTS/ORDER
15/9	72
WEEKLY GROSS	TOTAL GROSS
762/492	3762.

FIGURE 8.3
Handwritten sales order. The station salesperson usually writes the order at the client's business for initial approval. Courtesy Marketron.

sometimes it's a matter of minutes. No station is unfamiliar with having to make up spots due to late copy. It's irritating but a reality that you have to deal with."

Jan Hildreth, traffic manager of WARA-AM, Attleboro, Massachusetts, says that holiday and political campaign periods can place added pressure on the traffic person. "The fourth quarter is the big money time in radio. The logs usually are jammed, and availabilities are in short supply. The workload in the traffic department doubles. Things also get pretty chaotic around elections. It can become a real test for the nerves. Of course, there's always the late order that arrives at 5 P.M. on Friday that gets the adrenalin going."

There are few station relationships closer than that of the traffic department with programming and sales. Programming relies on the traffic manager for the logs that function as scheduling guides for on-air personnel. Sales depends on the traffic department to inform it of existing availabilities and to process orders onto the air. "It is crucial to the operation that traffic have a good relationship with sales and programming. When it doesn't, things begin to happen. The PD has to let traffic know when something changes; if not, the system breaks down. This is equally true of sales. Traffic is kind of the heart of things. Everything passes through the traffic department. Cooperation is very important," observes Barbara Kafulas, traffic manager, WNRI-AM, Woonsocket, Rhode Island.

THE TRAFFIC MANAGER'S CREDENTIALS

A college degree is not a criterion for the job of traffic manager. This is not to imply that skill and training are not necessary. Obviously the demands placed on the traffic manager are formidable, and not everyone is qualified to fill the position. "It takes a special kind of person to effectively handle the job of traffic. Patience, an eye for detail, plus the ability to work under pressure and with other people are just some of the qualities the position requires," notes WMJX-FM general manager Bill Campbell.

Typing or keyboard skills are vital to the job. A familiarity with computers and word processing has become necessary, as most stations have given up the manual system of preparing logs in favor of the computerized method.

Many traffic people are trained in-house and come from the administrative or clerical ranks. It is a position that traditionally has been filled by women. While traffic salaries generally exceed that of purely secretarial positions, this is not an area noted for its high pay. Although the traffic manager is expected to handle many responsibilities, the position generally is perceived as more clerical in nature than managerial.

Traffic managers frequently make the transition into sales or programming. The considerable exposure to those particular areas provides a solid foundation and good springboard for those desiring to make the change.

COMMERCIAL MATTER TYPE
CM COMMERCIAL MATTER
MRA MECHANICAL REPRODUCTION ANNOUNCEMENT
PA PUBLIC AFFAIRS
PAC PROMOTION ANNOUNCEMENT COMMERCIAL
PSA PUBLIC SERVICE ANNOUNCEMENT
SPA STATION PROMOTION ANNOUNCEMENT
/ ANNOUNCED AS SPONSORED

PROGRAM SOURCE
L LOCAL
R RECORDED
N SATELLITE NET
N ABC ENT

"C"
PROGRAM LOG

RADIO STATION WGNG
PAWTUCKET, RHODE ISLAND
ROGER WILLIAMS BROADCASTING COMPANY, INC

PROGRAM TYPE
A AGRICULTURE
E ENTERTAINMENT
EDIT EDITORIALS
I INSTRUCTIONAL
N NEWS
O OTHERS
PA PUBLIC AFFAIRS
POL POLITICAL
R RELIGIOUS
S SPORTS

PAGE ________ DAY ________________ DATE ________________

Announcer or Operator (Full Name) ____________________ Time On __________ Time Off __________
____________________ Time On __________ Time Off __________

STATION IDENT	PROGRAM TIME		PROGRAM TITLE-SPONSOR	MECH	COMM'L MAT'R			PROGRAM		
	ON	OFF			DURATION	TYPE	✓	SOURCE	TYPE	ANNC MNT
			STATION ID							
	:00	:04	LOCAL NEWS					L	N	
1	:04	:05								
	:05	:06	SPORTS					L	S	
			WEATHER					L	N	
			WGNG MUSIC					R	E	
6	:10	:12								
			WEATHER					L	N	
			PSA/		10	PSA				
3	:20	:22								
7	:25	:26								
	:30	:32	ABC/NEWS					ABC	N	
	:32	:34	LOCAL/NEWS					L	N	
2	:34	:35								
	:35	:37	SPORTS					L	S	
			WEATHER					L	N	
			PSA/		10	PSA				
5	:40	:42								
			WEATHER					L	N	
			PSA/		10	PSA				
4	:50	:52								
8	:55	:56								

FIGURE 8.4 Manual log master. Computers have not taken over traffic entirely. Many small stations still prepare logs using a typewriter. Courtesy WGNG.

FIGURE 8.4
continued

"D"

COMMERCIAL MATTER TYPE
CM COMMERCIAL MATTER
MRA MECHANICAL REPRODUCTION ANNOUNCEMENT
PA PUBLIC AFFAIRS
PAC PROMOTION ANNOUNCEMENT COMMERCIAL
PSA PUBLIC SERVICE ANNOUNCEMENT
SPA STATION PROMOTION ANNOUNCEMENT
✓ ANNOUNCED AS SPONSORED

PROGRAM SOURCE
L LOCAL
R RECORDED
N SATELLITE NET
N ABC ENT

PROGRAM LOG

RADIO STATION WGNG
PAWTUCKET, RHODE ISLAND
ROGER WILLIAMS BROADCASTING COMPANY, INC

PROGRAM TYPE
A AGRICULTURE
E ENTERTAINMENT
EDIT EDITORIALS
I INSTRUCTIONAL
N NEWS
O OTHERS
PA PUBLIC AFFAIRS
POL POLITICAL
R RELIGIOUS
S SPORTS

PAGE ______ DAY ______________ DATE ______________

Announcer or Operator (Full Name) ______________________ Time On __________ Time Off __________
______________________ Time On __________ Time Off __________

STATION IDENT	PROGRAM TIME ON	PROGRAM TIME OFF	PROGRAM TITLE-SPONSOR	MECH	COMM'L MAT'R DURATION	COMM'L MAT'R TYPE	✓	PROGRAM SOURCE	PROGRAM TYPE	ANNC UNIT
			STATION ID							
			WGNG MUSIC					R	E	
6	:03	:04								
			WEATHER					L	N	
			PSA/		10	PSA				
4	:10	:12								
2	:20	:22								
	:30	:34	ABC/NEWS					N	N	
			WEATHER					L	N	
			PSA/		10	PSA				
1	:40	:42								
3	:50	:52								
5	:55	:56								

"E"

COMMERCIAL MATTER TYPE

- CM COMMERCIAL MATTER
- MRA MECHANICAL REPRODUCTION ANNOUNCEMENT
- PA PUBLIC AFFAIRS
- PAC PROMOTION ANNOUNCEMENT COMMERCIAL
- PSA PUBLIC SERVICE ANNOUNCEMENT
- SPA STATION PROMOTION ANNOUNCEMENT
- ✓ ANNOUNCED AS SPONSORED

PROGRAM SOURCE

- L LOCAL
- R RECORDED
- N SATELLITE NET
- N ABC ENT

PROGRAM LOG

RADIO STATION WGNG
PAWTUCKET, RHODE ISLAND
ROGER WILLIAMS BROADCASTING COMPANY, INC.

PROGRAM TYPE

- A AGRICULTURE
- E ENTERTAINMENT
- EDIT EDITORIALS
- I INSTRUCTIONAL
- N NEWS
- O OTHERS
- PA PUBLIC AFFAIRS
- POL POLITICAL
- R RELIGIOUS
- S SPORTS

PAGE ______ DAY ______ DATE ______

Announcer or Operator (Full Name) ______ Time On ______ Time Off ______
______ Time On ______ Time Off ______

STATION IDENT	PROGRAM TIME		PROGRAM TITLE-SPONSOR	MECH	COMM'L MAT'R			PROGRAM		
	ON	OFF			DURATION	TYPE	✓	SOURCE	TYPE	ANNC INIT
			STATION ID							
	:00	:08	WGNG/NEWS					L	N	
1	:08	:09								
	:09	:10	SPORTS					L	S	
	:10		WEATHER					L	N	
			WGNG MUSIC					R	E	
3	:20	:22								
			WEATHER					L	N	
			PSA/		10	PSA				
5	:30	:34	ABC/NEWS					N	N	
2	:40	:42								
			WEATHER					L	N	
			PSA/		10	PSA				
4	:50	:52								
6	:55	:56								

FIGURE 8.4
continued

FIGURE 8.5 Spot schedule order form. Courtesy WMJX-FM.

Traffic Order

Date ______________

CODE GROUP				BUYER INFORMATION	CREDIT INFORMATION
Station	Product No.	Agency Estimate No.		Name	Name
Agency Number	Category	Rate Card		Telephone	Telephone
Advertiser Number	Billing Period	Co-Op Invoice	YES NO SORT	Account Type	Credit Check YES NO INITIALS
Contract Number	Invoice Times YES NO OPT	Dealer Invoice	YES NO # OF SHARES	X Y	Credit Limit
Salesman Number	Agency Commission %	Discount %		Z	Contract Year START END
Salesman Name		Annct. Type		Special Instructions:	
Advertiser		Advertiser Avail. Code			
To Be Logged As		Contract Avail. Code			

Option	Line No.	Group	Type	Minum. Separ. Prd	Minum. Separ. Adv	Broadcast Dates Start	Broadcast Dates End	Flight	Section	Class	Spons/Pgm Duration	Annct Length	Time Schedule Start	Time Schedule End	Spots Per Week	Mon	Tue	Wed	Thur	Fri	Sat	Sun	Control	Rate	Plan	Unit	Misc. Charges	Live	Video	Audio	Agency Copy Number	Rack Number	Cr. Code	Comp	M/G For

Agency Name and Address		PACKAGE PLAN	FREQ.	RATE	B'casts	Rate	Totals By Rate	B'casts	Rate	Totals By Rate
	Zip/P.C.									

Billing Name and Address		Contract Totals	Gross
			Agency Comm.
			Net
	Zip/P.C.	X To Cancel Order Before Start	

PRINTED IN U.S.A.

COMPUTERIZED TRAFFIC

There are three ways to computerize traffic at a radio station: *on-line systems* (also called "shared-time") in which a computer is linked by telephone lines to a central computer point, *multiterminal, in-house systems* whereby the traffic department is tied into the station's mainframe computer, and *microcomputer systems* (also called personal computers) that primarily address the functions of a single area. There are pros and cons to each approach. Nevertheless, computers vastly enhance the speed and efficiency of the traffic process. Computers store copious amounts of data, retrieve information faster than humanly possible, and schedule and rotate commercials with precision and equanimity, to mention only a few of the features that make the new technology especially adaptive for use in the traffic area.

FIGURE 8.6 Traffic managers check logs for accuracy before placing them in control room. Courtesy WHJJ-AM.

Computers are an excellent tool for inventory control, contends broadcast computer consultant Vicki Cliff. "Radio is a commodity not unlike a train carload of perishables, such as tomatoes. Radio sells time, which is progressively spoiling. The economic laws of supply and demand are classically applied to radio. Computers can assist in plotting that supply and demand curve in determining rates to be charged for various dayparts at any given moment. Inventory control is vital to any business. Radio is limited in its availabilities and seasonal in its desirability to the client. In a sold-out state, client value priorities must be weighted to optimize the station's billing. All things being equal, the

Marketron

WDMO-AM WEEKLY SPOT REPORT AS OF 5/24/82 PAGE 1
INCLUDED: SLOTS- ALL SLOTS
: PRIORITIES-ALL PRIORITIES

7A	5/24	5/25	5/26	5/27	5/28	5/29	5/30
02:00	0/ 0 502 # 2 60 HOUSEHO	0/ 0 529 # 1 60 ATLANTI	0/ 0 540 # 1 60 GENERAL	0/ 0 543 # 1 60 PLAYBOY	0/ 0 548 # 1 60 COCA CO	1/ 60	0/ 0 540 # 1 60 GENERAL
05:00	1/ 60	0/ 0 545 # 1 60 FIRESTO	0/ 0 545 # 1 60 FIRESTO	0/ 0 537 # 1 60 I H O P	0/ 0 557 # 1 60 NOXZELL	1/ 60	0/ 0 520 # 2 60 XEROX
06:00	1/ 10	1/ 10	1/ 10	1/ 10	1/ 10	1/ 10	1/ 10
10:00	1/ 0 518 # 2 60 NOXZELL	1/ 0 546 # 2 60 AMERICA	1/ 30 514 # 2 30 MRS PAU	1/ 0 546 # 2 60 AMERICA	1/ 30 535 # 1 30 CAPITAL	2/ 60	1/ 0 548 # 2 60 COCA CO
14:00	0/ 0 548 # 1 60 COCA CO	0/ 0 527 # 1 60 CARNATI	0/ 0 527 # 1 60 CARNATI	0/ 0 530 # 1 60 GREYHOU	0/ 0 527 # 1 60 CARNATI	0/ 0 502 # 2 60 HOUSEHO	1/ 60
16:00	0/ 0 519 # 2 30 AMERICA	0/ 0 510 # 3 30 HOLIDAY	0/ 0 508 # 2 30 FOTOMAT	1/ 30	1/ 30	1/ 30	1/ 30
22:00	0/ 0 504 # 6 60 DODGE	1/ 60	1/ 60	1/ 60	1/ 60	1/ 60	0/ 0 504 # 6 60 DODGE
23:00	0/ 0 551 # 1 60 COLGATE	0/ 0 1 # 1 60 SEARS	0/ 0 555 # 1 60 BIC COR	1/ 60	0/ 0 516 # 2 60 ROCKWEL	1/ 60	1/ 60
28:00	0/ 0 554 # 1 60 BASKIN	0/ 0 553 # 1 60 GENERAL	0/ 0 520 # 2 60 XEROX	0/ 0 555 # 1 60 BIC COR	0/ 0 512 # 2 60 FIRESTO	0/ 0 553 # 1 60 GENERAL	1/ 60
29:00	1/ 30 514 # 2 30 MRS PAU	1/ 0 534 # 1 60 HOWARD	1/ 30 519 # 2 30 AMERICA	1/ 0 553 # 1 60 GENERAL	1/ 0 518 # 2 60 NOXZELL	2/ 60	0/ 0 514 # 2 30 MRS PAU 508 # 2 30 FOTOMAT
32:00	0/ 0 512 # 2 60 FIRESTO	0/ 0 522 # 1 60 ANHEUSE	0/ 0 504 # 3 60 DODGE	0/ 0 509 # 2 60 CANADA	0/ 0 522 # 1 60 ANHEUSE	0/ 0 517 # 2 60 SCHLITZ	1/ 60
35:00	0/ 0 552 # 1 60 ELEKTRA	0/ 0 515 # 2 60 PRUDENT	0/ 0 518 # 2 60 NOXZELL	0/ 0 549 # 1 60 AMERICA	0/ 0 520 # 2 60 XEROX	1/ 60	0/ 0 509 # 2 60 CANADA

FIGURE 8.7
Manual and computerized traffic boards permit the traffic department to keep tabs on avails and rotation schedules. Courtesy Marketron.

credit rating of the client should be the deciding factor. Computers can eliminate the human subjectivity in formulating the daily log."

The cost of computerizing traffic has kept some stations from converting from the manual system, notes Cliff. "Purchasing a personal computer for the traffic function is okay, but if a station desires to perform several functions it may find itself out of luck. There are limits as to what can be done with a personal computer. On the other hand, a larger capacity, mainframe computer can be a major expense, although the functions it can perform are extraordinary. An on-line, real-time system can be costly also. Line expenses can really add up. Those station managers considering computerizing must gather all possible data to determine if the system they're considering will cost-justify itself."

A variety of traffic and billing software are

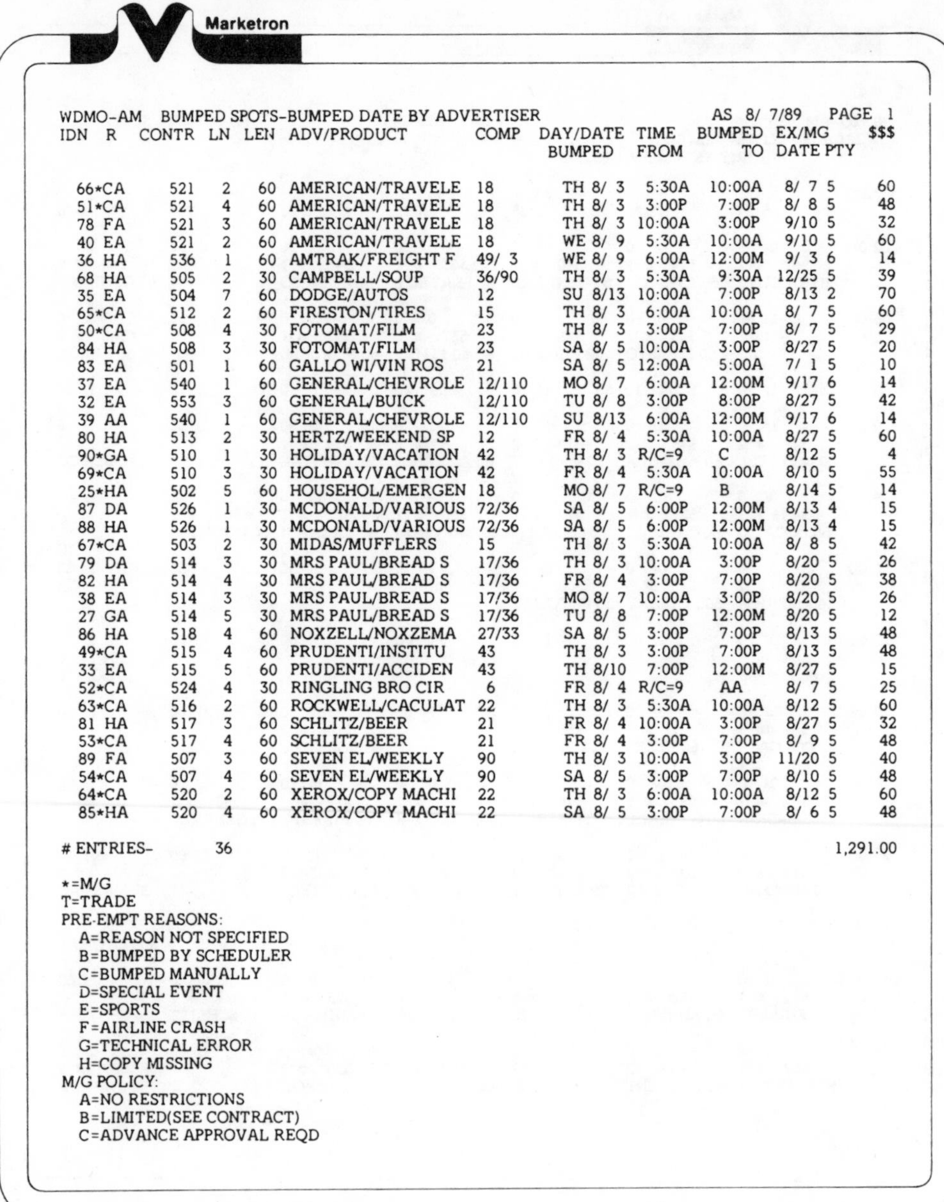

Marketron

WDMO-AM BUMPED SPOTS-BUMPED DATE BY ADVERTISER AS 8/ 7/89 PAGE 1

IDN R	CONTR	LN	LEN	ADV/PRODUCT	COMP	DAY/DATE BUMPED	TIME FROM	BUMPED TO	EX/MG DATE	PTY	$$$
66*CA	521	2	60	AMERICAN/TRAVELE	18	TH 8/ 3	5:30A	10:00A	8/ 7	5	60
51*CA	521	4	60	AMERICAN/TRAVELE	18	TH 8/ 3	3:00P	7:00P	8/ 8	5	48
78 FA	521	3	60	AMERICAN/TRAVELE	18	TH 8/ 3	10:00A	3:00P	9/10	5	32
40 EA	521	2	60	AMERICAN/TRAVELE	18	WE 8/ 9	5:30A	10:00A	9/10	5	60
36 HA	536	1	60	AMTRAK/FREIGHT F	49/ 3	WE 8/ 9	6:00A	12:00M	9/ 3	6	14
68 HA	505	2	30	CAMPBELL/SOUP	36/90	TH 8/ 3	5:30A	9:30A	12/25	5	39
35 EA	504	7	60	DODGE/AUTOS	12	SU 8/13	10:00A	7:00P	8/13	2	70
65*CA	512	2	60	FIRESTON/TIRES	15	TH 8/ 3	6:00A	10:00A	8/ 7	5	60
50*CA	508	4	30	FOTOMAT/FILM	23	TH 8/ 3	3:00P	7:00P	8/ 7	5	29
84 HA	508	3	30	FOTOMAT/FILM	23	SA 8/ 5	10:00A	3:00P	8/27	5	20
83 EA	501	1	60	GALLO WI/VIN ROS	21	SA 8/ 5	12:00A	5:00A	7/ 1	5	10
37 EA	540	1	60	GENERAL/CHEVROLE	12/110	MO 8/ 7	6:00A	12:00M	9/17	6	14
32 EA	553	3	60	GENERAL/BUICK	12/110	TU 8/ 8	3:00P	8:00P	8/27	5	42
39 AA	540	1	60	GENERAL/CHEVROLE	12/110	SU 8/13	6:00A	12:00M	9/17	6	14
80 HA	513	2	30	HERTZ/WEEKEND SP	12	FR 8/ 4	5:30A	10:00A	8/27	5	60
90*GA	510	1	30	HOLIDAY/VACATION	42	TH 8/ 3	R/C=9	C	8/12	5	4
69*CA	510	3	30	HOLIDAY/VACATION	42	FR 8/ 4	5:30A	10:00A	8/10	5	55
25*HA	502	5	60	HOUSEHOL/EMERGEN	18	MO 8/ 7	R/C=9	B	8/14	5	14
87 DA	526	1	30	MCDONALD/VARIOUS	72/36	SA 8/ 5	6:00P	12:00M	8/13	4	15
88 HA	526	1	30	MCDONALD/VARIOUS	72/36	SA 8/ 5	6:00P	12:00M	8/13	4	15
67*CA	503	2	30	MIDAS/MUFFLERS	15	TH 8/ 3	5:30A	10:00A	8/ 8	5	42
79 DA	514	3	30	MRS PAUL/BREAD S	17/36	TH 8/ 3	10:00A	3:00P	8/20	5	26
82 HA	514	4	30	MRS PAUL/BREAD S	17/36	FR 8/ 4	3:00P	7:00P	8/20	5	38
38 EA	514	3	30	MRS PAUL/BREAD S	17/36	MO 8/ 7	10:00A	3:00P	8/20	5	26
27 GA	514	5	30	MRS PAUL/BREAD S	17/36	TU 8/ 8	7:00P	12:00M	8/20	5	12
86 HA	518	4	60	NOXZELL/NOXZEMA	27/33	SA 8/ 5	3:00P	7:00P	8/13	5	48
49*CA	515	4	60	PRUDENTI/INSTITU	43	TH 8/ 3	3:00P	7:00P	8/13	5	48
33 EA	515	5	60	PRUDENTI/ACCIDEN	43	TH 8/10	7:00P	12:00M	8/27	5	15
52*CA	524	4	30	RINGLING BRO CIR	6	FR 8/ 4	R/C=9	AA	8/ 7	5	25
63*CA	516	2	60	ROCKWELL/CACULAT	22	TH 8/ 3	5:30A	10:00A	8/12	5	60
81 HA	517	3	60	SCHLITZ/BEER	21	FR 8/ 4	10:00A	3:00P	8/27	5	32
53*CA	517	4	60	SCHLITZ/BEER	21	FR 8/ 4	3:00P	7:00P	8/ 9	5	48
89 FA	507	3	60	SEVEN EL/WEEKLY	90	TH 8/ 3	10:00A	3:00P	11/20	5	40
54*CA	507	4	60	SEVEN EL/WEEKLY	90	SA 8/ 5	3:00P	7:00P	8/10	5	48
64*CA	520	2	60	XEROX/COPY MACHI	22	TH 8/ 3	6:00A	10:00A	8/12	5	60
85*HA	520	4	60	XEROX/COPY MACHI	22	SA 8/ 5	3:00P	7:00P	8/ 6	5	48

ENTRIES- 36 1,291.00

*=M/G
T=TRADE
PRE-EMPT REASONS:
A=REASON NOT SPECIFIED
B=BUMPED BY SCHEDULER
C=BUMPED MANUALLY
D=SPECIAL EVENT
E=SPORTS
F=AIRLINE CRASH
G=TECHNICAL ERROR
H=COPY MISSING
M/G POLICY:
A=NO RESTRICTIONS
B=LIMITED(SEE CONTRACT)
C=ADVANCE APPROVAL REQD

FIGURE 8.8
Spots that are not aired when scheduled must be rescheduled ("made good") according to the conditions of the purchase order. Courtesy Marketron.

available. Dozens of companies, most notably Marketron, Columbine, Bias Radio, Custom Business Systems, and The Management, specialize in providing broadcasters with software packages. Prices for computer software vary depending on the nature and content of the program. Some companies permit stations to purchase software on a time-payment basis. For example, The Management, based in Fort Worth, Texas, offers traffic and billing software for between $225 and $410 per month, and stations may take up to five years to pay in full. The Management offers several traffic and billing software options, ranging in price from fifteen hundred dollars for their basic "Simple Log" program to nine thousand dollars for their most elaborate—"Super Log."

Compatible hardware is specified by the software manufacturer. Most software is IBM-PC compatible, however. Companies and consultants specializing in broadcast computerization are listed in *Broadcasting Yearbook*.

ACTUAL TIME	SCHEDULED TIME	LENGTH	PROGRAM TITLES — ANNOUNCEMENTS	CART NUMBER	ANNCT TYPE	PROG. S	PROG. T	CONTRACT & LINE NUMBER

ON	OPERATOR	OFF	ON	OPERATOR	OFF

COLUMBINE SYSTEMS, INC. CSI-498 MADE IN U.S.A. 3-83 78741-1

FIGURE 8.9
Computerized log form. Courtesy WMJX-FM.

BILLING

At most stations, advertisers are billed for the airtime they have purchased after a portion or all of it has run. Few stations require that sponsors pay in advance. It is the job of the billing department to notify the advertiser when payment is due. Al Rozanski, business manager of WMJX-FM, Boston, explains the process involved once a contract has been logged by the traffic department. "We send invoices out twice monthly. Many stations bill weekly, but we find doing it every two weeks cuts down on the paperwork considerably. The first thing my billing

FIGURE 8.10
Computers were first installed in station traffic departments. Courtesy WMJX-FM.

FIGURE 8.11
Computerized log. Courtesy Marketron.

Marketron

PROGRAM LOG

PAGE 8
AM TUE 8/ 8/89
PACIFIC DAYLIGHT TIME

ACTUAL	SCHEDULED TIME	LENGTH	PROGRAM OR SPONSOR		COPY	SOURCE	TYPE
	7:00:00– 8:00:00		THE MORNING SHOW		LIVE	L	E
	7:00:00– 7:05:00		NEWS ON THE HOUR		LIVE	L	N
	02:00	60	ATLANTIC RECORDS/FATS DOMINO		CART #459		CM
	05:00	60	FIRESTONE/RADIALS		....		CM
	06:00	10	PSA/CONSERVE ENERGY		LIVE		PSA
	10:00	60	AMERICAN MOTORS/PACER	(ADJ)	CART #878		CM
	7:13:00– 7:16:00		SPORTS REPORT		LIVE	L	S
		5	SPONSOR BB		OPEN		
	14:00	60	CARNATION/YOGURT		CART #567		CM
		5	SPONSOR BB		CLOSE		
	16:00	30	HOLIDAY INN/VACATION		LIVE		CM
	7:20:00– 7:23:00		LOCAL NEWS			L	N
	22:00	60					
	23:00	60	SEARS/ALLSTATE		CART #304		CM
	28:00	60	(NET SPOT) GM/BUICK	(DB)	CART #5678		CM
	29:00	60	HOWARD JOHNSON/GET AWAY WEEKEND		CART #8966		CM
	7:30:00– 7:33:00		HEADLINES & NEWS UPDATES		LIVE	L	N
	32:00	60	ANHEUSER BUSCH/BUDWEISER	(FP)	CART #523		CM
	35:00	60	PRUDENTIAL LIFE/ACCIDENT INS		CART #784		CM
	36:00	60	AMERICAN AIRLINES/KANSAS CITY		CART #201		CM
	40:00	60	YES ON ONE COMMITTEE/POLITICAL		CART #682		CM
			LIVE DISCLAIMER		4		
	7:43:00– 7:46:00		BUSINESS NEWS			L	N
	44:00	60	GALLO WINES/VIN ROSE		CART #246 LT		CM
	46:00	60					
	52:00	10	TEMPERATURE & TIME		LIVE		OA
	52:00	5					
	53:00	30	HERTZ/WEEKEND SPECIAL		CART #778		CM
	53:30	30	KODAK/FILM	*MG	CART #668		CM
	57:00	60	BIC CORP/RAZORS	*TR	CART #345		CM
	58:00	30	RINGLING BRO CIRCUS		CART #990		CM
	58:30	30					
	59:55	5	STATION ID		LIVE		ID

TIME ON	LOGKEEPER'S SIGNATURE	TIME OFF	TIME ON	LOGKEEPER'S SIGNATURE	TIME OFF

person does is check the logs to verify that the client's spots ran. We don't bill them for something that wasn't aired. Occasionally a spot will be missed for one reason or another, say a technical problem. This will be reflected on the log because the on-air person will indicate this fact. Invoices are then generated in triplicate by our computer. We use an IBM System 34 computer and Columbine software. This combination is extremely versatile and efficient. The station retains a copy of the invoice and mails two to the client, who then returns one with the payment. The client also receives an affidavit detailing when spots were aired. If the client requests, we will notarize the invoice. This is generally necessary for clients involved in co-op contracts." The billing procedure at WMJX-FM is representative of that at many stations.

Not all radio stations have a full-time business manager on the payroll. Thus the person who handles billing commonly is responsible for maintaining the station's financial records or books as well. In this case, the services of a professional accountant may be contracted on a regular periodic basis to perform the more complex bookkeeping tasks and provide consultation on other financial matters.

Accounts that fail to pay when due are turned over to the appropriate salesperson for collections. If this does not result in payment, a station may use the services of a collection agency. Should its attempt also fail, the station likely would write the business off as a loss at tax time.

THE FCC AND TRAFFIC

Program log requirements were eliminated by the FCC in the early 1980s as part of the era's formidable deregulation movement. Before then the FCC expected radio stations to maintain a formal log, which—in addition to program titles, sponsor names, and length of elements—reflected information pertaining to the nature of announcements (commercial material, public service announcement), source of origination (live, recorded, network), and the type of program (entertainment, news, political, religious, other). The log in figure 8.4 reflects these requirements. Failing to include this information on the log could have resulted in punitive actions against the station by the FCC.

Although stations no longer must keep a program log under existing rules, some sort of document still is necessary to inform programming personnel of what is scheduled for broadcast and to provide information for both the traffic and billing departments pertaining to their particular functions. A log creates accountability. It is both a programming guide and a document of verification.

Stations are now at liberty to design logs that serve their needs most effectively and efficiently. The WMJX log form in figure 8.9 is an example of a log that has been designed to meet all of its station's needs in the most economical and uncomplicated way.

There are no stipulations regarding the length of time that logs must be retained. Before the elimination of the FCC program log regulations, stations were required to retain logs for a minimum of two years. Today most stations still hold onto logs for that amount of time for the sake of accountability.

ANA / RAB FORM FOR SCRIPT (IF TAPE IS USED, PREPARE SCRIPT FROM TAPE)

wmjx 106 fm
The magic is the music.®

ANA / RAB RADIO "TEAR-SHEET"
FORM AT BOTTOM OF SCRIPT PERMITS KNOWING HOW MANY TIMES THIS SCRIPT RAN, AT WHAT COST.

Client:	For:	
Begin:	End:	Date:

STATION DOCUMENTATION STATEMENT APPROVED BY THE CO-OPERATIVE ADVERTISING COMMITTEE OF THE ASSOCIATION OF NATIONAL ADVERTISERS

This announcement was broadcast________times, as entered in the station's program log. The times this announcement was broadcast were billed to this station's client at our invoice(s) number/dated ________at his earned rate of:

$________each for________announcements, for a total of $________
$________each for________announcements, for a total of $________
$________each for________announcements, for a total of $________

____
Signature of station official

________ (Notarize above) ________ Typed name and title ________ Station

FIGURE 8.12
Standard form used for clients involved in co-op agreements. Courtesy WMJX-FM.

THE FCC AND BILLING

Whereas the FCC has deregulated program logs, it takes a keen interest in station billing practices. Penalties for failure to adhere to FCC guidelines can result in fines of up to ten thousand dollars.

Section 73.1205 of the FCC's rules and regulations outlines the commission's position on fraudulent billing:

FIGURE 8.13
Example of station invoice sent to an ad agency. Notice that the commission is deducted from the total gross. Courtesy Marketron.

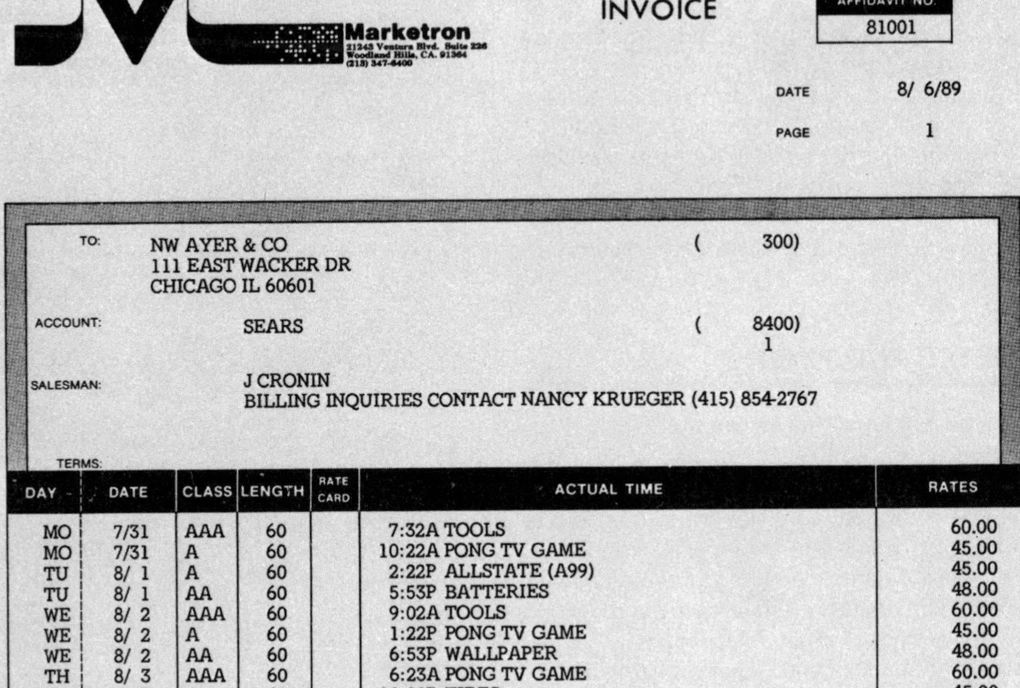

Marketron
21243 Ventura Blvd. Suite 226
Woodland Hills, CA. 91364
(213) 347-8400

INVOICE

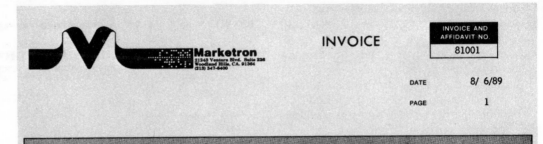

INVOICE AND AFFIDAVIT NO.
81001

DATE 8/ 6/89
PAGE 1

TO: NW AYER & CO (300)
111 EAST WACKER DR
CHICAGO IL 60601

ACCOUNT: SEARS (8400)
1

SALESMAN: J CRONIN
BILLING INQUIRIES CONTACT NANCY KRUEGER (415) 854-2767

TERMS:

DAY	DATE	CLASS	LENGTH	RATE CARD	ACTUAL TIME		RATES
MO	7/31	AAA	60		7:32A TOOLS		60.00
MO	7/31	A	60		10:22A PONG TV GAME		45.00
TU	8/ 1	A	60		2:22P ALLSTATE (A99)		45.00
TU	8/ 1	AA	60		5:53P BATTERIES		48.00
WE	8/ 2	AAA	60		9:02A TOOLS		60.00
WE	8/ 2	A	60		1:22P PONG TV GAME		45.00
WE	8/ 2	AA	60		6:53P WALLPAPER		48.00
TH	8/ 3	AAA	60		6:23A PONG TV GAME		60.00
TH	8/ 3	A	60		12:22P TIRES		45.00
TH	8/ 3	AA	60		4:02P BATTERIES		48.00
FR	8/ 4	AAA	60		7:29A PONG TV GAME		60.00
FR	8/ 4	A	60		11:22A PAINT		45.00
SU	8/ 6	A	60		10.22A PONG TV GAME		45.00
TU	8/ 1	AAA	60		5:30A TIRES	MISSED M/G TO RUN 8/ 8/89	N/C
MO	7/31	AA	60		3:00P BATTERIES	MISSED	N/C

NOTARIZED AFFIDAVITS REQUIRED

TOTAL SPOTS 13

TOTAL GROSS 654.00
LESS AGENCY COMMISSION 98.10

PAY THIS AMOUNT 555.90

THIS RADIO STATION WARRANTS THAT THE PROGRAM/ANNOUNCEMENTS INDICATED ABOVE WERE BROADCAST IN ACCORDANCE WITH OFFICIAL STATION LOG.
ALL TIMES ARE APPROXIMATE WITHIN 15 MINUTES AND ARE WITHIN THE TIME CLASSIFICATION ORDERED

§73.1205 Fraudulent billing practices.

(a) No licensee of an AM, FM, or television broadcast station shall knowingly issue or knowingly cause to be issued to any local, regional or national advertiser, advertising agency, station representative, manufacturer, distributor, jobber, or any other party, any bill, invoice, affidavit or other document which contains false information concerning the amount actually charged by the licensee for the broadcast advertising for which such bill, invoice, affidavit or other document is issued, or which misrepresents the nature or content of such advertising, or which misrepresents the quantity of advertising actually broadcast (number or length of advertising messages) or which substantially and/or materially misrepresents the time of day at which it was broadcast, or which misrepresents the date on which it was broadcast.

Marketron
21243 Ventura Blvd. Suite 226
Woodland Hills, CA. 91364
(212) 347-6400

STATEMENT OF ACCOUNT

STATEMENT NO 201

TO: ATTN: ACCOUNTS PAYABLE
NW AYER & CO
111 EAST WACKER DR
CHICAGO IL 60601

ADVERTISER: SEARS

ACCOUNT EXECUTIVE: J CRONIN

PREPARED: 8/ 7/78

DATE	TRANSACTION			CHARGE	CREDIT	BALANCE
3/26	CONTR#	200	INV# 30004 INV	768.00		
7/28			INV# 00000 O/A		720.00	
8/ 7			INV# 00000 RET CK	720.00		
4/30	CONTR#	300	INV# 40080 INV	564.00.		
7/19			INV# 40080 PAYM		400.00	
7/30			INV# 77080 ADJ		164.00	
5/28			INV# 50106 INV	783.75		
6/25			INV# 60132 INV	685.30		
7/30			INV# 70028 INV	423.60		
7/30	CONTR#	400	INV# 70022 INV	1,547.00		

CURRENT	30 DAYS	60 DAYS	90 DAYS	120 + DAYS	BALANCE
1,970.60	685.30	783.75		768.00	4,207.65

This is supplied as an aid in reconciling your records to ours. If not in agreement, call our Business Office. If your records do agree, PAYMENT IS DUE for the amounts outstanding.

Marketron

FIGURE 8.14
Statement of accounts payable. Maintaining accurate records is the responsibility of the station's billing person. Courtesy Marketron.

(b) Where a licensee and any program supplier have entered into a contract or other agreement obligating the licensee to supply any document providing specified information concerning the broadcast of the program or program matter supplied, including noncommercial matter, the licensee shall not knowingly issue such a document containing information required by the contract or agreement that is false.

(c) A licensee shall be deemed to have violated this section if it fails to exercise reasonable diligence to see that its agents and employees do not issue documents containing the false information specified in paragraphs (a) and (b) of this section.

Note: Commission interpretations of the rule may be found in a separate Public Notice issued June 10, 1976, entitled "Applicability of the Fraudulent Billing Rule," FCC 76–489 FR Doc. 76–17014, page 23673 (41 FR 23673).

[41 FR 23677, June 11, 1976]

CHAPTER HIGHLIGHTS

1. Each commercial slot on a station is called an availability. Availabilities constitute a station's salable inventory.

2. The traffic manager (or traffic director) controls and is accountable for the broadcast time inventory.

3. The traffic manager prepares a log to inform the deejays of what to broadcast and at what time.

4. The traffic manager is also responsible for ensuring that an ad order is logged as specified, that a record of when each client's spots are aired is maintained, and that copy and production tapes are in on time.

5. Programming relies on the traffic manager for the logs that function as scheduling guides for on-air personnel; the sales department depends upon the traffic manager to inform them of existing availabilities and to process orders onto the air.

6. Although most traffic people are trained in-house and are drawn from the administrative or clerical ranks, they must possess patience, an eye for detail, the ability to work under pressure, and keyboarding skills.

7. Many traffic departments have been computerized to enhance speed and efficiency. Therefore, traffic managers must be computer-friendly.

8. Based upon the spots aired, as recorded and verified by the traffic department, the billing department sends invoices weekly or biweekly to each client. Invoices are notarized for clients with co-op contracts.

9. Since the FCC eliminated program log requirements in the early 1980s, stations have been able to design logs that inform programming personnel of what is scheduled for broadcast and that provide necessary information for the traffic and billing departments.

10. FCC regulations pertaining to station billing practices are contained in Section 73.1205.

SUGGESTED FURTHER READING

Diamond, Susan Z. *Records Management: A Practical Guide*. New York: AMACOM, 1983.

Doyle, Dennis M. *Efficient Accounting and Record Keeping*. New York: David McKay and Company, 1977.

Heighton, Elizabeth J., and Cunningham, Don R. *Advertising in the Broadcast and Cable Media*, 2nd ed. Belmont, Calif.: Wadsworth Publishing, 1984.

Murphy, Jonne. *Handbook of Radio Advertising*. Radnor, Pa.: Chilton, 1980.

Slater, Jeffrey. *Simplifying Accounting Language*. Dubuque, Iowa: Kendall-Hall Publishing, 1975.

Zeigler, Sherilyn K., and Howard, Herbert H. *Broadcast Advertising: A Comprehensive Working Textbook*, 2nd ed. Columbus, Ohio: Grid Publishing, 1984.

9 Production

A SPOT RETROSPECTIVE

A typical radio station will produce thousands of commercials, public service announcements, and promos annually. Initially, commercials were broadcast live, due to a lack of recording technology. In the 1920s, most paid announcements consisted of lengthy speeches on the virtues of a particular product or service. Perhaps the most representative of the commercials of the period was the very first ever to be broadcast, which lasted over ten minutes and was announced by a representative of a Queens, New York, real estate firm. Aired live over WEAF in 1922, by today's standards the message would sound more like a classroom lecture than a broadcast advertisement. Certainly no snappy jingle or ear-catching sound effects accompanied the episodic announcement.

Most commercial messages resembled the first until 1926. On Christmas Eve of that year the radio jingle was introduced, when four singers gathered for a musical tribute to Wheaties cereal. It was not for several years, however, that singing commercials were commonplace. For the most part, commercial production during the medium's first decade was relatively mundane. The reason was twofold. The government had resisted the idea of blatant or direct commercialism from the start, which fostered a low-key approach to advertising, and the medium was just in the process of evolving and therefore lacked the technical and creative wherewithal to present a more sophisticated spot.

Things changed by 1930, however. The austere, no-frills pitch, occasionally accompanied by a piano but more often done a cappella, was gradually replaced by the dialogue spot that used drama or comedy to sell its product. A great deal of imagination and creativity went into the writing and production of commercials, which were presented live throughout the 1930s. The production demands of some commercials equaled and even exceeded those of the programs they interrupted. Orchestras, actors, and lavishly constructed sound effects commonly were required to sell a chocolate-flavored syrup or a muscle liniment. By the late 1930s, certain commercials had become as famous as the favorite programs of the day. Commercials had achieved the status of pop art.

Still, the early radio station production room was primitive by today's standards. Sound effects were mostly improvised show by show, commercial by commercial, in some cases using the actual objects sounds were identified with. Glass was shattered, guns fired, and furniture overturned as the studio's on-air light flashed. Before World War II, few sound effects were available on records. It was just as rare for a station to broadcast prerecorded commercials, although 78 rpm's and wire recordings were used by certain major advertisers. The creation of vinyl discs in the 1940s inspired more widespread use of electrical transcriptions for radio advertising purposes. Today sound effects are taken from specially designed LPs.

FIGURE 9.1
Pioneer announcer H.W. Arlin in the early 1920s. Courtesy Westinghouse Electric.

FIGURE 9.2
KDKA's pre–World War II control center. Courtesy Westinghouse Electric.

The live spot was the mainstay at most stations into the 1950s, when two innovations brought about a greater reliance on the prerecorded message. Magnetic recording tape and 33⅓ LPs revolutionized radio production methods. Recording tape brought about the greatest transformation and, ironically, was the product of Nazi scientists who developed acetate recorders and tape for espionage purposes.

The adoption of magnetic tape by radio stations was costlier and thus occurred at a slower pace than 33⅓ rpm's, which essentially required a turntable modification.

Throughout the 1950s advertising agencies grew to rely on LPs. By 1960, magnetic tape recorders were a familiar piece of studio equipment. More and more commercials were prerecorded. Some stations, especially those automated, did away with live announcements entirely, preferring to tape everything to avoid on-air mistakes.

Commercials themselves became more sophisticated sounding since practically anything could be accomplished on tape. Perhaps no individual in the 1960s more effectively demonstrated the unique nature of radio as an advertising medium than did Stan Freberg. Through skillful writing and the clever use of sound effects, Freberg transformed Lake Michigan into a basin of hot chocolate crowned by a seven-hundred-foot-high mountain of whipped cream, and no one doubted the feat.

Today the sounds of millions of skillfully prepared commercials trek through the ether and into the minds of practically every man, woman, and child in America. Good writing and production are what make the medium so successful.

FORMATTED SPOTS

In the 1950s the medium took to formatting in order to survive and prosper. Today listeners are offered a myriad of sounds from which to choose. There is something for practically every taste. Stations concentrate their efforts on delivering a specific format, which may be defined as Adult Contemporary, Country, Easy Listening, or any one of a dozen others. As you will recall from the discussion in chapter 3, each format has its own distinctive sound, which is accomplished through a careful selection and arrangement of compatible program elements. To this end, commercials also must reflect a station's format. What follows is a reference listing containing the observations of several radio producers concerning the effect that five key formats have on the production of commercials, from the copy stage through final mixdown.*

*Reprinted, with permission, from Michael C. Keith, *Production in Format Radio Handbook* (Washington, D.C.: University Press of America, 1984).

ADULT CONTEMPORARY

Copy

"As an Adult Contemporary station we make every effort to relate to the 25 to 49 year old in our copy."—John King, program director, WING-AM, Dayton, Ohio.

"The background and lifestyle of the station's audience must always be kept in mind when preparing copy. If the spot doesn't relate, the listener won't connect. Then you don't sell anything. Irrelevant copy may even result in audience tuneout."—Peter Fenstermacher, production director, WMJX-FM, Boston.

"Talk to people. Use short sentences, words with three or less syllables, and use common speech patterns. Remain on the level of your audience. Keep in mind who the A/C format is designed to attract and write from that perspective."—Bill Towery, production director, WFYR-FM, Chicago.

"The key word is 'adult.' We're going after the twenty-five plus crowd, so we must talk 'adult.' If I'm reading a spot and I come across a word or phrase that strikes me as wrong for our intended audience, I'll inform the person who wrote it and ask that it be changed. Just some good common sense can keep a spot adult. Write creatively, but stay cognizant of your audience, the station's format, and the nature of the account."—Paul Douglas, program director, WCIT-AM, Lima, Ohio.

Delivery

"Announcer delivery on A/C stations is not too punched-up or subdued. No screaming announcers on spots. Nothing too exaggerated. Not hyped, but not sleepy either. Somewhere in between is where you want to be in this format."—Jim Murphy, program director, WHDH-AM, Boston.

"The best way to reach the A/C listener in a spot is by being yourself, natural. The announcer should strive for realism. A one-on-one delivery does the job here. It fits the overall flow and pacing of the programming. Warmth and friendliness are important qualities in A/C spots."—Mike Scalzi, program director, WHBQ-AM, Memphis.

"The A/C format is actually less restrictive than some other formats. It's much looser in many respects. Sure, you can't scream a spot, but a lot of latitude exists in the way a spot can be read. In general, the easy, relaxed, unpretentious announcer style is what A/C stations are after."—Bill Towery, WFYR-FM.

Mixdown

"The production elements in an A/C spot should be congruous with the overall sound. For example, bed music should not deviate greatly from what the listener is accustomed to hearing on the station. For instance, A/C stations would not use music by "AC/DC" behind a commercial. Using hard rock as bed music would be like breaking format, and the result would be audience dissatisfaction and probable tune-out. You can't be too abrasive."—Mike Nutzger, program director, WGAR-AM, Cleveland.

"For the most part we stick with beds that are bright, pleasant, and adult, nothing with a real pronounced beat."—Jim Murphy, WHDH-AM.

"Moderation and balance are particularly important in Adult Contemporary. You've got to avoid overdoing production. Spots should flow. Too many sound effects and beds can clutter the sound and be distracting. This is a medium-intensity format. Moderation is the key."—Mike Scalzi, WHBQ-AM.

CONTEMPORARY HIT

Copy

"In this format the copywriter must be aware of current jargon and make an attempt to incorporate it in copy, if possible. Of course, it shouldn't be overdone. A copywriter can use colloquialisms to good advantage. You have to keep the teen audience in mind when creating copy. The key is speaking the language of the young without sounding pretentious."—Jim Cook, program director, WJET-AM, Erie, Pennsylvania.

"Colloquialisms can be effective when used properly, but be careful. Used incorrectly they can make you sound awfully foolish to the hip conscious teenager."—Michael Lowe, program director, WCKS-FM, Cocoa Beach, Florida.

"Keep the language simple and direct. Remember that you're dealing with young people, mostly of school age. Using words with ten syllables and elaborate phrasing will be lost."—Skip Hansen, program director, KZOZ-AM, Arroyo Grande, California.

Delivery

"This is a high-intensity format. You've got to convey energy and enthusiasm, although it is not necessary to scream at the listener."—Dave Taylor, program director, KAAY-AM, Little Rock, Arkansas.

"A wide variety of announcer deliveries work in hit radio. Voice schlock is not necessary, though. High energy and honesty work."—Jim Cook, WJET-AM.

Mixdown

"Hit radio is very production oriented. Spots can be pretty elaborate. In this format, use whatever hooks the listener. What it comes down to is selecting compatible Top 40 production values and mixing them with imagination and skill."—Mark Adams, program editor, WKLR-FM, Toledo, Ohio.

"When it comes to choosing music beds you have to use contemporary stuff. Hokey, dated music detracts from the total air sound. Commercials have to be in synch with the station's program sound or they cause product damage."—Jim Cook, WJET-AM.

"Sound effects are a popular element of Contemporary Hit commercials. You can really capture the attention of the listener by using a special effects sound from a popular movie or video game. Sound effects can be excellent grabbers and can really heighten the effectiveness of a spot."—Michael Lowe, WCKS-FM.

"Contemporary Hit requires production that is sharp, concise, and above all 'tight.' This format demands a good working knowledge of mixing and editing."—Dave Taylor, KAAY-AM.

EASY LISTENING

Copy

"You've got to work with the flow of words in EL. No interjections and harsh repetition. Lead the listener gently from point to point. Use more vowel sounds, with only limited use of consonants."—Enzo DeDominicis, vice president, WRCH/WRCQ, Hartford, Connecticut.

"The right words keep copy fluid. The wrong ones make it sluggish and less listenable, and 'listenability' is what this format is all about."—Dick Ellis, programmer, Peters Productions, San Diego.

"Write with intelligence and marry the copy to the EL sound. Avoid ultra hip and trendy jargon and stay away from cuteness. Maturity is the buzzword."—Phyllis Moore, general manager, WEZI-FM, Memphis.

Delivery

"Avoid sounding stilted. The one-on-one approach works better than the stiff shirt, even in a conservative format."—Dick Ellis, Peters Productions.

"Warmth and friendliness with authority and credibility. Delivery must fit the mood of the format. In EL the mood is set, so working with it is the best way to keep the listener with you."—Enzo DeDominicis, WRCH/WRCQ.

Mixdown

"No pronounced beats in beds. Match the flow. Light jazz is okay, as in Adult Contemporary, and even semi-classical. We keep a complete library of music appropriate for spot production within our format."—Phyllis Moore, WEZI-FM.

"EL is not big on sound effects. Some stations in this format don't even bed spots. Production is pretty direct and simple in keeping with the mood and tenor of the programming."—Dick Ellis, Peters Productions.

"Busy spots, those with several production elements, are fairly uncommon. Two-voicers are rare, too."—Enzo DeDominicis, WRCH/WRCQ.

COUNTRY

Copy

"Country stations pretty much let the nature of the account determine the writing style. We have that luxury over many other formats. Of course, we make certain that copy is keyed to our audience."—Dave Ross, production director, WGNG-AM, Providence, Rhode Island.

"Write intelligently. Don't insult anyone. The Country audience doesn't exclusively consist of 'good ol' boys' in pickup trucks. A lot of hip business people and professionals enjoy Country radio."—Steve Brelsford, program director, WQHK-AM, Ft. Wayne, Indiana.

"Here, whatever it takes to get the job done, we do. Hard-sell, soft-sell, you name it. We just want to give the client a good piece of copy. Creativity is the key objective. Humor can be very effective in Country."—Gary Tolman, program director, KAST-AM, Astoria, Oregon.

Delivery

"Delivery can be pretty varied in Country. If the copy calls for hard-sell, you crank it up and vice-versa."—Dave Ross, WGNG-AM.

"Much of Country is up-tempo, so delivery has to conform to the velocity of the music to achieve flow."—Roger Heinrick, program director, KCLS-AM, Flagstaff, Arizona.

"We don't talk 'country' or 'twang' here. Southern accents just aren't necessary. Announcer delivery is kept pretty free of regionalisms."—J.H. Johnson, program director, KRNR-AM, Roseburg, Oregon.

Mixdown

"Country music beds are not used in much abundance. We use a host of sounds, including pop, light rock, and others. Here we avoid the 'rural' sound in our spot beds. It breaks the sameness and keeps things fresh."—Jarrett Day, KSO-AM, Des Moines, Iowa.

"The Country format doesn't really impose many strictures on the use of effects. Use what sells the spot. Discretion should be employed, however."—Steve Brelsford, WQHK-AM.

"This is a fairly production heavy format. Country is comparable to hit stations in terms of the level of mixdown elements included in spots."—Dave Ross, WGNG-AM.

ALBUM-ORIENTED ROCK

Copy

"One-on-one writing is important so that the listener feels the message is just for him or her. Stiff or stuffy writing doesn't work in AOR."—Mel Myers, program director, KSNE-FM, Tulsa, Oklahoma.

"The perceptions of the AOR listener must be echoed in the choice of wording used. As a copywriter, you should know how your audience thinks and speaks. You write for lifestyle in this format."—Don Davis, program director, WWDC-FM, Washington, D.C.

Delivery

"The one-on-one style fits well. No real hyped or burn-out deliveries are acceptable."—Dennis Constantine, program director, KBCO-FM, Boulder, Colorado.

"AOR's on-air approach is less aggressive than most contemporary formats, and this same relaxed approach is necessary. Delivery should be natural, hip, and conversational, rather than contrived or forced."—Don Davis, WWDC-FM.

"Despite the fact that album rockers may be very heavy at times, the punched-up, super-slam spot is avoided here. It doesn't set well. Hard music is one thing, hard spots are another."—Mel Myers, KSNE-FM.

Mixdown

"Beds need to be generic and up-tempo. We find fusion and Jazz work best. Familiar music is avoided, except on concert spots. Edits using uptempo riffs, drumbeats, etc., are effective."—Dennis Constantine, KBCO-FM.

"Because of the attention paid to audio quality by AOR listeners, production values in commercials are important, too. The mixdown must meet the highest standards. Scratchy records, second or third generation dubbing, and improper levels defeat the purpose of production."—Don Davis, WWDC-FM.

THE PRODUCTION ROOM

Generally speaking, metro market stations employ a full-time production person (known variously as production director, production manager, or production chief). This individual's primary duties are to record voice-tracks and mix commercials and PSAs (public service announcements). Other duties involve the maintenance of the bed and sound effects library and the mixdown of promotional material and special programs, such as public affairs features, interviews, and documentaries.

Stations that do not have a slot for a full-time production person divide work among the on-air staff. In this case, the program director often oversees production responsibilities, or a deejay may be assigned several hours of production duties each day and be called the production director.

At most medium and small outlets on-air personnel take part in the production process. Production may include the simple dubbing of an agency spot onto a cartridge, a mixdown that requires a single bed (background music) under a thirty-second voicer, or a multi-element mixdown of a sixty-second two-voicer with sound effects and several bed transitions. Station production can run from the mundane and tedious (the dubbing of fifteen sixty-second Preparation H spots onto several different cartridges) to the exciting and challenging (a commercial without words conveyed through a confluence of sounds).

FIGURE 9.3 WMJX-FM Production Director Peter Fenstermacher voices a spot. Later he will mix the voice with other production elements, such as bed music and sound effects.

Most production directors are recruited from the on-air ranks, having acquired the necessary studio dexterity and know-how to meet the demands of the position. In addition to the broad range of mixdown skills required by the job, a solid knowledge of editing is essential. The production director routinely is called upon to make rudimentary splices or perform more complex editing chores, such as the rearrangement of elements in a sixty-second concert promo. Editing is covered in more detail later in this chapter.

The production director works closely with many people, but perhaps most closely with the program director. It is expected that the person responsible for production have a complete understanding of the station's programming philosophy and objective. This is necessary since commercials constitute an element of programming and therefore must fit in. A production person must be able to determine when an incoming commercial clashes with the station's image. When a question exists as to the spot's appropriateness, the program director will be called upon to make the final judgment, since it is he who is ultimately responsible for what gets on the air. In the final analysis, station production is a product of programming. In most broadcast organizations, the production director answers to the program director. It is a logical arrangement given the relationship of the two areas.

The production director also works closely with the station copywriter. Their combined efforts make or break a commercial. The copywriter conceives of the concept, and the producer brings it to fruition. The traffic department also is in close and constant contact with production, since it is one of its primary responsibilities to see that copy gets processed and placed in the on-air studio where it is scheduled for broadcast.

FIGURE 9.4
WHJJ/WHJY production studios. Although each studio is unique unto itself, the basics of layout are fairly consistent from station to station. For the sake of accessibility most studios are developed in a U shape, or a variation thereof.

FIGURE 9.4
continued

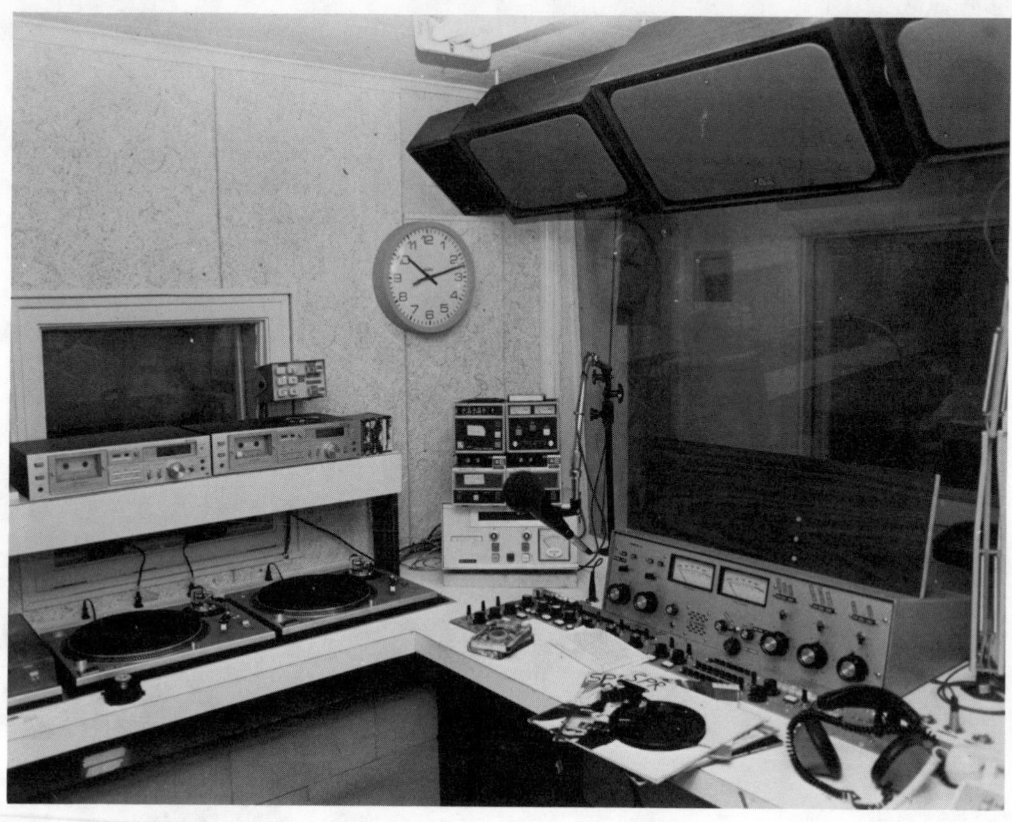

THE STUDIOS

A radio station has two kinds of studios: on-air and production. Both share basic design features and have comparable equipment. For ease of movement and accessibility, audio equipment commonly is set-up in a U shape within which the operator or producer is seated.

The standard equipment found in radio studios includes an audio console (commonly referred to as the "board"), reel-to-reel tape machines, cartridge (cart) machines, cassette decks, turntables, compact disk players (becoming a standard piece of equipment), and a patch panel.

Audio Console

The audio console is the centerpiece, the very heart of the radio station. Dozens of manufacturers produce audio consoles, and although design characteristics vary, the basic components remain relatively constant. Consoles come in all different sizes and shapes and may be monaural, stereo, or multitrack, but all contain *inputs* that permit audio energy to enter the console, *outputs* through which audio energy is fed to other locations, *VU meters* that measure the amount or level of sound, *pots (faders)* that control gain or the quantity of sound, *monitor gains* that control in-studio volume, and *master gains* for the purpose of controlling general output levels.

Since the late 1960s, the manufacture of consoles equipped with vertical faders has surpassed those with rotary faders. While vertical faders perform the same function as the more traditional pots, they are easier to read and handle.

Cue Mode. A low-power amplifier is built into the console so that the operator may hear audio from various sources without it actually being distributed to other points. The purpose of this is to facilitate the setup of certain sound elements, such as records and tapes, for eventual introduction into the mixdown sequence.

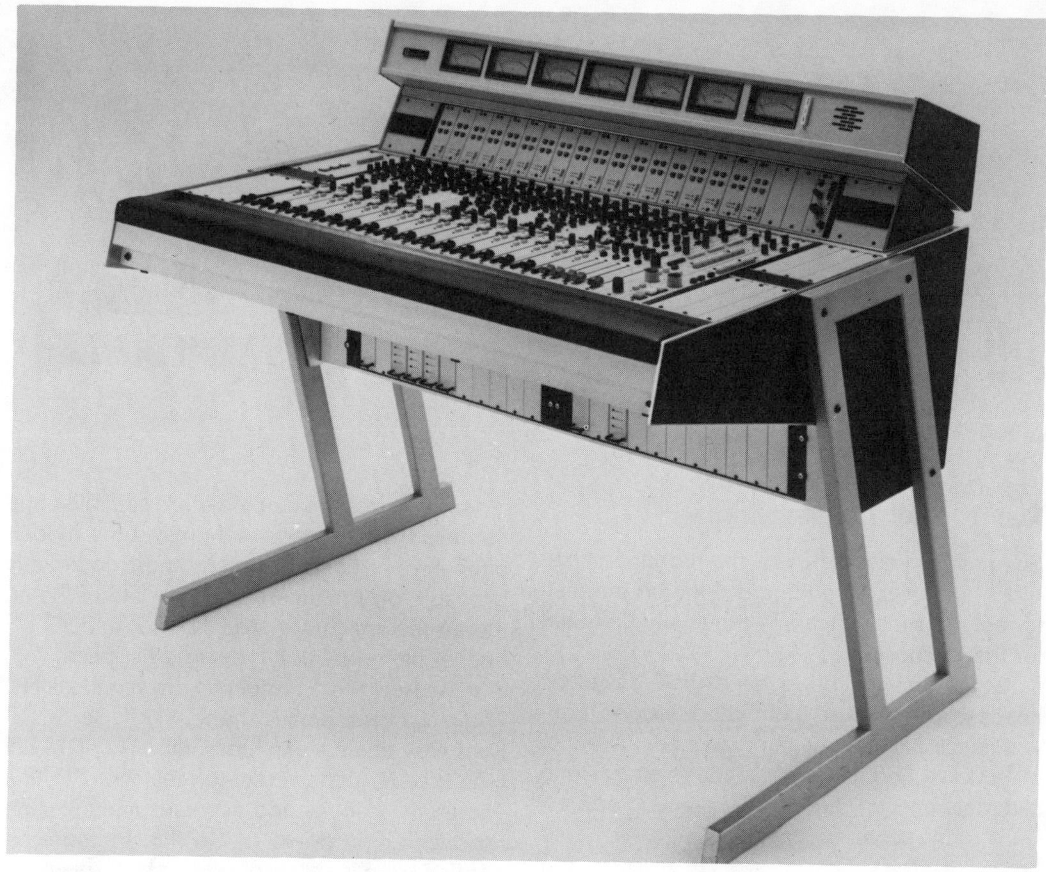

FIGURE 9.5
Multitrack audio console. Courtesy Auditronics.

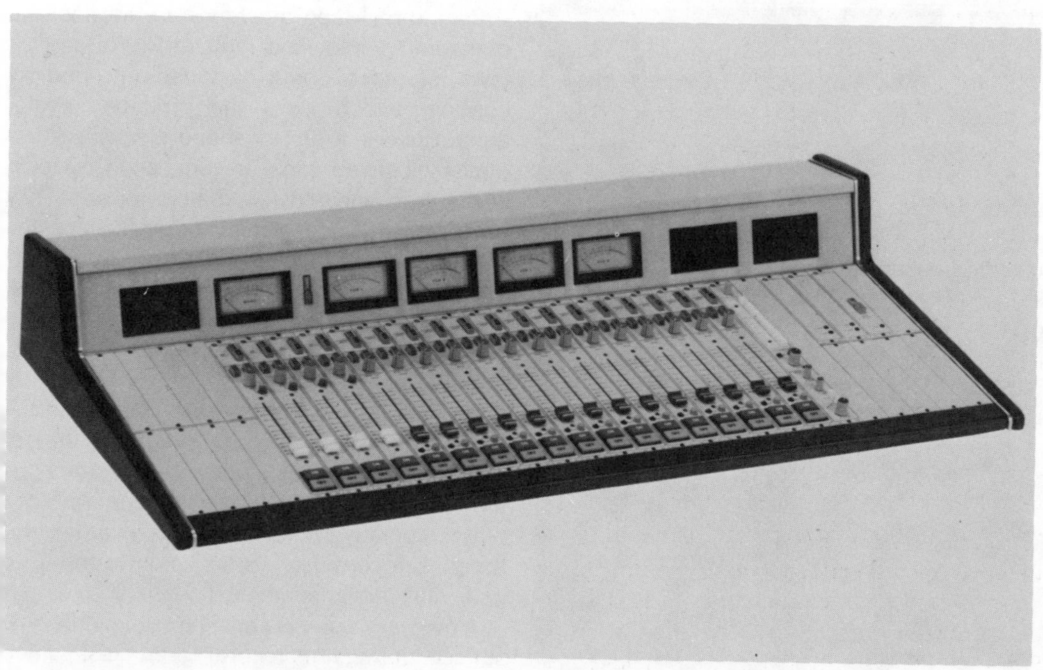

FIGURE 9.6
Audio console with vertical faders, popularly referred to as a "slide" board. Courtesy Auditronics.

FIGURE 9.7
Eight-mixer ("pot") monaural console. Courtesy Broadcast Electronics.

Reel-to-Reel Tape Machines

The reel-to-reel machine is the production studio workhorse. All editing is done on the reel-to-reel machine, since it is especially designed for that purpose.

Reel-to-reel machines have three magnetic heads whose purpose it is to record sound (convert electrical energy into magnetic energy), playback sound (convert magnetic energy into electrical energy), and erase magnetic impressions from recording tape.

FIGURE 9.8
Cue speaker within console permits producer to set up elements for audio processing. Courtesy LPB.

Most reel-to-reel machines are capable of recording at two speeds, although some models offer three. The tape speeds most commonly available on broadcast quality, state-of-the-art reel-to-reel machines are 3¾, 7½, and 15 IPS (inches per second). It is the middle speed, 7½, that is used most frequently by broadcasters. The reason is a practical one. While some stations may possess machines with 3¾ or 15 IPS tape speeds, all have reel-to-reels with 7½ IPS. Therefore, stations and agencies making dubs for distribution do so at 7½ IPS. In-house recording often is accomplished at high speed (15 IPS) because the sound quality is better and editing is easier.

Reel-to-reel tape machines are available in monaural, stereo, and multitrack. The last allows for overdubbing or sound-on-sound recording, which gives the producer greater control over the mix of sound elements. Multitrack recorders come in four, eight, sixteen, twenty-four, and thirty-six channel formats. They have become increasingly popular since the 1970s, although they can be extremely costly—prohibitively so for many small market stations.

Cart Machines

Cartridge tape machines came into use in the late 1950s and drastically simplified the recording and playback process. A continuous loop of quarter-inch magnetic tape within a plastic container (cart) is used to record everything from commercials and promos to music and short public affairs features.

A magnetic pulse is impressed against the tape surface during the recording process, which permits the cart to recue on its own. Carts come

FIGURE 9.9 (left) Pot in cue mode. Courtesy Broadcast Electronics.

FIGURE 9.10 (right) Magnetic head configuration—Erase, Record, and Playback.

FIGURE 9.11 Four-track reel-to-reels are the most common multitrack recorders at radio stations. Courtesy WMJX-FM.

in various lengths—10, 40, 70 seconds—depending on need. They consist of a hub (around which tape is loosely wound), guides, and pressure pads, which keep the tape against the heads of the machine. When the cart machine is activated, a pinch roller presses against a capstan to move the tape.

Cassette Tape Machines

Cassette machines are popular because of their ease of handling. The small tape cassettes used can be inserted and removed from the machines without the customary rethreading and rewinding that is required by reel-to-reel machines. Adopted by broadcasters in the mid-1960s, cassette recorders employ ⅛-inch tape that is moved at 1⅞ IPS. Cassette tapes can hold up to three hours of material. While not as integral to the production mixdown as the cart or reel-to-reel machine, cassettes are nonetheless an important and necessary piece of studio equipment. On-air cassette machines are generally used to

FIGURE 9.12 (right and far right)
Standard broadcast cartridges (carts) play a key role in the production process at most stations.

FIGURE 9.13
Platform-type magnetic tape eraser, commonly referred to as a "bulk" eraser.

FIGURE 9.14
Cartridge tape builder. Rather than do this themselves, the majority of stations simply purchase new carts or send existing ones out to be refilled. Courtesy Broadcast Electronics.

FIGURE 9.15 (far right)
Triple-deck cartridge machine. Courtesy Broadcast Electronics.

play music, features, and actualities, as well as for aircheck purposes.

Cassette tape recorders come in stereo and monaural and are available in everything from pocket-sized portables to permanent rack-mounted models.

Audio Tape

Magnetic tape is chemically treated with either polyester or acetate, primarily for the sake of preservation. Polyester-backed tape generally is preferred because of its greater durability. It is less affected by extreme temperatures and humidity. Audio tape is produced in varying thicknesses: .5, 1.0, and 1.5 millimeters (mm). The thinner the tape, the more that can be contained on a reel. For instance, a 7-inch reel of .5 mm holds 2400 feet of tape, whereas the same size reel of 1.5 mm only holds half as much, 1200 feet. The industry standard is 1.5-mm tape because it is stronger and more dependable.

The width of audio tape also varies. The most common tape found in a production studio is ¼ inch in width. This tape is used on standard mono and stereo reel-to-reel and cartridge machines. Multitrack recorders use anywhere from ½- to 2-inch-wide tape, depending on the number of tracks a machine possesses.

Head Cleaning. The oxide particles from the magnetic tape leave a residue on tape heads. Dust and dirt also accumulate on heads. This can result in a deterioration of sound quality. Therefore, it is necessary to clean a tape machine's heads frequently. This is generally accomplished with cotton swabs and a liquid head cleaner. Many stations use isopropyl alcohol because it is effective and inexpensive.

The procedure for cleaning heads is simple. Moisten the cotton swab with the head cleaner and run it across each head. When the swab becomes slightly discolored, use a fresh one. It is best to use a new swab for each head. Light pressure may be applied to the heads to ensure that surface particles are removed.

Compact Disks

Compact disk players (CDs) and disks are an innovation of the 1980s. Their appeal stems from a number of factors, but what makes them particularly attractive to radio broadcasters is the improved sound quality they offer compared to the conventional stylus turntable and vinyl LP. CDs are more likely to be found in a station's control room, but they are slowly being integrated into the production environment. By the mid-1980s, nearly 10 percent of the nation's

FIGURE 9.16 **Five-deck cartridge machine. Courtesy Broadcast Electronics.**

FIGURE 9.17 **Equipment rack containing cassette tape machines, cartridge machines, and computer terminal. Rack mounted for convenience and ease of handling. Courtesy WMJX-FM.**

radio stations had purchased CD equipment. Among the first to do so were classical music stations, which jumped at the opportunity to offer their listeners improved fidelity. Pop music stations have begun to follow their lead.

Because CD players use a laser beam rather than a stylus to decode a disk's surface, much of the extraneous noise, such as hissing, clicking, and popping, has been eliminated. Since a laser light is the only thing to touch a disk, the potential for damage is slight. A special coating is applied to the disks, which further ensures a noise- and trouble-free recording.

Another feature that broadcasters find appealing is the size of the CD. Less than five inches across, CDs are more easily stored and the CD player itself takes up far less room than does a turntable. Compact disk players will become a more integral part of the audio studio inventory when they can be used for recording purposes as well. CDs with recording capability are being developed, and this will have a dramatic impact on the audio studio. "Rerecording on CDs will eliminate the use of cartridges, enabling a station to record all of its music, commercials, promos, and any other 'canned' programs right to disk. No more carts. No more bulky cart racks. Everything will be on disk and will fit in a box sitting next to the board," predicts WMJX's production director, Peter Fenstermacher.

FIGURE 9.18
Dirty heads can create numerous problems, not the least of which is a deterioration in sound quality.

FIGURE 9.19
Compact disks signal a new era in sound reproduction. Courtesy Sony.

FIGURE 9.20
Standard broadcast turntable. Courtesy Broadcast Electronics.

Turntables

No audio studio is complete without turntables. This is especially true of the production room, even though many stations have phased out regular on-air use of turntables in favor of carted music and compact disks. Nonetheless, turntables usually are present in the control room for backup purposes and for special features that use LPs.

Broadcast turntables are designed for cueing purposes. Turntables are covered with felt to allow records to move freely when being cued, and the tone arm is carefully balanced to prevent record and stylus damage during this process.

Two methods of record cueing are used. One involves a technique called "dead-rolling," in which an LP is cued and then activated from the turntable's stop position. The other, known as "slip-cueing," requires that the record be held in place and released as the turntable rotates. Using the former technique, an LP must be backtracked approximately one-eighth of a turn from the start of the audio and a 45 rpm one-fourth of a turn. This permits the turntable to achieve its proper speed before the sound is reached. When a turntable does not achieve its proper speed before engaging the recording, distortion will occur in the form of a "wow."

FIGURE 9.21
In the 1980s record cueing goes on more in the production room than it does in the control room.

Patch Panel

A patch panel consists of rows of inputs and outputs connected to various external sources—studios, equipment, remote locations, network lines, and so forth. Patch panels essentially are routing devices that allow for items not directly wired into an audio console to become a part of a broadcast or production mixdown.

Microphones

Microphones are designed with different pickup patterns. *Omnidirectional* microphones are sensitive to sound from all directions (360 degrees), whereas *bidirectional* microphones pick up sound from two directions (180 degrees). The *unidirectional* microphone draws sound from only one path (90 degrees), and, because of its highly directed field of receptivity, extraneous sounds are not amplified. This feature has made the unidirectional microphone popular in both the control and production studios, where generally one person is at work at a time. Most studio consoles possess two or more microphone inputs so that additional voices can be accommodated when the need arises.

Omnidirectional and bidirectional microphones often are used when more than one voice is involved. For instance, an omnidirectional may be used for the broadcast or recording of a roundtable discussion, and the bidirectional during a one-on-one interview.

Announcers must be aware of a microphone's directional features. Proper positioning in relation to a microphone is important. Being outside the path of a microphone's pickup (off-mike) affects sound quality. At the same time, being too close to a microphone can result in distortion, known as popping and blasting. Keeping a hand's length away from a microphone will usually prevent this from occurring. Windscreens and blast filters may be attached to a microphone to help reduce distortion.

EDITING

Tape editing can range from a simple repair to a complicated rearrangement of sound elements. In either case, the ability to splice tape is essential. Figure 9.26 illustrates splicing steps.

FIGURE 9.22
The station's patch panel extends the reach and capacity of the control and production room consoles.

FIGURE 9.23
The look of microphones in the 1920s. Courtesy Jim Steele.

COPYWRITING

Poet Stephen Vincent Benét, who wrote for radio during its heyday, called the medium the "theater of the mind." Indeed the person who tunes radio gets no visual aids but must manufacture images on his or her own to accompany the words and sounds broadcast. The station employee who prepares written material is called a copywriter. What he or she writes primarily consists of commercials, promos, and PSAs, with the emphasis on the first of the three.

Not all stations employ a full-time copywriter. This is especially true in small markets where economics dictate that the salesperson write for his own account. Deejays also are called upon to pen commercials. At stations with bigger operating budgets, a full-time copywriter often will handle the bulk of the writing chores.

Copywriters must possess a complete understanding of the unique nature of the medium, a familiarity with the audience for which the commercial message is intended, and knowledge of the product being promoted. A station's format will influence the style of writing in a commercial; thus, the copywriter also must be thoroughly acquainted with the station's particular programming approach. Commercials must be compatible with the station's sound. For instance, copy written in Easy Listening usually is more conservative in tone than that written for Contemporary Hit stations, and so on.

WXXX

"Home of the Hits"

(SFX: Bed in)

TJ'S ROCKHOUSE, MARK STREET, DOWNTOWN BOISE, PRESENTS CLEO AND THE GANG ROCKING OUT EVERY FRIDAY AND SATURDAY NIGHT. AT TJ'S THERE'S NEVER A COVER OR MINIMUM, JUST A GOOD TIME. SUNDAY IDAHO'S MONARCHS OF ROCKABILLY, JOBEE LANE, RAISE THE ROOF AT TJ'S. YOU BETTER BE READY TO SHAKE IT, BECAUSE NOBODY STANDS STILL WHEN JOBEE LANE ROCKS. THURSDAY IS HALF PRICE NIGHT, AND LADIES ALWAYS GET THEIR FIRST DRINK FREE AT BOISE'S NUMBER ONE CLUB FOR FUN AND MUSIC. TAKE MAIN TO MARK STREET, AND LOOK FOR THE HOUSE THAT ROCKS, TJ'S ROCKHOUSE.

(SFX: Stinger out)

WYYY

"Soothing Sounds"

ELEGANT DINING IS JUST A SCENIC RIDE AWAY. (SFX: Bed in and under) THE CRITICALLY ACCLAIMED VISCOUNT (VY-COUNT) INN IN CEDAR GLENN OFFERS PATRONS AN EXQUISITE MENU IN A SETTING WITHOUT EQUAL. THE VISCOUNT'S 18TH CENTURY CHARM WILL MAKE YOUR EVENING OUT ONE TO REMEMBER. JAMISON LONGLEY OF THE WISCONSIN REGISTER GIVES THE VISCOUNT A FOUR STAR RATING FOR SERVICE, CUISINE, AND ATMOSPHERE. THE VISCOUNT (SFX: Royal fanfare) WILL SATISFY YOUR ROYAL TASTES. CALL 675-2180 FOR RESERVATIONS. TAKE ROUTE 17 NORTH TO THE VISCOUNT INN, 31 STONY LANE, CEDAR GLENN.

There are some basic rules pertaining to the mechanics of copy preparation that should be observed. First, copy is typed in UPPER CASE and double-spaced for ease of reading. Next, left and right margins are set at one inch. Sound effects are noted in parentheses at that point in the copy where they are to occur. Proper punctuation and grammar are vital, too. A comma in the wrong place can throw off the meaning of an entire sentence. Be mindful, also, that commercials are designed to be heard and not read. Keep sentence structure as uncomplicated as possible. Maintaining a conversational style will make the client's message more accessible.

Timing a piece of copy is relatively simple. There are a couple of methods: one involves counting words, and the other counting lines. Using the first approach, twenty-five words would constitute ten seconds, sixty-five words thirty seconds, and one hundred and twenty-five words one minute. Counting lines is an easier and quicker way of timing copy. This method is based on the assumption that it takes on the average three seconds to read one line

FIGURE 9.24
Production director setting levels for dubbing purposes. The foam windscreen on the microphone helps prevent distortion.

FIGURE 9.25
Production studio rack containing a variety of special equipment—harmonizer, compressor, equalizer, and reverb unit—designed to enhance the sound of commercials, PSAs, and promos. Courtesy WMJX-FM.

of copy from margin to margin. Therefore, nine to ten lines of copy would time out to thirty seconds, and eighteen to twenty lines to one minute. Of course, production elements such as sound effects and beds must be included as part of the count and deducted accordingly. For example, six seconds worth of sound effects in a thirty-second commercial would shorten the amount of actual copy by two lines.

Since everything written in radio is intended to be read aloud, it is important that words with unusual or uncommon pronunciations be given special attention. Phonetic spelling is used to convey the way a word is pronounced. For instance: "DINNER AT THE FO'C'SLE (FOKE-SIL) RESTAURANT IN LAITONE (LAY-TON) SHORES IS A SEA ADVENTURE." Incorrect pronunciation has resulted in more than one cancelled account. The copywriter must make certain that the announcer assigned to voice-track a commercial is fully aware of any particulars in the copy. In other words, when in doubt spell it out.

Excessive numbers and complex directions are to be avoided in radio copy. Numbers, such as an address or telephone number, should be repeated and directions should be as simple as possible. The use of landmarks ("ACROSS FROM CITY HALL . . .") can reduce confusion. Listeners seldom are in a position to write down something at the exact moment they hear it. Copy should communicate, not confuse or frustrate.

Of course, the purpose of any piece of copy is to sell the client's product. Creativity plays an important role. The radio writer has the world of the imagination to work with and is only limited by the boundaries of his own.

ANNOUNCING TIPS

Over forty thousand men and women in this country make their living as radio announcers. In few other professions is the salary range so broad. A beginning announcer may make little more than minimum wage, while a seasoned professional in a major market may earn a salary in the six-figure range.

While announcer salaries can be very good in smaller markets, the financial rewards tend to be far greater at metro market stations, which can afford to pay more. Of course, competition for positions is keener and expectations are higher. "You have to pay your dues in this profession. No one walks out of a classroom and into WNBC. It's usually a long and winding road. It takes time to develop the on-air skills that the big stations want. It's hard work to become really good, but you can make an enormous amount of money, or at least a very comfortable income, when you do," says Mike Morin, WZOU-FM, Boston.

The duties of an announcer vary depending on the size or ranking of a station. In the small station, announcers generally fill news and/or production shifts as well. For example, a mid-day announcer at WXXX, who is on the air from ten A.M. until three P.M., may be held responsible for the four and five P.M. newscasts, plus any production that arises during that same period. Meanwhile, the larger station may require nothing more of its announcers than the taping of voiceovers. Of course, the preparation for an airshift at a major market station can be very time consuming.

An announcer must, above all else, possess the ability to effectively read copy aloud. Among other things, this involves proper enunciation and inflection, which are improved through practice. WFYR's Bill Towery contends that the more a person reads for personal enjoyment or enrichment the easier it is to communicate orally. "I'd advise anyone who aspires to the microphone to read, read, read. The more the better. Announcing is oral interpretation of the printed page. You must first understand what is on the page before you can communicate it aloud. Bottom line here is that if you want to become an announcer, first become a reader."

Having a naturally resonant and pleasant-sounding voice certainly is an advantage. Voice quality still is very important in radio. There is an inclination toward the voice with a deeper register. This is true for female announcers as well as male. However, most voices possess considerable range and with training, practice, and experience even a person with a high-pitched voice can develop an appealing on-air sound. Forcing the voice into a lower register to achieve a deeper sound can result in injury to the vocal chords. "Making the most of what you already have is a lot better than trying to be something you're not. Perfect yourself and be natural," advises WZOU's Morin.

Relaxation is important. The voice simply is at its best when it is not strained. Moreover, announcing is enhanced by proper breathing,

FIGURE 9.26A
Mark tape to be cut against playback head using a light-colored grease pencil.

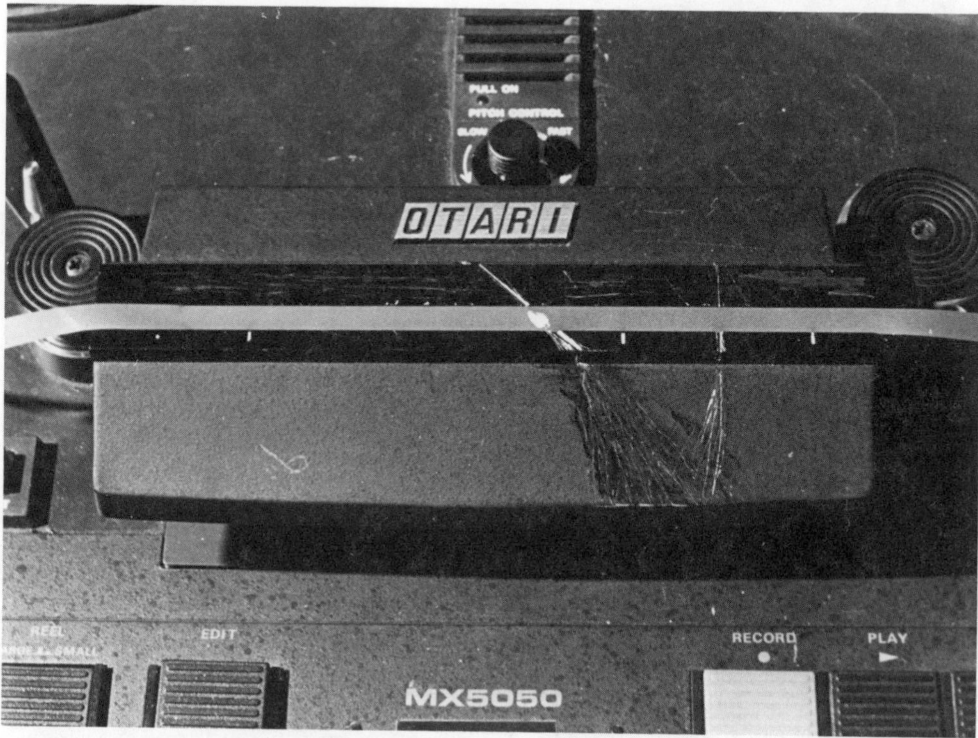

FIGURE 9.26B
Place magnetic tape in tracks of cutting block.

FIGURE 9.26C
Make a diagonal cut with a single-edge razor. Use a sharp razor to avoid uneven cut.

FIGURE 9.26D
Butt ends snugly together. There should be no gap or overlap.

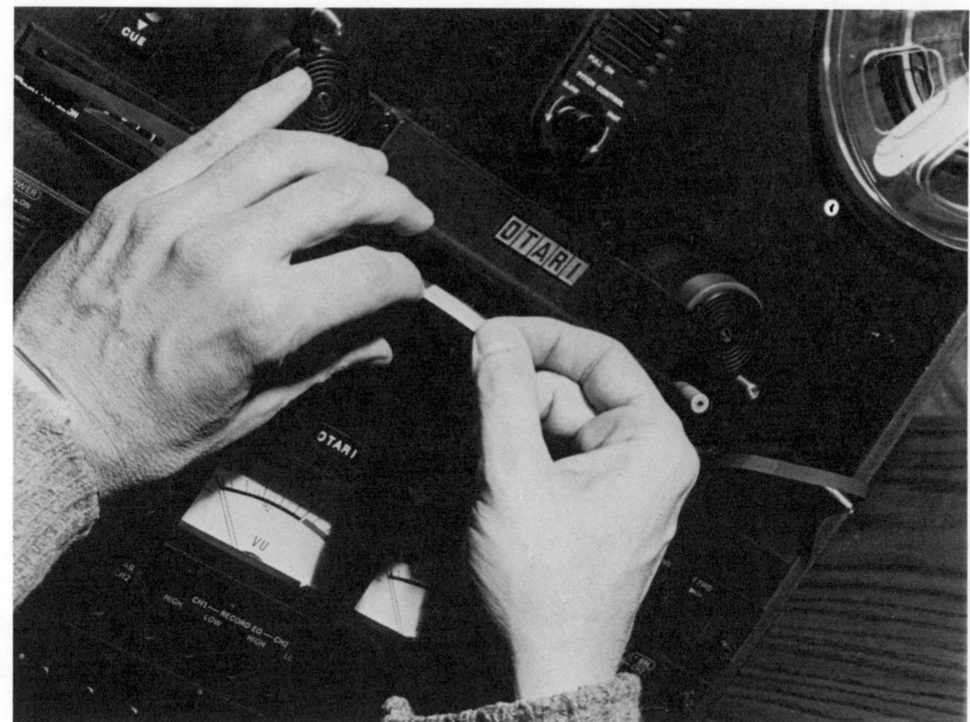

FIGURE 9.26E
Place a 3/4-inch-long piece of splicing tape against the nonrecording surface of the magnetic tape.

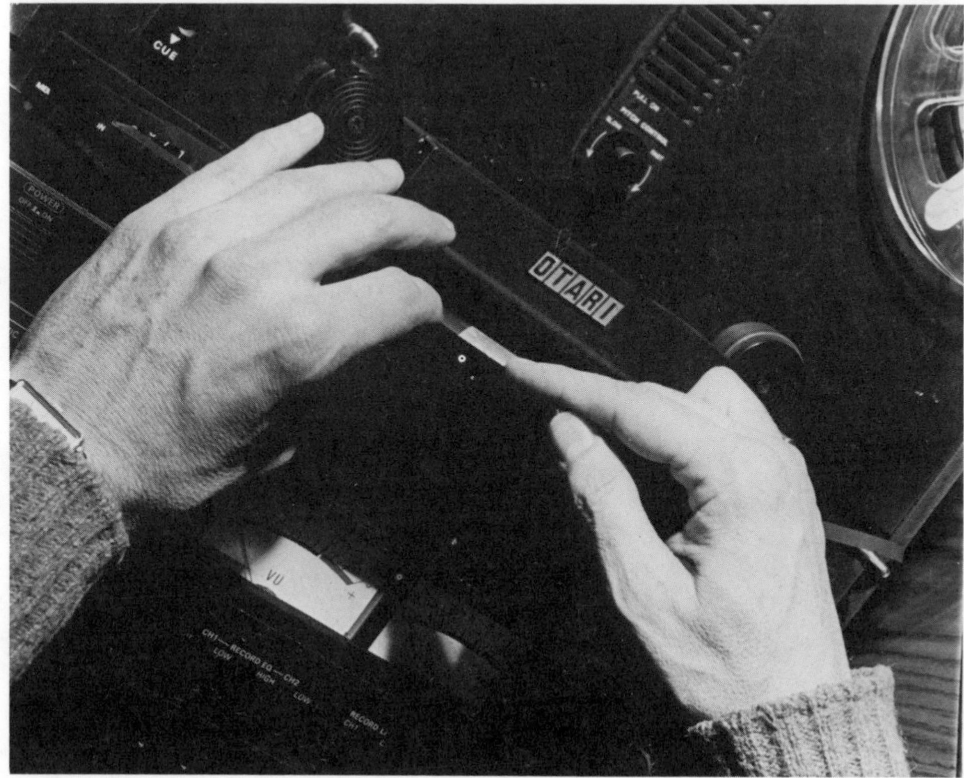

FIGURE 9.26F
Remove the air bubbles trapped between the splicing and magnetic tapes by gently rubbing the surface. Tape must adhere completely to ensure permanent grip.

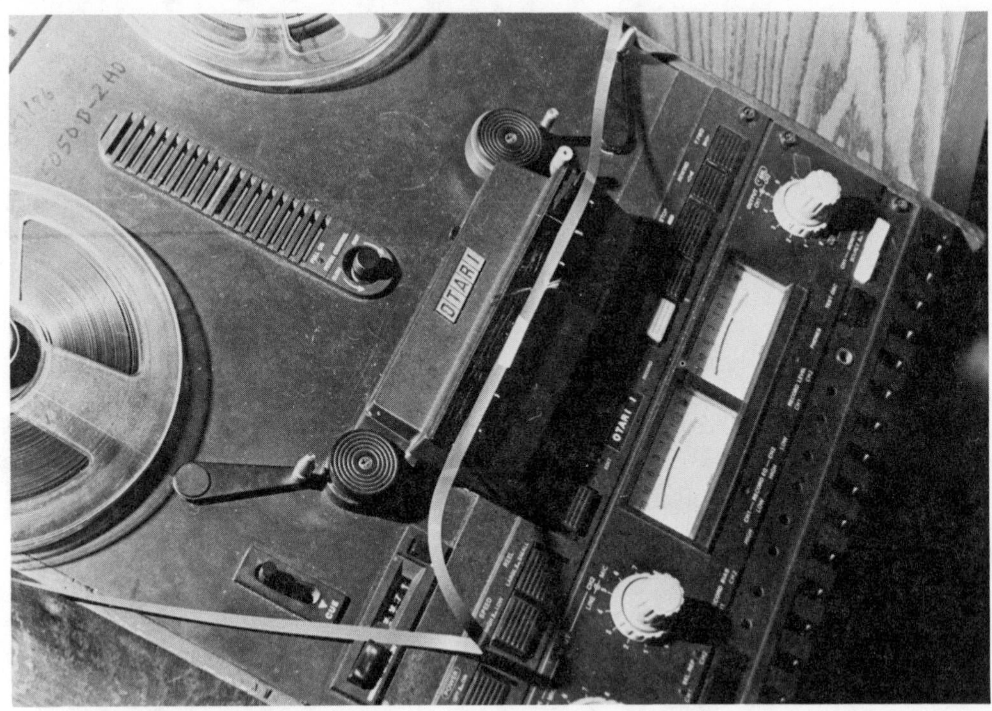

FIGURE 9.26G
Splicing tape must be properly aligned. Protruding splicing tape can adhere to surfaces and result in tape breakage. Adhesive also gums up tape heads.

FIGURE 9.26H
A too-small splice may come apart, and too much decreases tape's suppleness and flexibility. Be sure to carefully remove tape from the tracks of the cutting block. An abrupt tug may cause the tape to break.

WSUB COMMERCIAL COPY ORDER

Client *Rosebud Hardware* Date Submitted *Nov. 14, '86.*
Spec? Yes ☐ No ☐ Date Due *Nov. 19, '86.*
Sold? Yes ☑ No ☐ Date Start *Nov. 21, '86.*
Co-op? Yes ☐ No ☐ Date End *Nov. 29, '86.*
Number of spots to be written *1* Account Executive: *J. Leo*
Length of each spot 60 ☑ 30 ☐ Account Executive gets:
Live? ☐ Recorded? ☑ Tag Req.? ☐ ☑ Copy
Update Required? Yes ☐ No ☐ ☐ Cart.
Add to present rotation? Yes ☐ No ☐ ☐ Tape
Replace all present spots? Yes ☐ No ☐ ☑ Cassette
New Account? Yes ☑ No ☐ ☐ Co-op Copy for Client
Current Acc't-Add. Buy? Yes ☐ No ☐

Details: (Production, comment notes, voice, etc.)

COPY	PRODUCTION REQUIREMENTS
Name of business	Music?
Location	Male or female voice?
Product or service offered	Style
Unique selling proposition	

Rosebud Hardware — 41 East St.
10% to 15% Black + Decker and Stanley Power tools.
Wants sexy delivery. Use Nicki.
Fitting mood music bed.
See copy notes and newspaper Ad attached.

FIGURE 9.27
Station copy order. Courtesy WSUB-AM.

which is only possible when free of stress. Initially, being "on-mike" can be an intimidating experience, resulting in nervousness that can be debilitating. Here are some things that announcers do to achieve a state of relaxation:

1. Read copy aloud before going on the air. Get the feel of it. This will automatically increase confidence, thus aiding in relaxation.

2. Take several deep breaths and slowly exhale while keeping your eyes closed.

3. Sit still for a couple of moments with your arms limp at your sides. Tune-out. Let the dust settle. Conjure pleasant images. Allow yourself to drift a bit, and then slowly return to the job at hand.

4. Stand and slowly move your upper torso in a circular motion for a minute or so. Flex your shoulders and arms. Stretch luxuriously.

5. When seated, check your posture. Do not slump over as you announce. A curved diaphragm impedes breathing. Sit erect, but not stiffly.

6. Hum a few bars of your favorite song. The vibration helps relax the throat muscles and vocal chords.

7. Give yourself ample time to settle in before going on. Dashing into the studio at the last second will jar your focus and shake your composure.

In most situations, an accent—regional or otherwise—is a handicap and should be eliminated. Most radio announcers in the South do not have a drawl, and the majority of announcers in Boston put the *r* in the word *car*. A noticeable or pronounced accent will almost always put the candidate for an announcer's job out of the running. Accents are not easy to eliminate, but with practice they can be overcome.

THE SOUND LIBRARY

Music is used to enhance an advertiser's message—to make it more appealing, more listenable. The music used in a radio commercial is called a "bed" simply because it backs the voice. It is the platform on which the voice is set. A station may bed thousands of commercials over the course of a year. Music is an integral component of the production mixdown.

FIGURE 9.28 Production tape storage area ("tape morgue") contains commercials aired over the past year, or in some cases several years. Courtesy WHJY/WHJJ.

Bed music is derived from a couple of sources. Demonstration albums (demos) sent by recording companies to radio stations are a primary source, since few actually make it onto playlists and into on-air rotations. These albums are particularly useful because the music is unfamiliar to the listening audience. Known tunes generally are avoided in the mixdown of spots because they tend to distract the listener from the copy. However, there are times when familiar tunes are used to back spots. Nightclubs often request that popular music be used in their commercials to convey a certain mood and ambiance.

Movie soundtrack albums are another good place to find beds because they often contain a variety of music, ranging from the bizarre to the conventional. They also are an excellent source for special audio effects, which can be used to great advantage in the right commercial.

On-air albums are screened for potential production use as well. While several cuts on an album may be placed in on-air rotation and thereby eliminated for use in the mixdown of commercials, there will be cuts not programmed and therefore available for production purposes.

Syndicated or canned bed music libraries are available at a price and are widely used at larger stations. *Broadcasting Yearbook* contains a complete listing of production companies offering bed music libraries. The majority of stations choose to lift beds from in-house LPs.

Music used for production purposes is catalogued so that it may be located and reused. A system widely employed uses index cards, which can be stored for easy access in a container or on a rotating drum. At certain stations, computers are used to store production library information. Using the manual system, when an account is assigned a certain bed, a card is made and all pertinent data are included on it. Should a card exist for a bed not in current use and the bed be appropriate for a new account, then either a fresh card will be prepared or the new information penciled onto the existing card.

No production studio is complete without a sound effects library. Unlike bed music, sound effects are seldom derived from in-house anymore. Sound effect libraries can be purchased for as little as one hundred dollars or can cost thousands. The quality and selection of effects vary accordingly. Specially tailored audio effects also can run into the thousands but can add a unique touch to a station's sound.

CHAPTER HIGHLIGHTS

1. The first radio commercial aired on WEAF New York in 1922.

2. Early commercials were live readings: no music, sound effects, or singing.

3. Dialogue spots, using drama and comedy to sell the product, became prominent in the 1930s. Elaborate sounds effects, actors, and orchestras were employed.

4. With the introduction of magnetic recording tape and 33⅓ LPs in the 1950s, live commercial announcements were replaced by prerecorded messages.

:60 Uptempo, piano.
Album: "Liberace's Favorites"
Cut 3, Side 1.
Acct: Dottie's Candies
Date Aired: Mar/Apr. '87

:30 Med. Tempo, Strings.
Album: "Bert Kaempfert's Hits"
Cut 4, Side 2.
Acct: Joe Paul's Paints
Date Aired: Jan. 85 - TFN.

FIGURE 9.29
Index cards contain pertinent information about bed music.

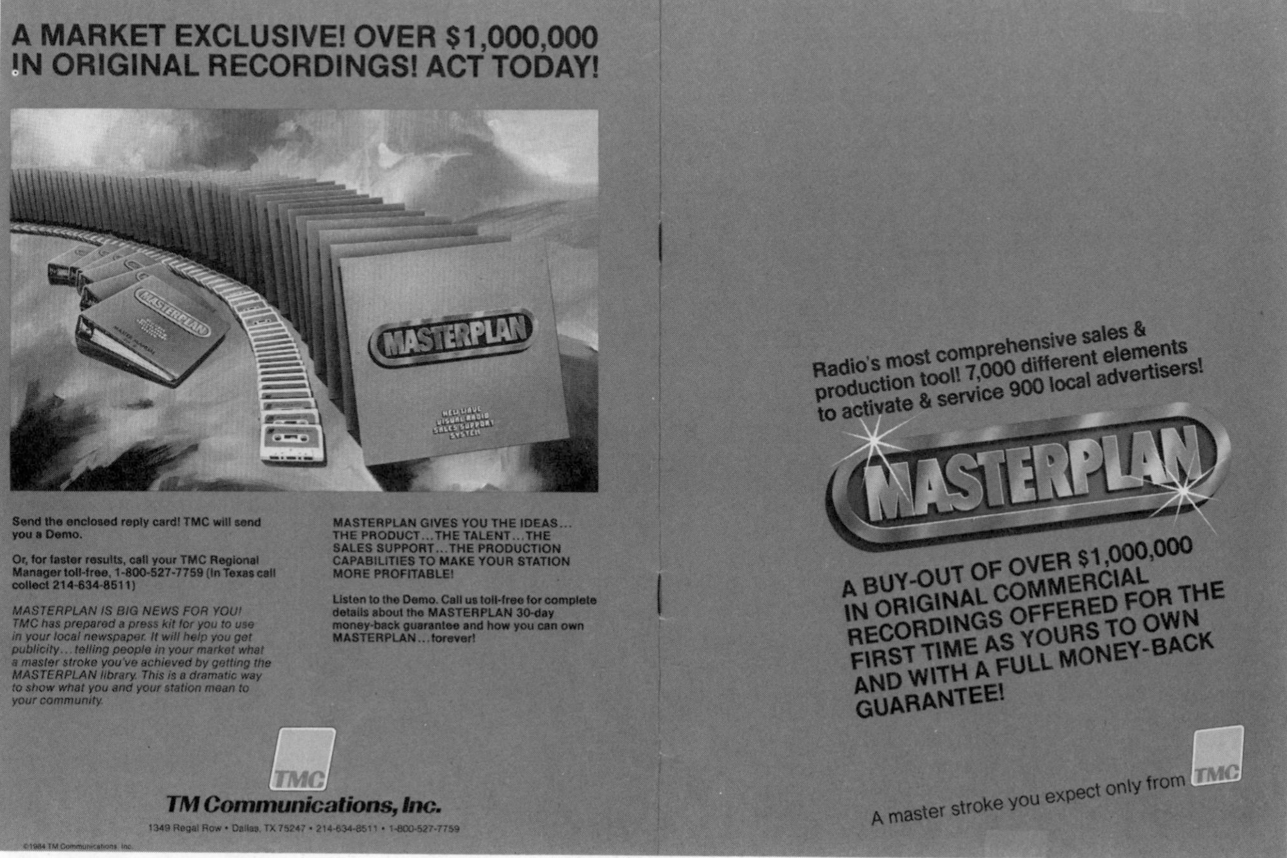

FIGURE 9.30
Elaborately produced sound libraries can be a valuable tool for a radio station. Courtesy TM Communications.

5. The copy, delivery, and mixdown of commercials must be adapted to match the station's format to avoid audience tune-out.

6. The production director (production manager, production chief) records voice-tracks, mixes commercials and PSAs, maintains the bed music and special effects libraries, mixes promotional material and special programs, and performs basic editing chores.

7. At smaller stations the production responsibilities are assigned part-time to on-air personnel or the program director.

8. The production director, who usually answers to the program director, also works closely with the copywriter and the traffic manager.

9. For ease of movement and accessibility, both on-air and production studio equipment are arranged in a U shape.

10. The audio console (board) is the central piece of equipment. It consists of inputs, which permit audio energy to enter the console; outputs through which audio energy is fed to other locations; VU meters, which measure the level of sound; pots (faders), which control the quantity (gain) of sound; monitor gains, which control in-studio volume; and master gains, which control general output levels.

11. When operating the console in "cue mode," the operator can listen to various audio sources without channeling them through an output.

12. Reel-to-reel tape machines are useful for recording at a variety of speeds and are necessary for editing.

13. Cartridge tape machines (cart machines) are used for recording and playback. They employ carts (plastic containers with continuous loops of magnetic tape), which are more compact and convenient than reels of tape.

14. Cassette tape machines, although not as integral to the mixdown process, are easy to handle and use convenient tape cassettes. They are often used for airchecks and actualities.

15. Audio tapes (magnetic tapes) come in a variety of thicknesses and widths. Acetate and polyester backings provide greater durability.

16. Because oxide particles from the magnetic tape, dust, and dirt accumulate on the heads of all types of tape machines, the heads should be cleaned regularly with a cotton swab and liquid head cleaner.

17. Compact disk players use a laser beam to decode the disk's surface, which eliminates stylus and turntable noises, distortion, and record damage.

18. Turntables, once the staple of control rooms, are being replaced by cart machines and compact disks. Turntables allow a record to be "cued," a process that allows the record to reach proper speed before the sound portion is engaged.

19. A patch panel is a routing device, consisting of inputs and outputs, connecting the audio console with various external sources.

20. Microphones are designed with different pickup patterns to accommodate different functions: omnidirectional (all directions), bidirectional (two directions), and unidirectional (one direction).

21. Tape editing ranges from simple repairs to complicated rearrangements of sound elements. The basic process to be mastered is called splicing.

22. The station copywriter, who writes the commercials, promos, and PSAs, must be familiar with the intended audience and the product being sold. The station's format and programming approach influence the style of writing. Copy should be typed in upper case, double-spaced, with one-inch margins. Sound effects are noted in parentheses, and phonetic spellings are provided for difficult words.

23. Aspiring announcers must be able to read copy aloud with proper inflection and enunciation. A naturally resonant and pleasant-sounding voice without a regional accent is an advantage.

24. Every station maintains a sound library for use in spot mixdowns. Commercially produced sound effects and bed-music collections, and unfamiliar cuts from LPs, are common source materials.

SUGGESTED FURTHER READING

Alten, Stanley R. *Audio in Media*. 2nd ed. Belmont, Calif.: Wadsworth Publishing, 1986.

Everest, Alton. *Handbook of Multichannel Recording*. Blue Ridge Summit, Pa.: Tab Books, 1975.

Guidelines for Radio Continuity. Washington, D.C.: NAB Publishing, 1982.

Hilliard, Robert L., ed. *Radio Broadcasting: An Introduction to the Sound Medium,* 3rd ed. New York: Longman, 1985.

Hoffer, Jay. *Radio Production Techniques.* Blue Ridge Summit, Pa.: Tab Books, 1974.

Hyde, Stuart W. *Television and Radio Announcing,* 3rd ed. Boston: Houghton Mifflin, 1979.

Keith, Michael C. *Production in Format Radio Handbook.* Washington, D.C.: University Press of America, 1984.

McLeish, Robert. *Techniques of Radio Production.* Boston: Focal Press, 1979.

Nisbet, Alec. *The Technique of the Sound Studio,* 4th ed. Boston: Focal Press, 1979.

Oringel, Robert S. *Audio Control Handbook,* 4th ed. New York: Hastings House, 1974.

Orlik, Peter B. *Broadcast Copywriting,* 2nd ed. Boston: Allyn and Bacon, 1982.

Runstein, Robert E. *Modern Recording Techniques.* Indianapolis: Howard Sams and Company, 1974.

10 Engineering

PIONEER ENGINEERS

Anyone who has ever spoken into a microphone or sat before a radio receiver owes an immense debt of gratitude to the many technical innovators who made it possible. Guglielmo Marconi, a diminutive Italian with enormous genius, first used electromagnetic (radio) waves to send a message. Marconi made his historical transmission, plus several others, in the last decade of the nineteenth century. Relying, at least in part, on the findings of two earlier scientists, James C. Maxwell and Heinrich Hertz, Marconi developed his wireless telegraph, thus revolutionizing the field of electronic communications.

Other wireless innovators made significant contributions to the refinement of Marconi's device. J. Ambrose Fleming developed the diode tube in 1904, and two years later Lee de Forest created the three-element triode tube called the Audion. Both innovations, along with many others, expanded the capability of the wireless.

On Christmas Eve of 1906, Reginald Fessenden demonstrated the transmission of voice over the wireless from his experimental station at Brant Rock, Massachusetts. Until that time, Marconi's invention had been used to send Morse code or coded messages. An earlier experiment in the transmission of voice via the electromagnetic spectrum also had been conducted. In 1892, on a small farm in Murray, Kentucky, Nathan B. Stubblefield managed to send voice across a field using the induction method of transmission, yet Fessenden's method of mounting sound impulses atop electrical oscillations and transmitting them from an antenna proved far more effective. Fessenden's wireless voice message was received hundreds of miles away.

FIGURE 10.1 Wireless transmitter in 1918. Courtesy Westinghouse Electric.

Few pioneer broadcast technologists contributed as much as Edwin Armstrong. His development of the regenerative and superheterodyne circuits vastly improved receiver efficiency. In the 1920s Armstrong worked at developing a static-free mode of broadcasting, and in 1933 he demonstrated the results of his labor—FM. Armstrong was a man ahead of his time. It would be decades before his innovation would be fully appreciated, and he would not live to witness the tremendous strides it would take.

Had it not been for these men, and many others like them, there would be no radio medium. Today's broadcast engineers and technologists continue in the tradition of their forebears. Without their knowledge and expertise there would be no broadcast industry because there would be no medium. Radio is first an engineer's medium. It is engineers who put the stations on the air and keep them there.

RADIO TECHNOLOGY

Radio broadcasters utilize part of the electromagnetic spectrum to transmit their signals. A natural resource, the electromagnetic spectrum is comprised of radio waves at the low frequency end and cosmic rays at the high frequency end. In the spectrum between may be

FIGURE 10.2
Early radio pioneer Dr. Frank Conrad. Conrad was responsible for putting station KDKA on the air. Courtesy Westinghouse Electric.

found infrared rays, light rays, x-rays, and gamma rays. Broadcasters, of course, use the radio wave portion of the spectrum for their purposes.

Electromagnetic waves carry broadcast transmissions (radio frequency) from station to receiver. It is the function of the transmitter to generate and shape the radio wave to conform to the frequency the station has been assigned by the FCC. Audio current is sent by a line from the control room to the transmitter. The current then modulates the carrier wave so that it may achieve its authorized frequency. A carrier wave that is undisturbed by audio current is called an unmodulated carrier.

The antenna radiates the radio frequency. Receivers are designed to pick up transmissions, convert the carrier into sound waves, and distribute them to the frequency tuned. Thus, in order for a station assigned a frequency of 950 kHz (kilohertz = one thousand hertz [Hz]) to reach a radio tuned to that position on the dial, it must alter its carrier wave 950,000 cycles (Hz) per second. The tuner counts the incoming radio frequency.

AM/FM

Several things distinguish AM from FM. To begin with, they are located at different points in the spectrum. AM stations are assigned frequencies between 535 and 1605 kHz on the Standard Broadcast band. FM stations are located between 88.1 and 107.9 MHz (megahertz = one million hertz) on the FM band.

Ten kilocycles (kc) separate frequencies in AM, whereas two hundred kilocycles is the distance between FM frequencies. FM broadcasters utilize thirty kilocycles for over-the-air transmissions and are permitted to provide subcarrier transmission (SCA) to subscribers on the remaining frequency. The larger channel width provides FM listeners a better opportunity to fine-tune their favorite stations as well as to receive broadcasts in stereo. To achieve parity, AM broadcasters developed a way to transmit

FIGURE 10.3
1940s "Air King" table model AM/FM receiver.

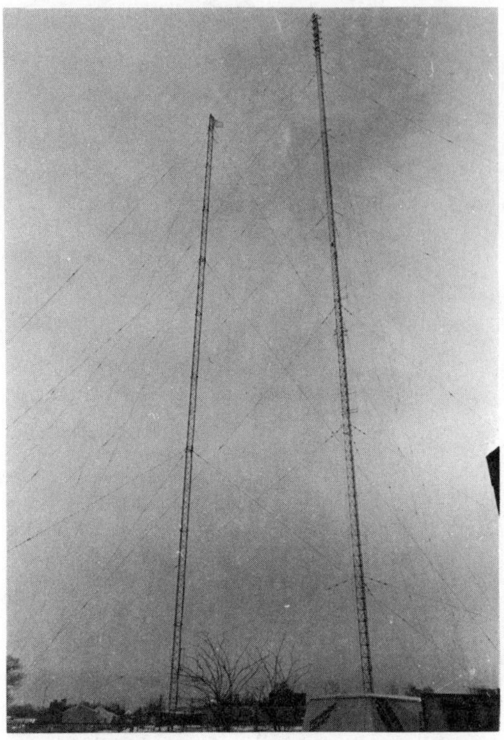

FIGURE 10.4 (left) Antennas (towers) propagate station signals.

FIGURE 10.5 (right) 30-kw FM transmitter. Courtesy Broadcast Electronics.

FIGURE 10.6 Lines leading from station transmitter to antenna base. FCC rules require fencing for safety purposes.

in stereo, and by the mid-1980s hundreds were doing so. The fine-tuning edge still belongs to FM, since its sidebands (15 kc) are three times wider than AM's (5 kc).

FM broadcasts at a much higher frequency (millions of cycles per second) compared to AM (thousands of cycles per second). At such a high frequency, FM is immune to low-frequency emissions, which plague AM. Whereas a car motor or an electric storm generally will interfere with AM reception, FM is static free. Broadcast engineers have attempted to improve the quality of the AM band, but the basic nature of the lower frequency makes AM simply more prone to interference than FM. FM broadcasters see this as a key competitive advantage and refer to AM's move to stereo as "stereo with static."

Signal Propagation

The paths of AM and FM signals differ from one another. *Ground waves* create AM's primary service area as they travel across the earth's surface. High-power AM stations are able to reach listeners hundreds of miles away during the day. At night AM's signal is reflected by the atmosphere (ionosphere), thus creating a *skywave* which carries considerably further, sometimes thousands of miles. Skywaves constitute AM's secondary service area.

In contrast to AM signal radiation, FM propagates its radio waves in a *direct* or *line-of-sight* pattern. FM stations are not affected by evening changes in the atmosphere and generally do not carry as far as AM stations. A high-power FM station may reach listeners within an eighty to one hundred mile radius since its signal weakens as it approaches the horizon. Since FM outlets radiate *direct waves,* antenna height becomes nearly as important as power. Generally speaking, the higher an FM antenna the further the signal travels.

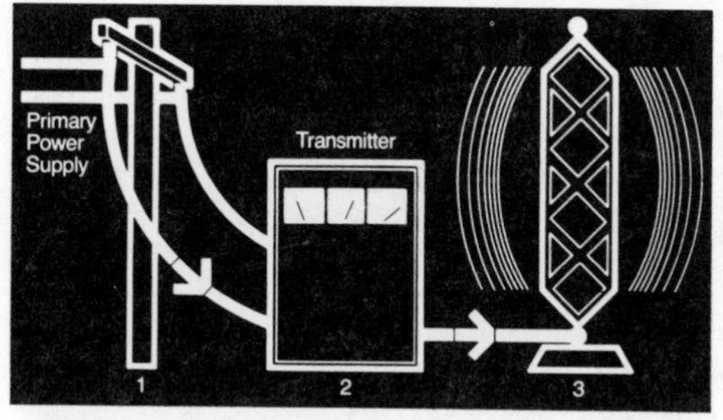

FIGURE 10.7
Stations receive their power from conventional utility companies. From FCC *Broadcast Operator's Handbook,* fig. 3–1.

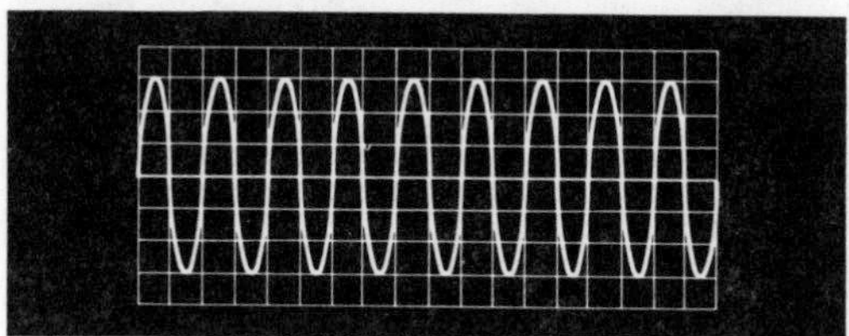

FIGURE 10.8
Unmodulated (undisturbed) carrier. From FCC *Broadcast Operator's Handbook,* fig. 5–1.

Skywave Interference

The fact that AM station signals travel greater distances at night is a mixed blessing. Although some stations benefit from the expanded coverage area created by the skywave phenomenon, many do not. In fact, over two thousand radio stations around the country must cease operation near sunset, while thousands more must make substantial transmission adjustments to prevent interference. For example, many stations must decrease power after sunset to ensure noninterference with others on the same frequency: WXXX-AM is five thousand watts (5 kw) days, but at night it must drop to one thousand watts (1 kw). Another measure designed to prevent interference requires that certain stations direct their signals away from stations on the same frequency. Directional stations require two or more antennas to shape the pattern of their radiation, whereas a nondirectional station that distributes its signal evenly in all directions only needs a single antenna. Because of its limited direct wave signal, FM is not subject to the postsunset operating constraints that affect most AM outlets.

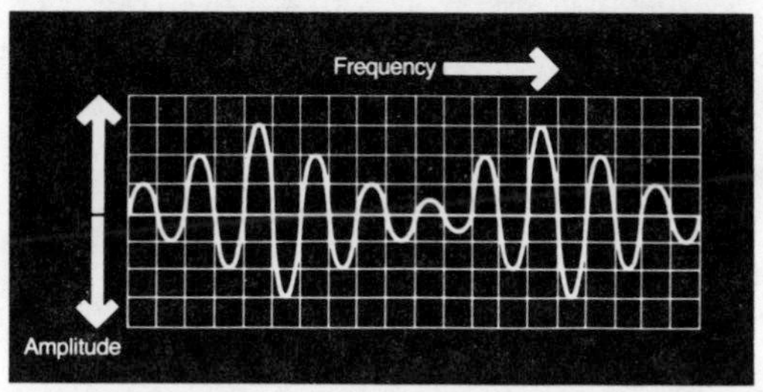

FIGURE 10.9
Amplitude modulated (AM) carrier. From FCC *Broadcast Operator's Handbook,* fig. 5–2.

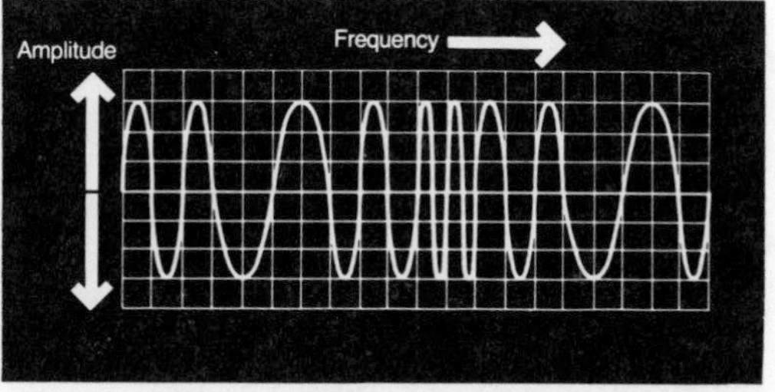

FIGURE 10.10
Frequency modulated (FM) carrier. From FCC *Broadcast Operator's Handbook,* fig. 5–4.

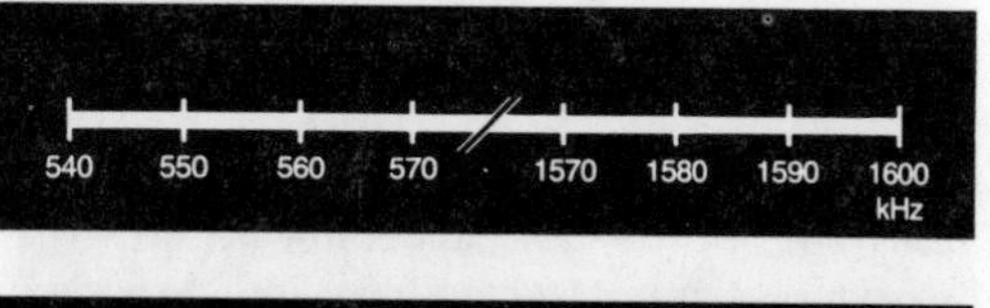

FIGURE 10.11
Standard AM band and FM band. From FCC *Broadcast Operator's Handbook,* figs. 4–7 to 4–8.

Station Classifications

To guarantee the efficient use of the broadcast spectrum, the FCC established a classification system for both AM and FM stations. Under this

RESEARCH

WLS FM COVERAGE

The map below indicates the area "covered" by the WLS FM signal. The inner circle represents 1 millivolt per meter. The outer circle represents the total reception range—50 microvolts per meter. Our transmitter move to the Sears Tower in June of 1983 provided a 22% increase in square mile coverage—the maximum allowable under present FCC rules.

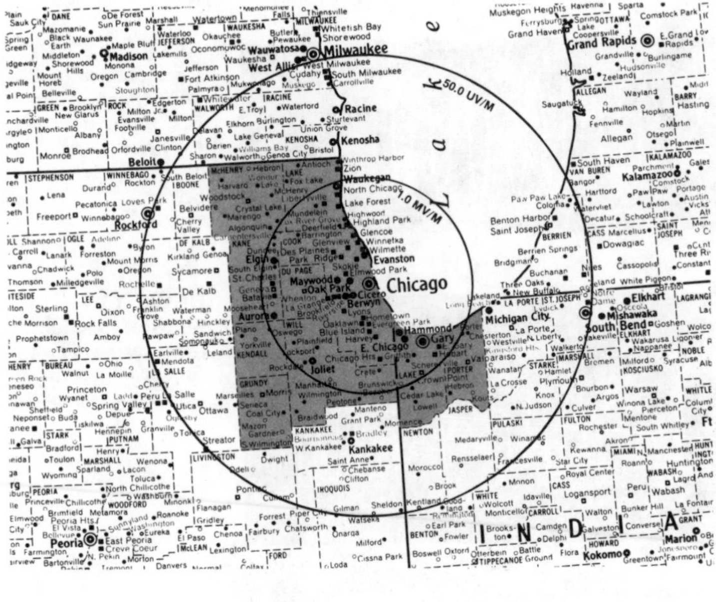

Eleven County Metro Area (effective Fall '84) includes: Cook, IL, DuPage, IL, McHenry, IL, Kane, IL, Lake, IL, Grundy, IL, Will, IL, Kendall, IL, Lake, IN Porter, IN and Kenosha, WI (AREA SHADED ON MAP).

FIGURE 10.12
Coverage maps show where a station's signal reaches. Courtesy WLS.

system the nation's ten thousand radio outlets operate free of the debilitating interference that plagued broadcasters prior to the Radio Act of 1927.

AM Classifications

Class I: Class I stations are clear channel stations with power not exceeding 50 kw. Their frequencies are protected from interference up to seven hundred and fifty miles. Among the pioneer or oldest stations in the country: KDKA, WBZ, WSM, WJR.

Class II: Stations in Class II are assigned power ranging from a minimum of 250 watts to a maximum of 50 kw. They must protect Class I outlets by altering their signals around sunset. As a Class II station, WINZ-AM in Miami is required to reduce power from 50 kw to 250 watts so as not to intrude upon other stations at 940 kHz.

Class III: Regional channel stations with less coverage area than Class I and II operations, Class III's are assigned up to 5 kw ERP (effective radiated power) and are intended to service the city or town and adjacent areas of their license. Regionals outnumber clear channel stations and are located all across the Standard band.

Class IV: Local channel Class IV stations operate with the lowest power (1 kw and under), often must sign off around sunset, and generally are found at the upper end of the dial—1200 to 1600 kHz. Class IV stations are particularly prevalent in small and rural markets.

FM Classifications

Class C: The most powerful FM outlets with the greatest service parameters are Class C's. Stations in this class may be assigned a maximum ERP of 100 kw and a tower height of up to 2000 feet. Class C radio waves carry on the average seventy miles from their point of transmission.

Class B: Class B stations operate with less power—up to 50 kw—than Class C's and are intended to serve smaller areas. The maximum antenna height for stations in this class is 500 feet. Class B signals generally do not reach beyond forty to fifty miles.

Class A: Class A stations are the least powerful of commercial FM stations; they do not exceed 3 kw ERP and 328 feet in antenna height. The average service contour for stations in this category is ten to twenty miles.

Class D: The Class D category is set aside for noncommercial stations with 10 watts ERP. This type of station is most apt to be licensed to a school or college.

In the 1980s the FCC introduced three new classes of FM stations under Docket 80–90 in an attempt to provide several hundred additional frequencies. They are as follows:

Class C1: Stations granted licenses to operate within this classification may be authorized to transmit up to 100 kw ERP with antennas not exceeding 984 feet. The maximum reach of stations in this class is about fifty miles.

Class C2: The operating parameters of stations in Class C2 are close to Class B's. The maximum power granted Class C2 outlets is 50 kw, and antennas may not exceed 492 feet. Class C2 stations reach approximately thirty-five miles.

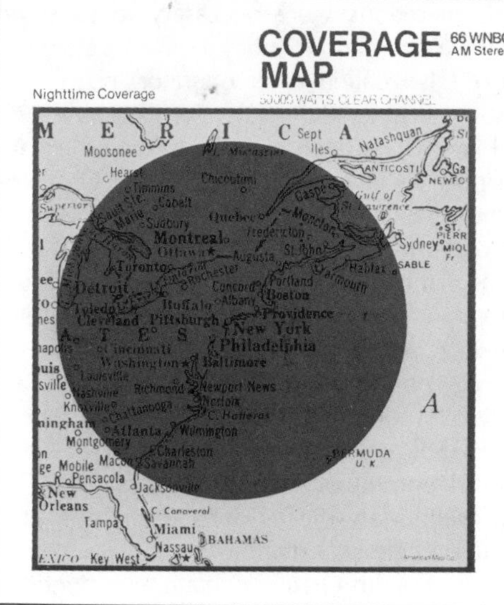

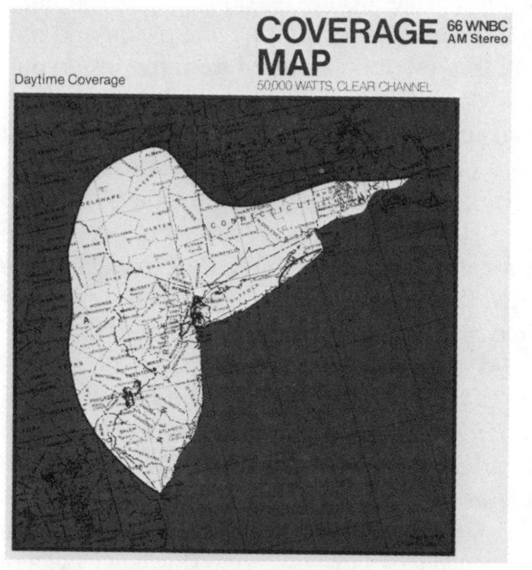

FIGURE 10.13
WNBC's coverage maps illustrate the difference between its daytime and nighttime reach. Courtesy WNBC-AM.

Class B1: The maximum antenna height permitted for Class B1 stations (328 feet) is identical to Class A's; however, Class B1's are assigned up to 25 kw ERP. Class B1 signals carry twenty-five to thirty miles.

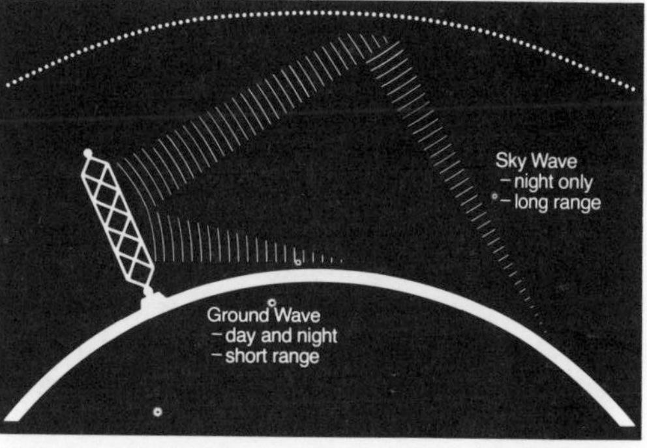

FIGURE 10.14
AM signal radiation. From FCC *Broadcast Operator's Handbook*, fig. 3–2.

BECOMING AN ENGINEER

Most station managers or chief engineers look for experience when hiring technical people. Formal training such as college ranks high but not as high as actual hands-on technical experience. "A good electronics background is preferred, of course. This doesn't necessarily mean ten years of experience or an advanced degree in electronic engineering, but rather a person with a solid foundation in the fundamentals of radio electronics, perhaps derived from an interest in amateur radio, computers, or another hobby of a technical nature. This is a good starting point. Actually, it has been my experience that people with this kind of a background are more attuned to the nature of this business. You don't need a person with a physics degree from MIT, but what you do want is someone with a natural inclination for the technical side. Ideally speaking, you want to hire a person with a tech history as well as some formal in-class training," contends Kevin McNamara, chief engineer, WMJX-FM.

Jim Puriez, chief engineer of WARA-AM, Attleboro, Massachusetts, concurs. "A formal ed-

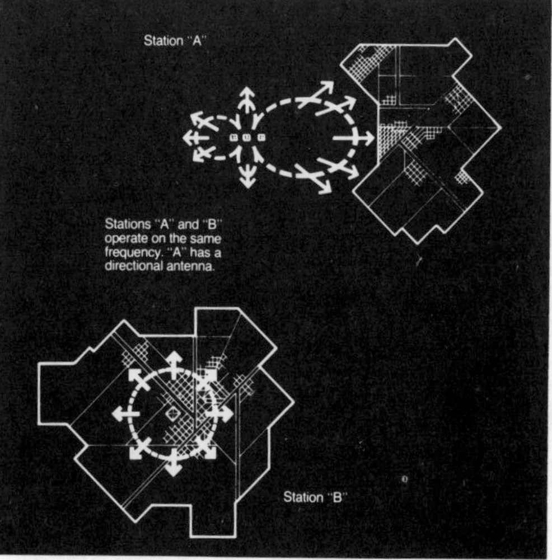

FIGURE 10.15
Nondirectional and directional antenna radiation. From FCC *Broadcast Operator's Handbook*, fig. 7–2.

ucation in electronics is good, but not essential. In this business if you have the desire and natural interest you can learn from the inside out. You don't find that many broadcast engineers with actual electronics degrees. Of course, most have taken basic electronics courses. The majority are long on experience and have acquired their skills on the job. While a college degree is a nice credential, I think most managers hire tech people on the basis of experience more than anything else."

Sid Schweiger, chief engineer of WVBF/WKOX, Framingham, Massachusetts, also cites experience as the key criterion for gaining a broadcast engineer's position. "When I'm in the market for a tech person, I'll check smaller market stations for someone interested in making the move to a larger station. This way, I've got someone with experience right from the start. The little station is a good place for the newcomer to gain experience."

Formal training in electronics is offered by numerous schools and colleges. The number shrinks somewhat when it comes to those institutions actually providing curricula in broadcast engineering. However, a number of technical schools do offer basic electronics courses applicable to broadcast operations.

Before August 1981 the FCC required that broadcast engineers hold a First Class Radiotelephone license. In order to receive the license, applicants were expected to pass an examination. An understanding of basic broadcast electronics and a knowledge of the FCC rules and regulations pertaining to station technical operations were necessary to pass the lengthy examination. Today a station's chief engineer (also called chief operator) need only possess a Restricted Operator Permit. Those who held First Class licenses prior to their elimination now receive either a Restricted Operator Permit or a General Radiotelephone license at renewal time.

FIGURE 10.16 Station engineer at the work bench. Courtesy WMJX-FM.

It is left to the discretion of the individual radio station to establish criteria regarding engineer credentials. Many do require a General Radiotelephone license or certification from associations, such as the Society of Broadcast Engineers (SBE) or the National Association of Radio and Telecommunications Engineers (NARTE), as a preliminary means of establishing a prospective engineer's qualifications. Appendices at the end of this chapter contain reproductions of SBE and NARTE membership application forms.

Communication skills rank highest on the list of personal qualities for station engineers, according to WMJX's McNamara. "The old stereotype of the station 'tech-head' in white socks, chinos, and shirt pocket pen holder weighed down by its inky contents is losing its validity. Today, more than ever, I think, the radio engineer must be able to communicate with members of the staff from the manager to the deejay. Good interpersonal skills are necessary. Things have become very sophisticated, and engineers play an integral role in the operation of a facility, perhaps more now than in the past. The field of broadcast engineering has become more competitive, too, with the elimination of many operating requirements."

Due to a number of regulation changes in the 1980s, most notably the elimination of upper-grade license requirements, the prospective engineer now comes under even closer scrutiny by station management. The day when a "1st phone" was enough to get an engineering job is gone. There is no direct "ticket" in anymore. As in most other areas of radio, skill, experience, and training are what open the doors the widest.

THE ENGINEER'S DUTIES

The FCC requires that all stations designate someone as chief operator. This individual is responsible for a station's technical operations. Equipment repairs and adjustments, as well as

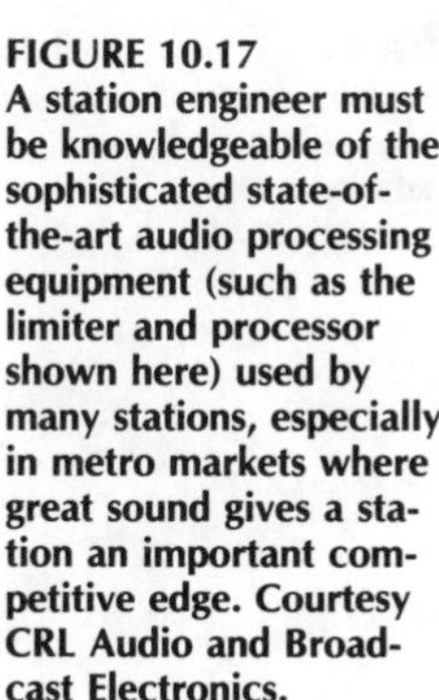
FIGURE 10.17
A station engineer must be knowledgeable of the sophisticated state-of-the-art audio processing equipment (such as the limiter and processor shown here) used by many stations, especially in metro markets where great sound gives a station an important competitive edge. Courtesy CRL Audio and Broadcast Electronics.

weekly inspections and calibrations of the station transmitter, remote control equipment, and monitoring and metering systems, fall within the chief's area of responsibility.

Depending on the makeup and size of a station, either a full-time or part-time engineer will be contracted. Many small outlets find they can get by with a weekly visit by a qualified engineer who also is available should a technical problem arise. Larger stations with more studios and operating equipment often employ an engineer on a full-time basis. It is a question of economics. The small station can less afford a day-to-day engineer, whereas the larger station usually finds that it can ill-afford to do without one.

WMJX's McNamara considers protecting the station's license his number one priority. "A station is only as good as its license to operate. If it loses it, the show is over. No other area of a station is under such scrutiny by the FCC as is the technical. The dereg movement in recent years has affected programming much more than engineering. My job is to first keep the station honest, that is, in compliance with the commission's rules. This means, keep the station operating within the assigned operating parameters, i.e., power, antenna phase, modulation, and so on, and to take corrective action if needed."

Steve Church, chief engineer of WFBQ/WNDE, Indianapolis, says that maintenance and equipment repairs consume a large portion of an engineer's time. "General repairs keep you busy. One moment you may be adjusting a pot on a studio console and the next replacing a part on some remote equipment. A broadcast facility is an amalgam of equipment that requires care and attention. Problems must be detected early or they can snowball. The proper installation of new equipment eliminates the chance of certain problems later on. The station's chief must be adept at a whole lot."

Other duties of the chief engineer include training techs, monitoring radiation levels, planning maintenance schedules, and handling a budget. Many stations hire outside engineering firms to conduct performance proofs, but it is ultimately the responsibility of the chief operator to ensure that the outlet meets its technical performance level. Proofs ascertain whether a station's audio equipment performance measurements fall within the prescribed parameters. A station's frequency response, harmonic distortion, FM noise level, AM noise level, stereo separation, crosstalk, and subcarrier suppression are gauged. If found adequate, the proof is passed. If not, the chief sees to it that necessary adjustments are made. Although the FCC no longer requires that Proof of Performance checks be undertaken, many stations continue to observe the practice.

The duties of a station engineer are wide ranging and demanding. It is a position that requires a thorough grasp of electronics relative to the broadcast environment, knowledge of FCC rules and regulations pertaining to station technical operations, and, especially in the case of the

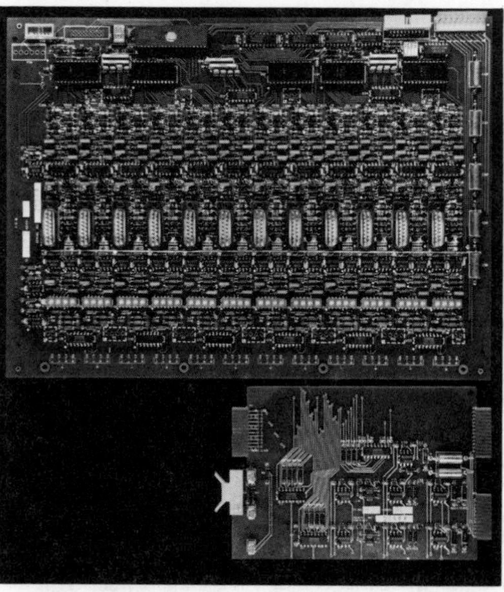

FIGURE 10.18
Solid state circuitry has become state-of-the-art in broadcasting. Courtesy IGM.

FIGURE 10.19 The luxury of solid-state electronics—pull out the bad, put in the good. Courtesy IGM.

chief engineer, the ability to manage finances and people.

STATION LOG

In 1983 the FCC dispensed with its requirement that radio stations keep maintenance and operating logs. In their place the commission created a new and considerably modified document called the Station Log, which stations must maintain. The new log requires that information pertaining to tower light malfunctions, Emergency Broadcast System (EBS) tests, and AM directional antenna systems be entered. Station Logs are kept on file for a period of two years.

Despite the fact that the FCC has eliminated the more involved logging procedures, some stations continue to employ the old system. "I like the accountability that Maintenance and Operating logs provide. We still use them here, and they are inspected daily. Despite the elimination of certain requirements, namely the tech logs, a station is still required to meet the operating stipulations of their license. Actually, enforcement action has been on the rise at the FCC, perhaps in reaction to the deregs. The commission is really interested in station technical operations. Keeping daily logs ensures compliance," says McNamara.

§ 73.1820 Station log.

(a) Entries must be made in the station log either manually by a properly licensed operator in actual charge of the transmitting apparatus, or by automatic devices meeting the requirements of paragraph (b) of this section. Indications of operating parameters that are required to be logged must be logged prior to any adjustment of the equipment. Where adjustments are made to restore parameters to their proper operating values, the corrected indications must be logged and accompanied, if any parameter deviation was beyond a prescribed tolerance, by a notation describing the nature of the corrective action. Indications of all parameters whose values are affected by the modulation of the carrier must be read without modulation. The actual time of observation must be included in each log entry. The following information must be entered:

(1) *All stations:* (i) Entries required by § 17.49 of this chapter concerning any observed or otherwise known extinguishment or improper functioning of a tower light:

(A) The nature of such extinguishment or improper functioning.

(B) The date and time the extinguishment or improper operation was observed or otherwise noted.

(C) The date, time and nature of adjustments, repairs or replacements made.

(ii) Any entries not specifically required in this section, but required by the instrument of authorization or elsewhere in this part.

(iii) An entry of each test of the Emergency Broadcast System procedures pursuant to the requirement of Subpart G of this part and the appropriate EBS checklist. All stations may keep EBS test data in a special EBS log which shall be maintained at any convenient location; however, such log should be considered a part of the station log.

(2) *Directional AM stations without an FCC-approved antenna sampling system* (See § 73.68): (i) An entry at the beginning of operations in each mode of operation, and thereafter at intervals not exceeding 3 hours, of the following (actual readings observed prior to making any adjustments to the equipment and an indication of any corrections to restore parameters to normal operating values):

(A) Common point current.

(B) When the operating power is determined by the indirect method, the efficiency factor F and either the product of the final amplifier input voltage and current or the calculated antenna input power. See § 73.51(e).

(C) Antenna monitor phase or phase deviation indications.

(D) Antenna monitor sample currents, current ratios, or ratio deviation indications.

(ii) Entries required by § 73.61 performed in accordance with the schedule specified therein.

(iii) Entries of the results of calibration of automatic logging devices (see paragraph (b) of this section), extension meters (see § 73.1550) or indicating instruments (see § 73.67) whenever performed.

(b) Automatic devices accurately calibrated and with appropriate time, date and circuit functions may be utilized to record entries in the station log *Provided:*

(1) The recording devices do not affect the operation of circuits or accuracy of indicating instruments of the equipment being recorded;

(2) The recording devices have an accuracy equivalent to the accuracy of the indicating instruments;

(3) The calibration is checked against the original indicators as often as necessary to ensure recording accuracy;

(4) Provision is made to actuate automatically an aural alarm circuit located near the operator on

WFBQ-FM operating log

DAY ______________

DATE ______________

Time	Plate Volts	Plate Amps	Power Out

1. All times are EST.
2. Unless otherwise noted, all operations are continued from previous day.
3. WFBQ leases its subcarrier to Comcast, Inc. for the operation of a background music service. Unless otherwise noted, subcarrier program material is background music provided by the Muzak Corp.,subcarrier is on continuously when modulation present, and, with the exception of short breaks between music segments modulation is continuous.
4. Responsibility for tower maintenance and light checks delegated to WRTV
5. Po = Ep x Ip x Eff
   Eff= 74.12% as ind. in mfrs instr.

Comments:

POWER TOLERANCE: Low Limit = 18,000 w / High Limit = 21,000 w

EBS TEST SENT	EBS TEST RECEIVED	EBS RECEIVER CHECK
	VIA ______	
______ time ______ initial	______ time ______ initial	______ time ______ initial

Operator on	Time	Operator off	Time

Ch.Op.Review:

____

Date ______

FIGURE 10.20 Operating log currently in use at WFBQ-FM. Courtesy WFBQ-FM.

duty if any of the automatic log readings are not within the tolerances or other requirements specified in the rules or station license;

(5) The alarm circuit operates continuously or the devices which record each parameter in sequence must read each parameter at least once during each 30 minute period;

(6) The automatic logging equipment is located at the remote control point if the transmitter is remotely controlled, or at the transmitter location if the transmitter is manually controlled;

(7) The automatic logging equipment is located in the near vicinity of the operator on duty and is inspected periodically during the broadcast day. In

R&R/Friday, November 9, 1984

THIRTEEN STATIONS HIT

FCC Issues $40,500 In Radio Station Technical Fines

FCC Chairman **Mark Fowler** apparently wasn't kidding when he vowed before the **Texas Association of Broadcasters** that the Commission intends to punish "spectrum slobs" and to "never be asleep at the switch when it comes to protecting the integrity of signals" (**R&R** 11-2).

In the month of October, the Commission issued a total of $40,500 in fines against 13 radio stations, including three in the top 50 markets, for violations of its technical rules. Several of the actions had been announced earlier, but were completed by the Commission staff after the stations were given a chance to respond to notices of "apparent liability for forfeiture."

Victims of the Mass Media Bureau actions were:

- **KWOD/Sacramento,** fined $10,000 for building and operating a tower at an unauthorized height, failing to get permission to move an auxiliary station, and for causing spurious emissions and overmodulation.
- **KFFB/Fairfield Bay, AR,** fined $10,000 for failing to cease operation during a remote control system malfunction, failing to install control and monitoring equipment, failing to calibrate instruments, and failing to ensure accurate entries in station logs.
- **WVAB/Virginia Beach, VA,** fined $10,000 for exceeding its authorized post-sunset power limit.
- **WLAC-FM/Nashville,** fined $4000 for unauthorized operation.
- **KCCN/Honolulu,** fined $2000 for failing to respond to an FCC notice of violation within ten days.
- **WBEN/Buffalo,** fined $600 for breaking rules that require a partial proof of performance measurement at least once each third calendar year.
- **KGYN/Guymon, OK,** fined $600 for breaking rules that require a partial and skeleton antenna proof of performance measurement.
- **WZXM/Gaylord, MI,** fined $500 for failing to make required equipment performance measurements.
- **KMED/Medford, OR,** fined $500 for failing to report an ownership change to the FCC within 30 days.
- **WOVO/Glasgow, KY** and **KLIK/Jefferson City, MO,** fined $500, and **KCLR/Ralls, TX,** fined $800, for failure to submit annual equipment performance measurements. KCLR was also charged with failure to maintain operable remote control equipment.
- **KKCC/Clinton, OK,** fined $500 for inaccurate logs and improperly-graded meters.

FIGURE 10.21
This article in an industry trade journal bears out the fact that the FCC is still out there. Reprinted with permission from *Radio and Records*.

the event of failure of malfunctioning of the automatic equipment, the employee responsible for the log shall make the required entries in the log manually at that time.

(8) The indicating equipment conforms to the requirements of § 73.1215 (Indicating instruments—specifications) except that the scales need not exceed 2 inches in length. Arbitrary scales may not be used.

(c) In preparing the station log, original data may be recorded in rough form and later transcribed into the log.

[43 FR 45854, Oct. 4, 1978, as amended at 44 FR 58735, Oct. 11, 1979; 47 FR 24580, June 7, 1982; 48 FR 38481, Aug. 24, 1983; 48 FR 44806, Sept. 30, 1983; 49 FR 33603, Aug. 23, 1984]

§ 73.1835 Special technical records.

The FCC may require a broadcast station licensee to keep operating and maintenance records as necessary to resolve conditions of actual or potential interference, rule violations, or deficient technical operation.

[48 FR 38482, Aug. 24, 1983]

§ 73.1840 Retention of logs.

(a) Any log required to be kept by station licensees shall be retained by them for a period of 2 years. However, logs involving communications incident to a disaster or which include communications incident to or involved in an investigation by the FCC and about which the licensee has been notified, shall be retained by the licensee until specifically authorized in writing by the FCC to destroy them. Logs incident to or involved in any claim or complaint of which the licensee has notice shall be retained by the licensee until such claim or complaint has been fully satisfied or until the same has been barred by statute limiting the time for filing of suits upon such claims.

(b) Logs may be retained on microfilm, microfiche or other data-storage systems subject to the following conditions:

(1) Suitable viewing—reading devices shall be available to permit FCC inspection of logs pursuant to § 73.1226, availability to FCC of station logs and records.

(2) Reproduction of logs, stored on data-storage systems, to full-size copies, is required of licensees if requested by the FCC or the public as authorized by FCC rules. Such reproductions must be completed within 2 full work days of the time of the request.

(3) Corrections to logs shall be made:

(i) Prior to converting to a data-storage system pursuant to the requirements of § 73.1800 (c) and (d), (§ 73.1800, General requirements relating to logs).

(ii) After converting to a data-storage system, by separately making such corrections and then associating with the related data-stored logs. Such corrections shall contain sufficient information to allow those reviewing the logs to identify where corrections have been made, and when and by whom the corrections were made.

(4) Copies of any log required to be filed with any application; or placed in the station's local public inspection file as part of an application; or filed with reports to the FCC must be reproduced in fullsize form when complying with these requirements.

[45 FR 41151, June 18, 1980, as amended at 46 FR 13907, Feb. 24, 1981; 46 FR 18557, Mar. 25, 1981; 49 FR 33663, Aug. 24, 1984]

THE EMERGENCY BROADCAST SYSTEM

The Emergency Broadcast System came into existence following World War II as the nation and the world entered the nuclear age. The system is designed to provide the president and heads of state and local government with a way to communicate with the public in the event of a major emergency.

This is how it works. Should the president deem it necessary, he is empowered to activate

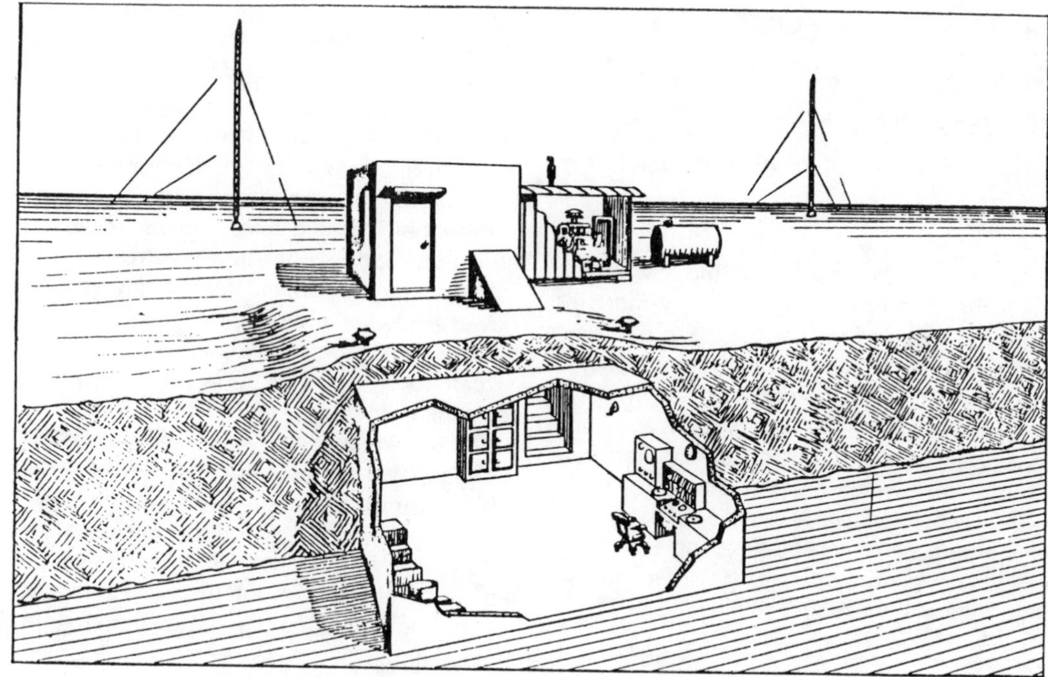

FIGURE 10.22
FEMA rendering of a protected underground shelter for an EBS station.

the national EBS. When this is done, all broadcast stations must take the following steps as outlined in Subpart G, Section 73.901, of the FCC's regulations.

1. Receive Emergency Action Notification (EAN) via AP/UPI teletype, network feed, or EBS monitor receiver. Continue to monitor for further instructions.
2. Refer to EBS Checklist. Each station has this folder on hand and must post it so that it is readily accessible to on-duty operators. The folder contains procedural information pertaining to a station's participation in the Emergency Action Notification System.
3. Authenticate EAN. This applies to AP/UPI subscribers and network affiliates only.
4. Discontinue normal programming and broadcast the first short announcement given in the Checklist.
5. Transmit attention signal.

Stations not designated to remain in operation in the event of an EAN then remove their carriers from the air after advising listeners where to tune for further information. Those participating stations continue to broadcast information as it is received from the nation's base of operations. Every radio station is required to install and operate an EBS monitor. Failure to do so can result in a substantial penalty imposed by the FCC.

Stations are also required to test the Emergency Broadcast System by airing both an announcement and an attention signal. Here is the text that must be broadcast once a week on a rotating basis between 8:30 A.M. and local sunset:

EBS TEST MESSAGE TEXT

THE FOLLOWING IS A TEST OF THE

EMERGENCY BROADCAST SYSTEM.

(ATTENTION SIGNAL) THIS IS A TEST OF THE

EMERGENCY BROADCAST SYSTEM. THE

BROADCASTERS OF YOUR AREA IN

VOLUNTARY COOPERATION WITH FEDERAL,

STATE, AND LOCAL AUTHORITIES

DEVELOPED THIS SYSTEM TO KEEP YOU

INFORMED IN THE EVENT OF AN

EMERGENCY. IF THIS HAD BEEN AN ACTUAL

EMERGENCY, THE ATTENTION SIGNAL YOU

JUST HEARD WOULD HAVE BEEN

FOLLOWED BY OFFICIAL INFORMATION, NEWS, AND INSTRUCTIONS. THIS STATION SERVES THE (OPERATIONAL AREA NAME) AREA. THIS CONCLUDES THIS TEST OF THE EMERGENCY BROADCAST SYSTEM.

EBS tests are documented in the Station Log when they are broadcast. The entry must include the time and the date of the test.

The Federal Emergency Management Agency (FEMA) makes funds available to stations designated to remain on the air during an authentic emergency through the Broadcast Station Protection Plan. Under this provision the government provides financial assistance to EBS stations for the purpose of constructing and equipping a shelter designed to operate for at least fourteen days under emergency conditions.

AUTOMATION

The FCC's decision in the mid-1960s requiring that AM/FM operations in markets with populations of more than one hundred thousand originate separate programming 50 percent of the time provided significant impetus to radio automation. Before then combo stations, as they were called, simulcast their AM programming on FM primarily as a way of curtailing expenses. FM was still the poor second cousin of AM.

Responding to the rule changes, many stations resorted to automation systems as a way to keep expenses down. Interestingly enough, however, automation programming, with its emphasis on music and deemphasis on chatter, actually helped FM secure a larger following, resulting in increased revenue and stature.

Today, over a fifth of all commercial stations are automated. Some are fully automated, while others rely on automation part of their broadcast day. Automation is far more prevalent on FM, but in the late 1970s and 1980s many AM outlets were employing automation systems to present Nostalgia and Easy Listening programming. The advent of AM stereo also has generated more use of automation on the Standard Broadcast band.

Although a substantial initial investment usually is necessary, the basic purpose of automation is to save a station money, and this it does by cutting staffing costs. Automation may also reduce the number of personnel problems. However, despite early predictions that automation eventually would replace the bulk of the radio work force, very few jobs have actually been lost. In fact, new positions have been created.

Automated stations employ operators as well as announcers and production people. The extent to which a station uses automation often bears directly on staffing needs. Obviously, a fully automated station will employ fewer programming people than a partially automated outlet.

A simple automation system consists of a small memory bank that directs the sequence of program elements, whereas a more complex system may be run by a computer that also produces logs, music sheets, invoices, affidavits, and so on. In either case, an automation system primarily consists of reel-to-reel tape decks, with ten- and/or twelve-inch reel capacity for longer play, and cartridge units, known as carousels or stack racks.

Programming elements are aired when a trip mechanism in a tape machine is activated by a cue tone. All reels and carted elements, such

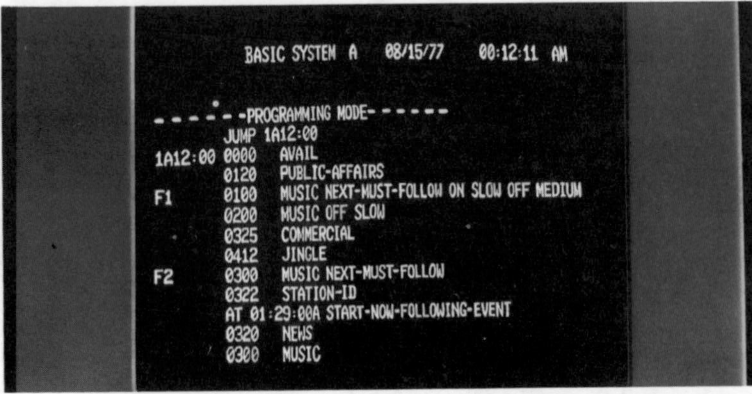

FIGURE 10.23 Standard computer keyboard and display screen. Many automated systems are run by computers. The operator simply types in the programming routine for the system to follow. Courtesy IGM.

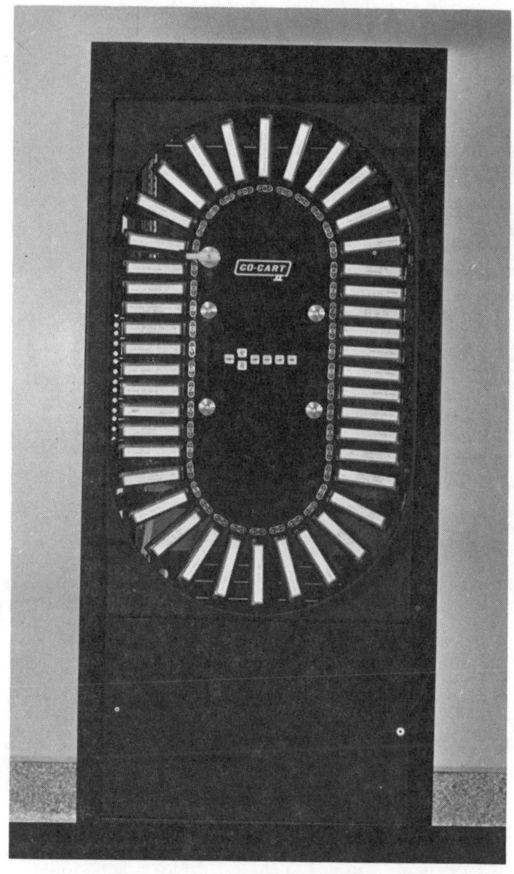

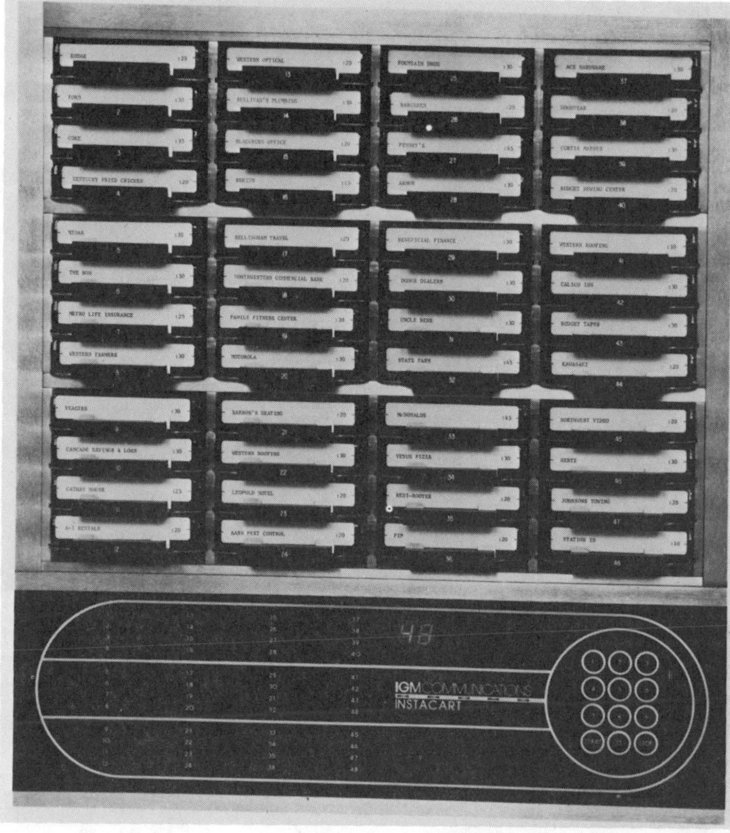

FIGURE 10.24 (above)
Automation stack cart system containing carted commercials. Courtesy IGM.

FIGURE 10.25 (above left)
Automation carousel unit containing carted music. Courtesy IGM.

as commercials, promos, PSAs, newscasts, features, and the weather must be impressed with a cue tone in order that the programming chain be maintained. A tape source without a cue tone will not signal the next program item in the sequence. To prevent extended periods of dead air, automation systems are equipped with silence sensors which, after a period of two to forty seconds depending on how they are set, automatically trigger the next element in the sequence.

Automation systems often are housed in a separate room that may be adjacent to the on-air studio. Remote start and stop switches located in the control studio permit the operator to go live whenever necessary. The engineer is responsible for the repair and maintenance of automation equipment, and at some stations he or she actually operates the system.

POSTING LICENSES AND PERMITS

The FCC requires that a station's license and the permits of its operators be posted. What follows are the rules pertaining to this requirement as outlined in Subpart H, Section 73.1230 of the FCC's regulations.

§ 73.1230 Posting of station and operator licenses.

(a) The station license and any other instrument of station authorization shall be posted in a conspicuous place and in such a manner that all terms are visible at the place the licensee considers to be the principal control point of the transmitter. At all other control or ATS monitoring and alarm points a photocopy of the station license and other authorizations shall be posted.

(b) The operator license of each station operator employed full-time or part-time or via contract, shall be permanently posted and shall remain posted so long as the operator is employed by the licensee. Operators employed at two or more stations, which are not co-located, shall post their operator license or permit at one of the stations, and a photocopy of the license or permit at each other station. The operator license shall be posted where the operator is on duty, either:

(1) At the transmitter; or

(2) At the extension meter location; or

(3) At the remote control point, if the station is operated by remote control; or

(4) At the monitoring and alarm point, if the station is using an automatic transmission system.

(c) Posting of the operator licenses and the sta-

FIGURE 10.26 Microprocessor program automation system. Courtesy Broadcast Electronics.

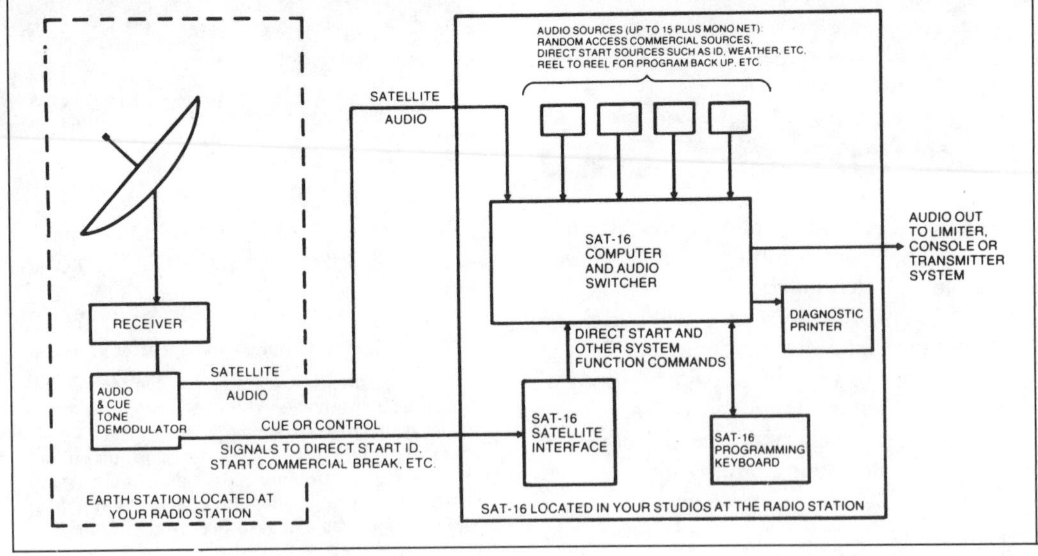

FIGURE 10.27 Satellite-linked automation systems are available in certain formats. Diagram shows how the system works. Courtesy Broadcast Electronics.

tion license and any other instruments of authorization shall be done by affixing the licenses to the wall at the posting location, or by enclosing them in a binder or folder which is retained at the posting location so that the documents will be readily available and easily accessible.
[43 FR 45847, Oct. 4, 1978, as amended at 49 FR 29069, July 18, 1984]

CHAPTER HIGHLIGHTS

1. Guglielmo Marconi first used electromagnetic (radio) waves to send a message at the end of the nineteenth century. Marconi used earlier findings by James C. Maxwell and Heinrich Hertz.

2. J. Ambrose Fleming developed the diode tube (1904), and Reginald Fessenden transmitted voice over the wireless (1906).

3. Edwin Armstrong developed the regenerative and superheterodyne circuits, and first demonstrated the static-free FM broadcast signal (1933).

4. Broadcast transmissions are carried on electromagnetic waves. The transmitter creates and shapes the wave to correspond to the "frequency" assigned by the FCC.

5. Receivers pick up the transmissions, converting the incoming radio frequency (RF) into sound waves.

6. AM stations are assigned frequencies between 535 and 1605 kHz, with ten kilocycle separations between frequencies. AM is disrupted by low-frequency emissions, can be blocked by irregular topography, and can travel hundreds (along surface-level ground waves) or thousands (along nighttime skywaves) of miles.

7. Because AM station signals travel greater distances at night, in order to avoid "skywave" interference, over two thousand stations around the country must cease operation near sunset. Thousands more must make substantial nighttime transmission adjustments (decrease power), and others (directional stations) must use two or more antennas to shape the pattern of their radiation.

8. FM stations are assigned frequencies between 88.1 and 107.9 MHz, with two-hundred-kilocycle separations between frequencies. FM is static free, with direct waves (line-of-sight) carrying up to eighty to a hundred miles.

9. To guarantee efficient use of the broadcast spectrum and to minimize station-to-station interferences, the FCC established four classifications for AM stations and seven classifications for FM. Lower-classification stations are obligated to avoid interference with higher-classification stations.

10. A station's chief engineer (chief operator) needs experience with basic broadcast electronics, as well as a knowledge of the FCC regulations affecting the station's technical operation. The chief must repair and adjust equipment, as well as perform weekly inspections and calibrations. Other duties may include installing new equipment, training techs, planning maintenance schedules, and handling a budget.

11. A Proof of Performance involves checking the station's frequency response, harmonic distortion, FM noise level, AM noise level, stereo separation, crosstalk, and subcarrier suppression.

12. Although the FCC dispensed with the maintenance and operating log requirements (1983), a Station Log must be maintained. The log lists information about tower light malfunctions, Emergency Broadcast System tests, and AM directional antenna systems.

13. The Emergency Broadcast System (EBS), implemented after World War II, provides the government with a means of communicating with the public in an emergency. Stations must follow rigid instructions both during periodic tests of the system and during an actual emergency.

14. Over one-fifth of today's commercial stations are fully or partially automated. More prevalent in FM stations, automation reduces staffing costs but requires a significant equipment investment. Automated programming elements are aired when a trip mechanism in a reel-to-reel tape deck is activated by a cue tone, which is impressed on all program material. Either an operator or a computer can maintain the programming chain.

15. The FCC requires that a station's license and the permits of its operators be accessible in the station area.

APPENDIX 10A: SBE CANONS OF ETHICS

CANONS OF ETHICS

Foreword

Honesty, justice and courtesy form a moral philosophy which, associated with mutual interest between men, constitutes the foundation of ethics. The broadcast engineer should recognize such a standard, not in passive observance, but as a set of dynamic principles guiding his conduct and way of life. It is his duty to practice his profession according to those Canons of Ethics.

As the keystone of professional conduct is integrity, the engineer will discharge his duties with fidelity to the public his employers and with fairness and impartiality to all. He should uphold the honor and dignity of his profession and also avoid association with any enterprise of questionable character. In his dealings with fellow engineers he should be fair and tolerant.

Professional Life

Section 1. The broadcast engineer will cooperate in extending the effectiveness of the engineering profession by interchanging information and experience with other broadcast engineers and students and by contributing to the work of engineering societies, schools and the scientific and engineering press.

Section 2. The broadcast engineer will avoid all conduct or practice likely to discredit or unfavorably reflect upon the dignity or honor of the profession.

Relations With The Public

Section 3. The broadcast engineer will endeavor to extend public knowledge of broadcasting engineering and will discourage the spreading of untrue, unfair, and exaggerated statements.

Section 4. He will have due regard for the safety of life and health of the public and employees who may be affected by the work for which he is responsible.

Section 5. He will express an opinion when it is founded on adequate knowledge and honest conviction while he is serving as a witness before a court, commission or other tribunal.

Section 6. He will not issue ex parte statements, criticisms or arguments on matters connected with public policy which are inspired or paid for by private interests, unless he indicates on whose behalf he is making the statement.

Section 7. He will refrain from expressing publicly an opinion on an engineering subject unless he is informed as to the facts relating thereto.

Relations With Clients And Employers

Section 8. The broadcast engineer will act in professional matters for his employer as a faithful agent or trustee.

Section 9. He will act with fairness and justice between his employer.

Section 10. He will make his status clear to his employer before undertaking an engagement if he may be called upon to decide on the use of inventions, apparatus, or any other thing in which he may have a financial interest.

Section 11. He will guard against conditions that are dangerous or threatening to life, limb or property on work for which he is responsible, or if he is not responsible, will promptly call such conditions to the attention of those who are responsible.

Section 12. He will present clearly the consequences to be expected from deviations proposed if his engineering judgment is overruled by non-technical authority in cases where he is responsible for the technical adequacy of engineering work.

Section 13. He will engage, or advise his employer to engage, and he will cooperate with, other experts and specialists whenever the employer's interests are best served by such service.

Section 14. He will disclose no information concerning the business affairs or technical processes of employers without their consent.

Section 15. He will not accept compensation, financial or otherwise, from more than one interested party for the same service, or for services pertaining to the same work, without the consent of all interested parties.

Section 16. He will not accept commissions or allowances, directly or indirectly, from contractors or other parties dealing with his employer in connection with work for which he is responsible.

Section 17. He will not be financially interested in the bids as or of a contractor on competitive work for which he is employed as an engineer unless he has the consent of his employer.

Section 18. He will promptly disclose to his employer any interest in a business which may compete with or affect the business of his employer. He will not allow an interest in any business to affect his decision regarding engineering work for which he is employed, or which he may be called upon to perform.

Relations With Engineers

Section 19. The broadcast engineer will endeavor to protect the broadcast profession collectively and individually from misrepresentation and misunderstanding.

Section 20. He will take care that credit for work is given to those to whom credit is properly due.

Section 21. He will uphold the principle of appropriate and adequate compensation for those engaged in broadcast work, including those in subordinate capacities, as being in the public interest and maintaining the standards of the profession.

Section 22. He will endeavor to provide opportunity for the professional development and advancement of personnel in his employ.

Section 23. He will not directly or indirectly injure the professional reputation, prospects or practice of another collegue. However, if he considers that an individual is guilty of unethical, illegal or unfair practice, he will present the information to the proper authority for action.

Section 24. He will exercise due restraint in criticizing another collegue's work in public, recognizing the fact that the engineering societies and engineering press provide the proper forum for technical discussions and criticism.

Section 25. He will not try to supplant another engineer in a particular employment after becoming aware that definite steps have been taken toward the other's employment.

Section 26. He will not use the advantages of a salaried position to compete unfairly with another engineer.

Section 27. He will not become associated in responsibility for work with engineers, who do not conform to ethical practices.

APPENDIX 10B: SBE APPLICATION

An invitation to join SBE

Who are we?

The Society of Broadcast Engineers, formed in 1963 as the Institute of Broadcast Engineers, is a non-profit organization serving the interests of broadcast engineers. We are the only society devoted to all levels of broadcast engineering.

Our membership, which is international in scope, is made up of studio and transmitter operators and technicians, supervisors, announcer-technicians, chief engineers of large and small stations and of commercial and educational stations, engineering vice presidents, consultants, field service and sales engineers, broadcast engineers from recording studios, schools, CCTV and CATV systems, production houses, advertising agencies, corporations, audio-visual departments, and all other facilities that utilize broadcast engineers.

What can we do for you?

Help you keep pace with our rapidly changing industry through educational seminars, and a look at new technology through industry tours and exhibits at monthly chapter meetings, regional conventions, and our national meeting held in conjunction with the National Association of Broadcasters (NAB).

Give you national representation. To serve as a voice for you in the industry; a liaison for you with governmental agencies as well as other industry groups.

To provide a forum for the exchange of ideas and sharing of information with other broadcast engineers and industry people.

To promote the profession of broadcast engineering.

To establish standards of professional education and training for broadcast engineering, and to recognize achievement of these standards.

In addition to the intangible benefits of membership in the SBE, the tangible benefits of an insurance program, communications through **The SBE Signal**, certification and re-certification opportunities, and a readily available network of specialized professionals.

All this adds up to an increase in your worth as a broadcast engineer to your employer.

Where does your money go?

A small office staff handling membership, certification, and the day-to-day business of the Society. Many duties of the SBE are handled by officers and board members who volunteer their time with no remuneration.

A library of videotape training material for loan from the national headquarters.

The production of our bi-monthly newsletter, **The SBE Signal**.

Allows SBE representation—through a professionally designed informational booth at state and regional meetings as well as NAB and NRBA.

A portion of your annual dues returns to subsidize the local chapter.

Supplements expenses for special events such as NAB, Chapter Chairman and Certification Chairman Meetings, and invitational opportunities to represent the SBE.

How to be one of us

Membership categories include Student, Associate Member, Member, Senior Member, Honorary Member, and Fellow.

Qualification for Member grade requires that the individual be actively engaged in broadcast engineering or have an academic degree in electrical engineering or its equivalent in scientific or professional experience in broadcast engineering or a closely related field or art.

The cost of membership is $20 annually for member and associate member grades, and $10 for student memberships.

What do we want from you?

First, we'd like to have your name on the SBE roster.

Strength in numbers gives us additional clout. And we want and need your participation and input at the regional and national levels as well as at the local level.

Certification Program

The program issued its first certificates on January 1, 1977, and now conducts tests at various times and places for those people, either members or non-members, who wish to have a certificate attesting to their competence as broadcast engineers. The certificates are issued for two different levels of achievement in either radio or TV and are valid for five years from date of issue. Recertification may be accomplished by earning professional credits for activities which maintain competence in the state-of-the-art or by re-examination.

Emphasis in the tests is on practical working knowledge rather than general theory. The tests are as valid for people in related industries as they are for broadcasters.

An entry-level certificate was added to the certification program in January 1982 to attract new technical talent to the broadcast industry and provide incentive for them to grow with technology.

The certification program is conducted by the SBE to benefit everyone in the industry. The program recognizes professional competence as judged by one's peers, and encourages participation in seminars, conventions, and meetings to help keep abreast of the constantly changing technology in broadcast engineering.

If you ever wanted to meet the people who design the equipment you use,

- to talk with your fellow engineers and technicians,
- to tour the many facilities that employ engineers and technicians,
- to keep abreast of the state-of-the-art equipment,
- to upgrade your skills for certification,
- here is the opportunity to become a member of the most prestigious society in its field.

SOCIETY OF BROADCAST ENGINEERS, INC.
P.O. Box 50844, Indianapolis, Indiana 46250

Group Insurance Program

When you join the SBE, you have the opportunity to participate in the Group Insurance Program for SBE members and their dependents, which offers a wide range of coverage to suit your individual needs. The low rates are made possible through the economics of group administration and by the fact that SBE does not profit from the insurance program. Please note that requests for coverage under some of the plans are subject to insurance company approval.

- Term Life Insurance Plan offers options of up to $195,000 for eligible members, with lesser amounts for dependents.
- High-Limit Accident Insurance provides protection wherever you go, 24 hours a day, and eliminates the need for special accident insurance every time you travel.
- Disability Income Plan protects your income by providing monthly benefit payments when you are unable to work due to a disabling illness or accident.
- Excess Major Medical Plan supplements your regular hospital/medical coverage in the event of a catastrophic illness or accident, paying up to $1,000,000 after you satisfy your deductible.
- In-Hospital Plan pays up to $100 per day for every day you spend in the hospital—up to 365 days—directly to you, to spend as you wish.
- Major Medical Expense Insurance is designed for members who have little or no basic medical coverage.

For further information concerning membership, certification, application, regional meetings and conventions, contact the Society of Broadcast Engineers, Inc., P.O. Box 50844, Indianapolis, IN 46250, (317) 842-0836.

APPENDIX 10B: continued

MEMBERSHIP APPLICATION

SOCIETY OF BROADCAST ENGINEERS
P.O. Box 50844 ● Indianapolis, Indiana 46250 ● 317/842-0836

(Please type or print)

Application For:
☐ New Member $20.00
☐ Associate Member $20.00
☐ Student Member $10.00
☐ Change in Grade:
☐ To Member
☐ Sr. Member
Reinstate:
☐ Former Member # ______

Name: ____________________

Full home Address: (don't abbreviate) ___ Receive SBE Mail here?

____

____________________ Home Phone ( ) ______

Full Company Name and Address: ___ or here?

____

____

____________________ Business Phone ( ) ______

If accepted, please consider me a member of ____________________ Chapter

SBE Certification # ______________ (if applicable) Date of Birth ______

Current Job Title: ______________ Date Employed: ______

Type of Facility: ____________________

Description of Duties: ____________________

____

Total years of responsible Engineering Experience: ______ Field of Activity: Radio ______ Television ______ Other ______

PROFESSIONAL LICENSES OR CERTIFICATES

____

____

Additional Information Requested on Reverse Side

ADMISSIONS COMMITTEE ACTION

Date: ______

Action deferred for more information ____________________

Admissions Committee Chairman's Signature: ______________

Approved for Grade ______

Candidate Notified ______

Entered in Records ______

EXPERIENCE RECORD

List in chronological order, beginning with the most recent, all formal experience in Broadcast Engineering or related employment. Indicate field or fields of specialization under 'Position.' Please do not limit yourself to the four spaces below. ATTACH A BRIEF DESCRIPTION OF DUTIES.

From Mo. Yr.	To Mo. Yr.	Company Name and Location	Position or Title	Type of Facility

EDUCATION

College, University, or Technical Institute	From Mo. Yr.	To Mo. Yr.	Credits or Yrs. Compl.	Course or Major	Degree

List Short Courses, Seminars Related to Broadcast-Communications Technology

____

____

SPECIAL ACHIEVEMENTS

List awards, patents, books, articles, etc.

____

____

REFERENCES

List two references - familiar with your work

Name	Company	Address	Phone

Have you ever been convicted of a violation of the Communications Act of 1934, as amended. Yes ☐ No ☐. If so, describe in full.
(Use additional space if necessary) ____________________

______ I have enclosed the required application fee.

Signed ____________________ Date ______ ______ 19__

I agree to abide by the By-Laws of the Society if admitted.

APPENDIX 10C: NARTE APPLICATION

APPLICATION FOR CERTIFICATION AND ENDORSEMENT
THE NATIONAL ASSOCIATION OF RADIO & TELECOMMUNICATIONS ENGINEERS, INC.

P.O. BOX 15029 • SALEM, OR 97309
503-581-3336

☐ CHECK HERE IF YOU ARE A NARTE MEMBER. NUMBER ____________
☐ CHECK HERE IF YOU WANT A NARTE MEMBERSHIP APPLICATION.

Date of Application ____________

FOR CERTIFICATION COMMITTEE ONLY
Date Recieved ____________ 19____
Date Reviewed ____________ 19____
Date Returned ____________ 19____
Date Granted ____________ 19____
Date Denied ____________ 19____
Certificate No. Assigned ____________

1. I, the undersigned, hereby apply for NARTE Certification and Endorsement under the **Grandfather** provisions outlined in the NARTE Administrative Rules Handbook on Certification and Endorsement, and submit the following information to substantiate my experience in the relevant disciplines. *My check for $________ is enclosed.*

2. FULL NAME ____________
LAST FIRST MIDDLE OR INIT (AS TO BE SHOWN ON CERTIFICATE)

3. DATE OF BIRTH ____________ PLACE OF BIRTH ____________
CITY STATE/PROVINCE COUNTRY

I AM A CITIZEN OF ____________
COUNTRY

PREFERRED MAILING ADDRESS ____________
P.O BOX OR STREET
CITY STATE ZIP

CHECK IF THIS IS
☐ HOME OR
☐ BUSINESS

TELEPHONE ____________

4. Existing professional licensing (include FCC)

ISSUING AGENCY	FIRST DATE OF ISSUE	LIC. OR CERT. NO.	TYPE	CHECK IF CURRENT OR LAPSED	

ATTACH COPY OF PRESENT LICENSES, IF ANY.

5. Education

A GIVE THE HIGHEST ELEMENTARY OR HIGH SCHOOL GRADE COMPLETED ________ B IF YOU COMPLETED HIGH SCHOOL, GIVE DATE ________

C NAME AND LOCATION OF COLLEGE OR UNIVERSITY	DATES ATTENDED		YEARS COMPLETED		CREDIT HOURS	DEGREES RECEIVED
	FROM	TO	DAY	NIGHT	SEMESTER	

D CHIEF GRADUATE COLLEGE SUBJECTS	CREDIT HOURS SEMESTER	E CHIEF UNDERGRADUATE COLLEGE SUBJECTS	CREDIT HOURS SEMESTER

USE ADDITIONAL SHEETS IF NECESSARY

FORM 002 6-84

APPENDIX 10C continued

6. Other schools, short courses, industrial courses, military, etc.

SCHOOL OR COURSE NAME	LOCATION	HRS.	DATES ATTENDED	CERTIFICATE, DEGREE, LICENSE, ETC.

7. Significant personal achievements or contributions in radio or telecommunications:

8.a. Grade of Certificate for which application is made:

☐ Technician

☐ Engineer (4th Class not available for Engineers)

☐ First Class ☐ Second Class ☐ Third Class ☐ Fourth Class

b. NON RF RADIATION ENDORSEMENT PROGRAM SUBCATEGORIES:

[ ] A. TELEPHONE INSIDE PLANT
[ ] B. TELEPHONE LOCAL OUTSIDE PLANT
[ ] C. TELEPHONE TOLL OUTSIDE PLANT
☐ D. SWITCHING SYSTEMS
[ ] E. TRAFFIC ENGINEERING
[ ] F. MULTIPLEX SYSTEMS
☐ G. WIRE TRANSMISSION SYSTEMS
☐ H. LIGHTWAVE SYSTEMS
☐ I. ADMINISTRATIVE/ REGULATORY
[ ] J. COMPUTER TELE-COMMUNICATIONS
[ ] K. CONTROL SYSTEMS
[ ] L. ENCRYPTION, VOICE AND DATA
[ ] M. VIDEO PRODUCTION TECHNOLOGY
☐ N. OPERATIONS
☐ O. CABLE TRANSMISSION SYSTEMS
[ ] P.
[ ] Y. MASTER ENDORSEMENT
☐ Z. SPECIAL - ON REQUEST AND NEED

c. RF RADIATION ENDORSEMENT PROGRAM SUBCATEGORIES:

☐ A. RADAR SYSTEMS
☐ B. SATELLITE SYSTEMS
☐ C. MICROWAVE SYSTEMS
☐ D. MILLIMETER WAVE SYSTEMS
☐ E. LAND MOBILE SYSTEMS
☐ F. SCATTER SYSTEMS
☐ G. INTERFERENCE ANALYSIS/SUPPRESSION
☐ H. CONTROL SYSTEMS (INCL SCADA)
☐ I. CELLULAR RADIO SYSTEMS
☐ J. LF, MF & HF RADIO (NON BROADCAST)
☐ K. INTERNATIONAL BROADCAST*
☐ L. BROADCAST AM
☐ M. BROADCAST FM
☐ N. BROADCAST TV
☐ O. FREQUENCY COORDINATION (AVAILABLE ONLY IN 1ST & 2ND CLASS ENGINEERING CATEGORY)
☐ P. TELEGRAPHY*
☐ Q. AERONAUTICAL/ MARINE*
☐ R. ANTENNA SYSTEMS
☐ S. INTERNATIONAL PUBLIC FIXED*
☐ T. ADMINISTRATIVE/ REGULATORY
☐ U. SPECIAL FIELD TEST*
☐ Y. MASTER ENDORSEMENT
☐ Z. SPECIAL - ON REQUEST AND NEED

* Supplements but does not replace FCC's continuing licensing program as required by International Treaty.

Holders of RF Radiation Certificates may apply for and receive endorsements in Non-RF Radiation subcategories.

APPENDIX 10C continued

EXPERIENCE AND QUALIFICATIONS STATEMENT OF ______________________

DATE OF EMPLOYMENT (month, year)	TITLE				
CLASSIFICATION GRADE	PLACE OF EMPLOYMENT	WAS THIS QUALIFYING EXPERIENCE?			
NAME AND ADDRESS OF EMPLOYER (firm, organization, etc.)	NAME & TITLE OF SUPERVISOR	ENGINEER		TECHNICIAN	
DESCRIPTION OF WORK		YES	NO	YES	NO

DATE OF EMPLOYMENT (month, year)	TITLE				
CLASSIFICATION GRADE	PLACE OF EMPLOYMENT	WAS THIS QUALIFYING EXPERIENCE?			
NAME AND ADDRESS OF EMPLOYER (firm, organization, etc.)	NAME & TITLE OF SUPERVISOR	ENGINEER		TECHNICIAN	
DESCRIPTION OF WORK		YES	NO	YES	NO

DATE OF EMPLOYMENT (month, year)	TITLE				
CLASSIFICATION GRADE	PLACE OF EMPLOYMENT	WAS THIS QUALIFYING EXPERIENCE?			
NAME AND ADDRESS OF EMPLOYER (firm, organization, etc.)	NAME & TITLE OF SUPERVISOR	ENGINEER		TECHNICIAN	
DESCRIPTION OF WORK		YES	NO	YES	NO

DATE OF EMPLOYMENT (month, year)	TITLE				
CLASSIFICATION GRADE	PLACE OF EMPLOYMENT	WAS THIS QUALIFYING EXPERIENCE?			
NAME AND ADDRESS OF EMPLOYER (firm, organization, etc.)	NAME & TITLE OF SUPERVISOR	ENGINEER		TECHNICIAN	
DESCRIPTION OF WORK		YES	NO	YES	NO

USE ADDITIONAL SHEETS IF NECESSARY

APPENDIX 10C continued

FEE SCHEDULE

The fee schedule for the NARTE certification and endorsement program is on a tiered structure basis as follows, payable in U.S. funds:

	BY APPLICATION DURING THE GRANDFATHER PERIOD (12/31/85)		BY EXAMINATION AFTER THE GRANDFATHER PERIOD (12/31/85)		RENEWALS
	TECH	ENGINEER	TECH	ENGINEER	
1st Class Certificate	$30.00	$60.00	––– TO BE DETERMINED –––		
2nd Class Certificate	$20.00	$50.00	––– TO BE DETERMINED –––		
3rd Class Certificate	$15.00*	$40.00**	––– TO BE DETERMINED –––		
4th Class Certificate	$10.00**	N/A	––– TO BE DETERMINED –––		

ENDORSEMENTS ARE $5.00 EACH DURING THE GRANDFATHER PERIOD.
MASTER ENDORSEMENTS ARE $30.00.
**Endorsements not available.

THE BASIC CERTIFICATE

During the grandfather period, all certificates must have at least one endorsement in a subcategory approved by the NARTE Board of Directors.

RENEWAL

The Basic Certificate is valid for two years from date of issue. Endorsements are valid from date of issue and renewal shall coincide with the Basic Certificate.

ENDORSEMENTS

Endorsements are $5.00 each for the established endorsements during the grandfather period. Special endorsements not covered in the NARTE categories are $50.00 each unless the Board of Directors feels interest in the special category is enough to warrant a lesser amount.

Endorsements are renewable at $2.00 each to a maximum of $30.00 for all endorsements in the Non-RF Radiation or RF Radiation categories.

Master Endorsements indicate a broad range of experience and require substantial documentation of at least 50% of all endorsement subcategories. It indicates applicant is capable of working in any phase of the Radio and Telecommunications industry.

GRACE PERIODS

A member of NARTE whose dues are current, but whose Basic Certificate and Endorsements have expired, has two years to apply for reinstatement, without examination, by application and the payment of the past two year's fees for the Basic Certificate and the appropriate endorsements.

Neither a member nor a non-member of NARTE may exercise the privileges of the Basic Certificates or endorsements during an expired period, unless application for renewal was made prior to expiration.

Any person who holds a Basic Certificate with Endorsements, from NARTE, may continue to exercise the privileges of that Certificate with Endorsements if that person has made application for renewal prior to expiration. A copy of the application for renewal must be posted with the original of the Basic Certificate with Endorsements until notice of renewal is received.

REFERENCES:

NAME ADDRESS TELEPHONE

1. ______________________________

2. ______________________________

3. ______________________________

Upon application, the Board of Directors, upon recommendation by the membership or certification committee, may waive any of the requirements set forth above if, in the opinion of the Board, conditions justify such a waiver.

I certify that the statements made by me on this application are true and correct to the best of my knowledge and belief and are made in good faith.

Date ______________ 19_____ ______________________________

SUGGESTED FURTHER READING

Antebi, Elizabeth. *The Electronic Epoch*. New York: Van Nostrand Reinhold, 1982.

Cheney, Margaret. *Tesla: Man Out of Time*. Englewood Cliffs, N.J.: Prentice-Hall, 1983.

Considine, Douglas M., ed. *Van Nostrand's Scientific Encyclopedia*. New York: Van Nostrand Reinhold, 1983.

Davidson, Frank P. *Macro: A Clear Vision of How Science and Technology Will Shape Our Future*. New York: William Morrow, 1983.

Ennes, Harold E. *AM–FM Broadcasting: Equipment, Operations, and Management*. Indianapolis: Howard Sams, 1974.

Evans, Walter H. *Introduction to Electronics*. Englewood Cliffs, N.J.: Prentice-Hall, 1962.

Finnegan, Patrick S. *Broadcast Engineering and Maintenance Handbook*. Blue Ridge Summit, Pa.: Tab, 1976.

Noll, Edward M. *Broadcast Radio and Television Handbook,* 6th ed. Indianapolis: Howard Sams, 1983.

Reitz, John R. *Foundations of Electromagnetic Theory*. Reading, Mass.: Addison-Wesley, 1960.

Roberts, Robert S. *Dictionary of Audio, Radio, and Video*. Boston: Butterworth, 1981.

Starr, William. *Electrical Wiring and Design: A Practical Approach*. New York: John Wiley, and Sons, 1983.

11 Consultants and Syndicators

RADIO AID

Two things directly contributed to the rise of radio consultants: more stations—from two thousand in the 1950s to ten thousand in the 1980s—and more formats—from a half dozen to several dozen during the same period. Broadcast consultants have been around almost from the start, but it was not until the medium set a new course following the advent of television that the field grew to real prominence. By the 1960s consultants were directing the programming efforts of hundreds of stations. In the 1970s over a third of the nation's stations enlisted the services of consultants. Today, radio consultants work with an even greater number of stations, and the growing level of competition in the radio marketplace should increase station involvement with consultants into the 1990s and beyond. In fact, some industry experts predict that the day will come when the overwhelming majority of stations, regardless of size, will solicit the aid of an outside consultant or consultancy firm in some way, shape, or form. Whether or not this comes to pass, consultants do play an important role in the shaping and management of the medium today.

Stations use consultants for various reasons. According to Dave Scott, president of Century 21 Programming, Dallas, Texas, a lack of research expertise on the local station level prompts many stations to use consultants. "We're well into the information age, the age of highly sophisticated research techniques and computerized data. It takes a lot of resources to assess a market and prescribe a course of action. Most stations do not have the wherewithal. At Century 21 Programming, each of our consultants goes through more ratings surveys and research data than most station owners, managers, or program directors will in a lifetime. The way the marketplace is today, using a consultant generally is a wise move. Radio stations that attempt to find their niche by trial and error make costly mistakes. A veteran consultant can accelerate a station's move on the road to success."

Boston-based radio consultant Donna Halper agrees with Scott and adds, "Consultants give their client stations an objective viewpoint and another experienced person's input. Consultants are support people, resource people, who bring to a situation a broader vision rather than the purely local perspective. Consultants, and not just out-of-work PDs who call themselves consultants but in reality aren't, have a lot of research, information, and expertise they can make available to a client with an ailing station."

Mikel Hunter of Mikel Hunter Broadcast Services, Las Vegas, says consultants help stations develop a distinctiveness that they need in order to succeed. "Unfortunately most station PDs are bandwagon riders. Many watch what other stations do around the country and clone them in their markets. Sometimes this works. Often it doesn't. It likely was a consultant who helped design the programming of that successful station being copied, and the consultant did so based on what was germane to that particular market, not one a thousand miles away. Therein lies the problem. Simply because a station in Denver is doing great book by programming a certain way does not guarantee that a station in Maryland can duplicate that success. A good consultant brings originality and creativity to each new situation, in addition to the knowledge and experience he possesses. The follow-the-leader method so prevalent among programmers actually creates a lot of the problems that consultants are called on to remedy."

There are over a hundred broadcast consultants listed in the various media directories around the country. More than half of this number specialize in radio. Generally, consultancy companies range in size from twenty to fifty employees, but may be comprised of as few as two or three. Many successful program directors also provide consultancy to stations in other markets in addition to their regular programming duties. A growing number of station rep companies provide their client stations consultancy services for an additional fee.

CONSULTANT SERVICES

Stations hire program consultants to improve or strengthen their standings in the ratings surveys. An outside consultant may share general program decisions with the station's PD or may be endowed with full control over all decisions affecting the station's sound, contends Donna Halper. "I have as little or as much involvement as the client desires. Depending on the case, I can hire and train staff (or fire staff), design or fine-tune a format, or simply motivate and direct deejays, which is actually anything but simple. Whatever a station wants, as a professional consultant I can provide. Usually, I make recommendations and then the owner or GM decides whether or not I will carry them out. At some of my stations, I've functioned as the acting PD, for all intents and purposes. At other client stations, I've been sort of the unofficial mother figure, providing support, encouragement, and sometimes a much-needed kick in the behind."

Most consultant firms are equipped to provide either comprehensive or limited support to stations. "In some cases, consultants offer a packaged 'system for success' in the same way a McDonald's hamburger franchise delivers a 'system for success' to an investor. The consultant gets control. In other instances, consultants deliver objective advice or research input to a station more on a one-to-one basis. This parallels the role of most accountants or attorneys in that the decisions are still made by the station management, not the consultant," notes Century 21's Dave Scott.

Program consultants diagnose the problems that impair a station's growth and then prescribe a plan of action designed to remedy the ills. For example, WXXX, located in a twenty-station market, is one of three that programs current hits, yet it lags behind both of its competitors in the ratings. A consultant is hired to assess the situation and suggest a solution. The consultant's preliminary report cites several weaknesses in WXXX's overall programming. The consultant's critique submitted to the station's general manager may be written like this:

Consultants Report Card

Here's a clear picture of how consultants' client stations fared in the spring book. In addition to percentages of up and down 12+ figures, you'll see how each radio doctor's clients performed in key demos.

Consulted stations are noted by *(A)* for **Burkhart/Abrams/Michaels/Douglas;** *(B)* **Gary Burns;** *(BP) Beau Phillips;* *(C)* **Rick Carroll;** *(F)* **Frank Felix;** *(DH)* **Donna Halper;** *(H)* **Bob Hattrik;** *(M)* **Mike McVay;** *(P)* **Jeff Pollack;** *(S)* **John Sebastian;** *(TW)* **Todd Wallace.** Co-consultations are also noted.

All stations indicated as such were consulted during the spring 1984 rating book. Stations may have terminated or begun consultation arrangements since.

*The Burkhart/Abrams organization serves **WMMR/Philadelphia** in a research-only capacity. The nature and extent of client/consultant relationships vary.

Consultant	Stations	12+	#1's Adults 18-34	#1's Men 18-34	#1's Men 25-34	#1's Teens
B/A/M/D	50	33 up (66%) 14 down (28%) 3 flat (6%)	24 (55%)	36 (77%)	26 (55%)	14 (30%)
Burns	3	3 down	—	—	—	1
Carroll	4	4 down	1	1	1	1
Felix	5	2 up 3 down	4 (80%)	3 (75%)	2 (50%)	1 (25%)
Halper	1	1 down	1	1	—	—
Hattrik	2	1 up 1 down	—	1	1	—
McVay	2	1 down 1 flat	—	—	—	—
Phillips	3	3 up	—	2	—	1
Pollack	28	18 up (64%) 8 down (29%) 2 flat (7%)	13 (48%)	18 (69%)	9 (35%)	3 (12%)
Sebastian	**6	3 up 2 down 1 flat	—	1	—	1
Wallace	1	1 up	—	—	—	—

FIGURE 11.1 Consultants' report card listing firms and their ratings achievements. Reprinted with permission from *Radio and Records*.

Dear GM:

Following a month-long analysis of WXXX's on-air product, here are some initial impressions. A more extensive report on each of the areas cited herein will follow our scheduled conference next week.

1. *Personnel:* Morning man Jay Allen lacks the energy and appeal necessary to attract and sustain an audience in this daypart. While Allan possesses a smoothness and warmth that would work well in other time slots, namely midday or evenings, he does not have the "wake-up and roll" sound, nor the type of humor listeners have come to expect at this time of day. The other "hot hit" stations in the market offer bright and lively morning teams. Allen does not stand up against the competition. His contrasting style is ill-suited for AM drive, whereas midday man Mike Curtis would be more at home during this period. His upbeat, witty, and casual style when teamed with newspeople Chuck Tuttle and Mark Fournier would strengthen the morning slot.

Tracy Jessick and Michelle Jones perform well

**BURKHART/ABRAMS/MICHAELS/DOUGLAS AND ASSOCIATES, INC.
BEGINS KEY PERSON
PERSONNEL RECRUITMENT SERVICE**

ATLANTA - As of February 1, 1982, Burkhart/Abrams/Michaels/Douglas and Associates will provide both clients and non-clients a key person recruitment service.

According to President Kent Burkhart, the service will search for presidents, managers, sales managers and sales people, engineers, business managers, program directors, and extraordinary air talent.

"We know, or know of, just about everyone in the radio business. . . small, medium, or large markets," said Burkhart. He continued, "We receive weekly calls, letters, and resumes from good people looking for new opportunities. We also receive a lot of inquiries from our clients and non-clients looking for new talent. It seems like this is a natural marriage. Of course, we plan to add additional personnel at B/A/M/D to increase our awareness of known and hidden talent. There is a general radio cry that there is not enough management, sales, engineering, or air talent to be found. We have and will continue to disprove that supposition."

For more information call Kent Burkhart at 404-955-1550.

BURKHART/ABRAMS/MICHAELS/DOUGLAS AND ASSOCIATES, INC.
6500 River Chase Circle, East
Atlanta, Georgia 30328

FIGURE 11.2
Consultant companies often hire personnel for client stations. Courtesy Burkhart, Abrams, Michaels, Douglas and Associates.

in their respective time periods. Overnight man Johnny Christensen is very adequate. Potential as midday man should Curtis be moved into morning slot.

Weekend personnel uneven. Better balance needed. Carol Mirando, two to seven P.M. Sunday, is the strongest of the part-timers. Serious pacing problems with Larry Coty in seven to midnite slot on Saturday. Can't read copy.

2. *Music:* Rotation problems in all dayparts. Playlist narrowing and updating necessary. Better definition needed. As stands, station verges on Adult Contemporary at certain times of the day, especially during AM drive. On Monday the fourteenth, during evening daypart, station abandoned currents and assumed Oldies sound. More stability and consistency within format essential. Computerized music scheduling possible solution. Separate report to follow.

3. *News Programming:* General revamping necessary. Too heavy an emphasis during both drive periods. Cut back by 20 to 30 percent in these two dayparts. Fifteen-minute "Noon News" needs to be eliminated. Tune-out factor in targeted demos. Same holds true for half-hour, five to five-thirty P.M., "News Roundup." Hourly five-minute casts reduced to minute headlines after seven P.M. Both content and style of newscasts presently inappropriate for demos sought. Air presentations need adjusting to better, more compatibly suit format. Tuttle and Fournier of morning show are strong, whereas P.M. drive news would benefit from a comparable team. Ovitt, Hart, and Lexis do not complement each other. Van Sanders is effective in evening slot. More sounders and actualities in hourly newscasts. Greater "local" slant needed, especially on sports events.

4. *General Programming:* Too much clutter! A log-jam in drive dayparts. Spots clustered four deep in spot sets, sometimes at quarter hour. So much for "maintenance." Rescheduling needed for flow purposes as well. "Consumer Call" at eight A.M., noon, and five P.M. not suitable for demos. "Band News" good, but too long. One-minute capsule versions scheduled through day would be more effective. Friday evening "Oldies Party" too geriatric, breaks format objective. Sends target demos off to competition by appealing to older listeners with songs dating back to 1960s. Public Affairs programs scheduled between nine A.M. and noon on Sundays delivers teens to competition that airs music during same time period. Jingles and promos dated. Smacks of decade ago. New package would add contemporary luster needed to sell format to target demo.

5. *Promotions:* "Bermuda Triangle" contest aimed at older demos. Contest prizes geared for twenty-five- to thirty-nine-year-old listener. Ages station. Concert tie-in good. Album "giveaway" could be embellished with other prizes. Too thin as is. Response would indicate lack of motivation. True also of "Cash Call." Larger sums need to be awarded. Curtis's "Rock Trivia" on target. Hits demos on the money. Expand into other dayparts. Bumper stickers and "X-100" calendar do not project appropriate image. New billboards and busboards also need adjusting. Paper ads focus on weak logo. Waiting to view TV promo. Competition promos are very weak. A good "X-100" TV promo would create advantage in this area. Opportunity.

6. *Technical:* Signal strong. Reaches areas that competition does not. Significant null in Centerville area. Competition's signals unaffected. Occasional disparity in levels. Spots sometimes very hot. Promos and PSAs, especially UNICEF and American Cancer Society, slightly muddy. In general, fidelity acceptable on music. Extraneous noise, possibly caused by scratches or dirt, on some power rotation cuts. Stereo separation good. Recommend compressor and new limiter. Further plant evaluation in progress. A more detailed report to follow.

Following an extensive assessment of a station's programming, a consultant may suggest a major change. "After an in-depth evaluation and analysis, we may conclude that a station is improperly positioned in its particular market and recommend a format switch. Sometimes station management disagrees. Changing formats can be pretty traumatic, so there often is resistance to the idea. A critique more often recommends that adjustments be made in an existing format than a change over to a different one. There are times when a consultant is simply called upon to assist in the hiring of a new jock or newsperson. Major surgery is not always necessary or desired," says Donna Halper.

10 WAYS TO WIN

1. IN-MARKET ANALYSIS AND STATION CRITIQUE
A top programmer from our firm comes to your market to meet face-to-face to discuss your changing competition, positioning of your product, promotions, and how well you are doing with your format.

2. RESEARCHED MUSIC LISTS
These music lists are broken down into categories and rotations. This national music base is used in conjunction with local input to improve product that is on the air.

3. WEEKLY MUSIC CALLS WITH ONE OF OUR NATIONALLY KNOWN PROGRAMMERS
Up-to-the-minute research findings on music. . . which songs are burning out, hot new **reaction** records we are having success with, and programming advice on problems your program director may have run into during the week.

4. THE FORMATICS
How to get the most out of your call letters, promotions, air talent (talent workshops), etc.

5. BI-MONTHLY LIFESTYLE RESEARCH REPORTS
These can help jock content, contests, station image, development of older, more profitable demos, etc.

6. DISC JOCKEY WRITTEN CRITIQUES EVERY 20 DAYS
Each station is asked to send tapes to Atlanta for analysis by the programmer working on your project.

7. MONTHLY PROGRAMMING MEMOS
These address the need for information on the best new radio innovations. . . sales/client tie-ins that have made money for stations, engineering devices, and general advice on what problem areas have been noticed in certain markets.

8. SALES PRESENTATIONS
Sales meeting presentations that will help educate your sales staff on the product they are selling.

9. NATIONAL EMPLOYEE SEARCH DIVISION
This division can find the key personnel, program directors, and jocks that do make a difference in product development.

10. ADVICE ON HOW TO DO LOCAL RESEARCH
This includes computer systems for call-out research, monthly market surveys, actual music programming, designing perception-oriented questionnaires, a complete focus group turnkey operation, and The Music Test.

Proven.

BURKHART / ABRAMS / MICHAELS / DOUGLAS (404) 955-1550 for more information.
6500 RIVER CHASE CIRCLE, EAST • ATLANTA, GEORGIA 30328

FIGURE 11.3 Consultant promo piece. Courtesy Burkhart, Abrams, Michaels, Douglas and Associates.

Today the majority of stations in major and medium markets switching formats do so with the aid of a consultant. According to the NAB, 10 to 15 percent of the nation's stations change formats each year. Consultant fees range from three hundred to over one thousand dollars a day, depending on the complexity of the services rendered and the size of the station.

D. Halper & Associates
Radio Programming Consultants
28 Exeter Street · #611
Boston, MA 02116
(617) 266-5666

HOW TO KEEP THE LISTENERS (and the consultant...) HAPPY-- SOME DO's AND DON'T's

DO: SAY THE CALL LETTERS EVERY TIME YOU OPEN THE MIKE. Believe it or don't, most folks have no idea in the world what station they are listening to, especially in a market with several stations that have similar call letters...so don't keep them guessing.

DO: SOUND FRIENDLY BUT NATURAL ON THE RADIO. Phony radio voices are DEAD. Natural communication is what the audience likes best. So don't try to imitate those jive, big-voiced top-40 jocks you heard as a child. You don't need to imitate some famous major market d.j.-- just be YOURSELF. Listeners think of a radio station as a friend, so give them the opportunity to feel comfortable with you.

DO: FOLLOW THE FORMAT. It isn't there to make your life miserable or ruin your creativity. It is there to make sure the listeners get what THEY need-- consistency and familiarity.

DO: Tape yourself every day without fail!!! GET USED TO BEING TAPED-- it's a fact of life in most larger markets, so if you have any plans to ever move up, you may as well learn not to be afraid of a skimmer. And, believe it or don't, it's there to help you improve...

DO: CONCENTRATE ON YOUR SHOW!!! That may sound like a cliche, but when you are spending all your time on the phone and then jumping to put on another record and open the mike, it SHOWS. Have some idea what you are going to say, even if it's just reading a liner. And be prepared...you owe it to yourself as well as your audience.

And then, a few DON'T's...

DON'T TAKE OUT YOUR MOOD ON THE AUDIENCE!!! It isn't their fault that cart machine 3 didn't work or that you think the format is stupid. Listeners don't even perceive a format, and they don't much care. If you just had a fight with the PD or the GM or with your mother, IT'S NOT THE AUDIENCE'S FAULT!!! You are being paid to be friendly and personable during your airshift. I know it isn't always easy, but that's a sign of professionalism...keep your personal problems OUT OF YOUR AIRSHIFT!!!

DON'T COME OUT OF NEWS WITH A SLOW RECORD!!! News or any other long non-music element slows the pace down anyway. So pick it back up again right away.

DON'T PLAY TWO SOUND-ALIKES IN A ROW. I can't stress this enough. Boring radio is not what adult radio should be. If you just played a slow ballad, have the good sense to play something different next, be it a medium or an up. If you just played a country-rock song, don't follow it with another one unless we are doing a country-rock special...VARIETY makes for a good radio show, so plan out your music so that you give the audience that variety.

DON'T AVOID SONGS THAT YOU PERSONALLY HATE. For every song that you are in love with, there are probably 50 in the format that you despise. Tough luck-- the listeners love those songs and you are showing them no respect by refusing to play what they want. You may think Air Supply is wimpy, but your audience is sitting there hoping you'll play it soon... And that brings up the next point---

DON'T PROGRAM YOUR SHOW TO THE ACTIVES. You are an active. Anybody who calls the radio station for a request is an active. Your friends are probably actives too. BUT THE VAST MAJORITY OF THE AUDIENCE IS PASSIVE!!! Actives make up at most about 5% of your audience, and if you let them, they will gladly run your show and drive away everybody else. So, be nice to them when they call and if they want a song that happens to be on the playlist, terrific. But remember-- the majority of your audience will NEVER call, and they tend to HATE the songs that the actives like...So try to appease the passives and you'll be a lot safer-- actives make a lot of noise, but there aren't a whole lot of them...

And one parting word-- don't worry about other radio stations. Just be friendly, consistent, and hit-oriented and people will prefer to listen to you. More news later...

DER KOMMISSAR

Donna

FIGURE 11.4 Consultant's memo to client stations. Courtesy Donna Halper.

CONSULTANT QUALIFICATIONS

Many consultants begin as broadcasters. Some successfully programmed stations before embarking on their own or joining consultancy firms. According to Rick Sklar, president of Sklar Communications, consultants who have a background in the medium have a considerable edge over those who do not. "The best way to fully understand and appreciate radio is to work in it. As you might imagine, radio experience is very helpful in this business." Not all consultants have extensive backgrounds in the medium. Most do possess a thorough knowledge of how radio operates on all its different levels, from having worked closely with stations and having acquired formal training in colleges offering research methodology, audience measurement, and broadcast management courses. "A solid education is particularly important for those planning to become broadcast consultants. It is a very complex and demanding field today, and it is becoming more so with each passing day. My advice is to load up. Get the training and experience up front. It is very competitive out there. You make your own opportunities in this profession," says Century 21's Dave Scott.

Both Halper and Scott rate people skills and objectivity highly. "Consulting requires an ability to deal with people. Decisions sometimes result in drastic personnel changes, for example, changing formats. A consultant must be adept at diplomacy but must act with conviction when the diagnosis has been made. Major surgery invariably is traumatic, but the idea is to make the patient, the station, healthy again. You can't let your own personal biases or tastes get in the way of what will work in a given market," observes Halper. Dave Scott shares Halper's sentiments. "A consultant, like a doctor, must be compassionate and at the same time maintain his objectivity. It is our intention and goal as a program consultant service to make our client stations thrive. As consultants, we're successful because we do what we have to do. It's not a question of being mercenary. It's a question of doing what you have to do to make a station prosper and realize its potential."

Consultant company executives also consider wit, patience, curiosity, sincerity, eagerness, competitiveness, and drive, not necessarily in that order, among the other virtues that the aspiring consultant should possess.

CONSULTANTS: PROS AND CONS

There are as many opponents of program consultants within the radio industry as there are advocates. Broadcasters who do not use consultants argue that local flavor is lost when an outsider comes into a market to direct a station's programming. Donna Halper contends that this may be true to some degree but believes that most professional consultants are sensitive to a station's local identity. "Some consultants do clone their stations. Others of us do not. In fact, I'd say most do not. For those of us who recognize local differences, there need not be any loss whatsoever as a consequence of consultant-recommended changes. But the hits are pretty much the hits, and good radio is something that Tulsa deserves as well as Rochester. So I do try to localize my music research and acquire a good feel for the market I'm working in. But as far as basic rules of good radio are concerned, those don't vary much no matter what the market is. It's important for a station to reflect the market it serves, and I support my clients in that. Because I work out of Boston doesn't mean that my AOR client in Duluth should sound like a Boston album rocker. It should sound like a solid AOR station that could be respected in any city but fits the needs of Duluth."

Dwight Douglas of Burkhart/Abrams/Michaels/Douglas and Associates says that localization is essential for any radio station and that consultants are amply aware of this fact. "It is an industry axiom that a station must be a part of its environment. An excellent station will be uniquely local in relating to its audience. That tends to take the form of news, weather, sports, public service, general information, and jock talk. A good consultant will free a station from music worries and allow it to concentrate on developing local identity. We work hard at customizing formats to suit the geo-demographics or lifestyles of the audiences of our client stations."

A station has an obligation to retain its sense of locality regardless of what a consultant may suggest, contends Mikel Hunter. "No station should simply turn itself over body and soul to a consultant. Local flavor does not have to be sacrificed if a station has a strong PD and a general manager who doesn't insist that the PD merely follow the consultant's suggestions. A station should not let itself become a local franchise. Consultants are a valuable resource, but both the station and the consultant must pool their wisdom to make the plan work."

The cost factor is another reason why some stations do not use consultants. "Consultants can be expensive, although most consultants scale their fees to suit the occasion, that is, the size of the market. A few hundred dollars a day can be exhorbitant for many smaller stations. But the cost of the research, analysis, and strategy usually is worth the money. I believe that a station, in most cases, gets everything it pays for when it uses a consultant. It's worth investing a few thousand to make back a million," contends Dwight Douglas.

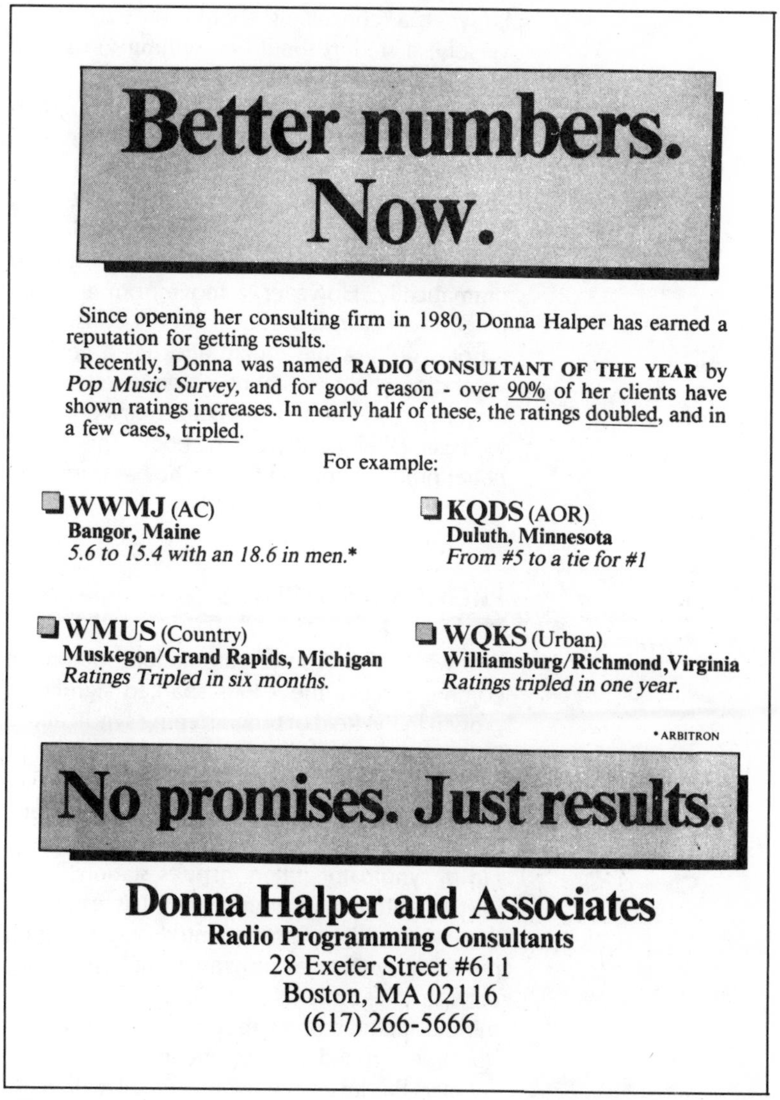

Century 21's Dave Scott believes that certain stations can become too dependent on consultants. "A consultant is there to provide support and direction when needed. If a station is infirm, it needs attention, perhaps extensive care. However, when a station regains its health, an annual or semi-annual checkup is usually sufficient. A checkup generally can prevent problems from recurring."

Mikel Hunter agrees with Scott, adding "A radio doctor needs the cooperation of his client. On the other hand, a station must insist that a consultant do more than diagnose or critique. Positive input, that is, a remedying prescription,

Better numbers. Now.

Since opening her consulting firm in 1980, Donna Halper has earned a reputation for getting results.

Recently, Donna was named **RADIO CONSULTANT OF THE YEAR** by *Pop Music Survey*, and for good reason - over 90% of her clients have shown ratings increases. In nearly half of these, the ratings doubled, and in a few cases, tripled.

For example:

- **WWMJ** (AC) **Bangor, Maine** *5.6 to 15.4 with an 18.6 in men.**
- **KQDS** (AOR) **Duluth, Minnesota** *From #5 to a tie for #1*
- **WMUS** (Country) **Muskegon/Grand Rapids, Michigan** *Ratings Tripled in six months.*
- **WQKS** (Urban) **Williamsburg/Richmond,Virginia** *Ratings tripled in one year.*

*ARBITRON

No promises. Just results.

Donna Halper and Associates
Radio Programming Consultants
28 Exeter Street #611
Boston, MA 02116
(617) 266-5666

FIGURE 11.5 Consultant promo piece proclaiming ratings success for clients. Courtesy Donna Halper.

is what a consultant should provide. Conversely, a station should be willing to use the aid that the consultant provides."

Statistically, those stations that use programming consultants more often than not experience improved ratings. In case after case consultants have taken their client stations from bottom to top in many of the country's largest markets. Of course, not all succeed quite so dramatically. However, a move from eleventh place to sixth in a metro market is considered a noteworthy achievement and has a very invigorating effect on station revenue. "The vast majority of consultants benefit their clients by increasing their position in the book. This means better profits," notes Halper, who has improved the ratings of 90 percent of her client stations.

PROGRAM SUPPLIERS

The widespread use of automation equipment commencing in the 1960s sparked significant growth in the field of programming syndication. Initially the installation of automation systems motivated station management to seek out syndicator services. Today the highly successful and sophisticated program formats offered by a myriad of syndicators often inspires stations to invest in automation equipment. It is estimated that nearly half of the country's radio outlets purchase syndicated programming, which may consist of as little as a series of one- or two-minute features or as much as a twenty-four-hour, year-round station format.

Peters Productions programmer Dick Ellis cites economics as the primary reason why stations resort to syndicators. "We supply high-quality programming and engineering at a relatively low cost. For instance, for a few hundred dollars a month a small market operator gets a successful program director, a highly skilled mastering engineer, all the music he'll ever need (no service problems with record companies) recorded on the highest-quality tapes available. It takes a programmer eight hours to program one twenty-four cut reel. It takes a mastering engineer eight hours to remove all the pops and clicks found on even brand-new records, plus place the automation tones. All of this frees the local operator to concentrate his efforts on promotion and, of course, sales."

William Stockman of Schulke Radio Productions (SRP) says that stations are attracted to syndicators because of the highly professional, major market sound they are able to provide. "By using SRP's unique programming service a smaller station with limited resources can sound as polished and sophisticated as any metro station."

Both economics and service motivate radio stations to contract syndicators, contends Larry C. Vanderveen, president, Radio Arts, Inc. "While a syndicator can be very cost efficient, of equal importance is the quality of the programming available through a particular house. A station can pay a lot and not get a whole lot. We concentrate on delivering a first-rate product at a reasonable price. That is why our list of client stations continues to expand."

The demand for syndicator product has paralleled the increase in the number of radio outlets since the 1960s. With the current projections indicating that there will be hundreds of additional AM and FM frequencies entering the marketplace over the next decade, this trend should continue.

SYNDICATOR SERVICES

The major program syndicators usually market several distinctive fully packaged radio formats. "We make available a complete format service with each of our format blends at Peters Productions. We're not merely a music service. Our programming goal is the emotional gratification of the type of person attracted to a particular format," says Dick Ellis, whose company offers nine different formats, including Beautiful Music, Easy Listening, Standard Country, Modern Country, Adult Contemporary, Standard MOR, Super Hits, Easy Contemporary, and a country and contemporary hybrid called Natural Sound.

Century 21 Programming also is a leader in format diversity, explains Dave Scott. "Our inventory includes everything from the most contemporary super hits sound to several Christian formats. We even offer a full-time Jazz format. We have programming to fit any need in any market."

Drake-Chenault Enterprises, Far West Communications, Radio Arts, Inc., Concept Productions, and TM Communications also are among the most successful of those syndicators marketing several program formats. Some syndicators prefer to specialize in one or two pro-

Eighteen Successful Sounds: One's Right For You

The Z Format. Since 1973, Century 21's contemporary hit Z Format has delivered the best track record in the business!

The Hot Z Format. Today's hottest hits give you the most popular music, so your station will be the most popular in your market!

Album Oriented Z. Century 21's album rock format is available either unannounced or uniquely custom-voiced.

The A-C Format. Adult contemporary music goes hand-in-hand with a big, responsive audience.

Good Ol' Rock & Roll. Here are the top hit oldies of the 50's, 60's & 70's that still sound good today.

You can choose from our *three* different formats: modern, traditional, or pop/crossover country.

The C-C Format. This Country-Crossover Format blends country with adult contemporary music.

The E-Z Format. Century 21's E-Z Format moves "middle-of-the-road" music into *your* lane.

MORE BEAUTIFUL. MORe Beautiful blends middle-of-the-road vocals with the finest instrumentals.

Simply Beautiful. Beautiful music is the favorite format of broadcasters seeking trouble-free, stable operations.

MEMORY MUSIC. Here's the ideal mixture of M-O-R and nostalgia. Music vintages are tailored to your station.

SACRED SOUNDS. Century 21 delivers four Christian formats: beautiful, adult contemporary, traditional & country gospel.

The Jazz-Z Format. As either a full- or part-time format, Century 21's Jazz-Z sound gives your station unique popularity.

FIGURE 11.6 Century 21's catalogue of syndicated formats. Courtesy Century 21.

gramming areas. For example, Bonneville Broadcasting System, Churchill Productions, and Schulke Radio Productions, which was acquired by Bonneville in 1984, primarily specialize in the adult Easy Listening format.

Each format is fully tested before it is marketed, explains SRP's Stockman. "At Schulke our strategy has been to reorient the music from essentially a producer-oriented to a consumer-oriented product. Music is tested on a cut-by-cut basis in several markets coast-to-coast. Using patented and proven methodology, music is carefully added or selectively deleted. By determining what songs the listeners like to hear

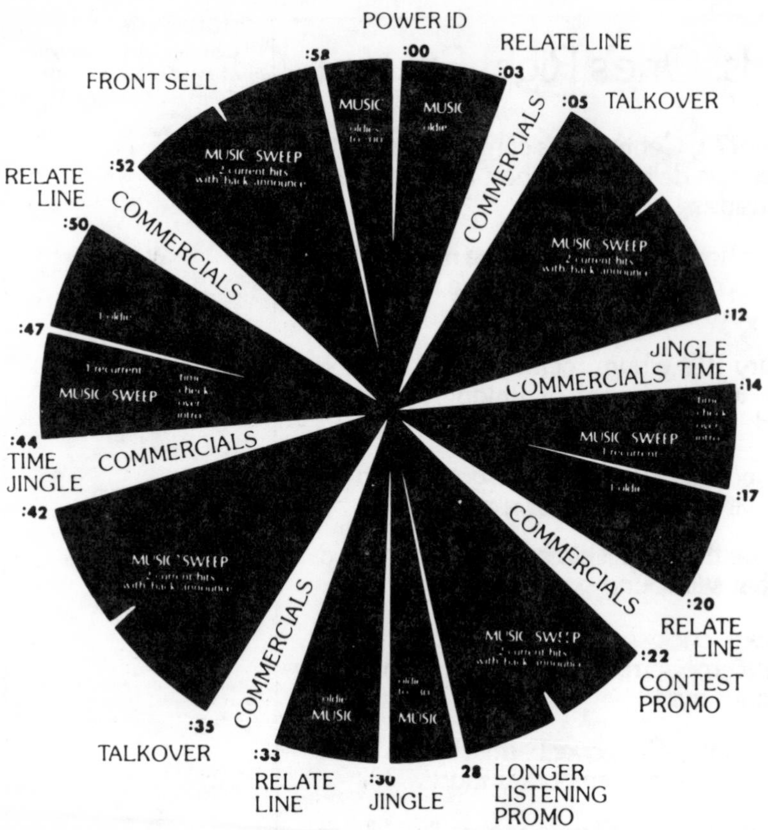

FIGURE 11.7 Sample customized sound hour. Courtesy Century 21.

and which songs they dislike, SRP has assembled a totally researched library that has been on the air via our subscriber stations since March 1983. Every song played on our stations has been rated by the listeners as a 'winner,' and all the 'stiffs' that have a high dislike factor have been eliminated altogether."

Customized sound hours are designed for each format to ensure consistency and compatibility on the local station level. "An exact clock is tailored for our client station after our market study. The format we provide will perfectly match the station in tempo, style, music mix, announcing, promos, news, weather, and commercial load," says Century 21's Dave Scott.

Audience and market research and analysis are conducted by syndicators before implementing a particular format. "Our clients receive comprehensive consulting services from our seasoned staff. We begin with a detailed study of our client station's market. We probe demographics, psychographics, and population growth trends of a station's available audience. We analyze a client's competition quantitatively through available ratings and qualitatively from airchecks. Then the programming Century 21 provides is professionally positioned to maximize our client's sales, ratings, and profits. All of our programming is solidly backed by systematic studies of the listening tastes of each format's target audience. Our research includes call-out and focus group studies, in-depth market analysis, attitudinal audience feedback, psychographic patterns and tests, and several in-house computers with ratings data on-line," says Dave Scott.

Format programming packages include hundreds of hours of music, as well as breaks, promos, and IDs, by seasoned metro market announcers. Customized identity elements, such as jingles and other special formatic features (taped time checks), are made available by the majority of syndicators. "We try to cover all bases to insure the success of our clients. We back each of our formats a dozen different ways. For example, image builders in the form of promotions, contests, and graphics also are an element of our programming service at Radio Arts," says Vanderveen.

To stay in step with the ever-changing marketplace, syndicators routinely update the programming they provide their subscribers. "When you want people to listen to a station a lot, you've got to keep them interested in it. To do so you have to air a sound that's always fresh and current. Tape updates from Century 21 are plentiful. We give stations the most extensive initial collection of music tapes available. Then we follow them up with hundreds more throughout the year. For instance, our CHR, AOR, and Country subscribers receive over one hundred updates annually. All categories have frequent updates, so our client's sound stays fresh and vital," says Dave Scott, who adds that the lines of communication are kept open between the client and syndicator long after the agreement has been signed. "Since the success of our clients is very important to us, continuing consultation and assistance via a toll-free hotline is always available twenty-four hours a day. Automation-experienced broadcasters are in our production studios around the clock, and consultants can be reached at work or home any time. Help is as close as the phone."

Syndicators assist stations during the installation and implementation stage of a format and provide training for operators and other station personnel. Comprehensive operations manuals

Rub yourself out of the RED and into the BLACK.

HIGH OPERATING COSTS?

Satellite becomes your air staff, your programming and research department. You don't have to worry about constant talent wage increases, benefit packages or fringes, so you get your programming payroll costs under control once and for all. A music library? You don't need one. Overhead goes down, so profits go up.

DAY-TO-DAY MANAGEMENT HASSLES?

With Satellite you have live professional programming 24 hours a day, 7 days a week. And you don't have to schedule vacations, juggle air shifts, search for part-timers, hire, fire, or sit through meetings. Finally . . . you'll have the time to concentrate on the important things, like planning, promotion, and most of all, increasing sales.

WONDERING WHICH FORMAT IS RIGHT FOR YOU?

You get your choice of the four most popular and successful music formats in radio: top 40 (CHR), adult contemporary, MOR, and country. They're all created and researched by Burkhart/Abrams to be consistent, competitive, and exciting. And they feature major-market caliber live personalities who really know how to deliver an audience.

CONCERNED ABOUT KEEPING A LOCAL IMAGE?

Satellite's always there, your basic, full-time live programming. You can add as much or as little local material as you like, whenever you like. Our Flex Clock[sm] concept lets you schedule local news, traffic, sports, features, and specials. And LocaLiners[sm] recorded individually for you by satellite personalities add extra local color!

WORRY ABOUT COMPLICATED SET-UP AND MAINTAINANCE?

It's a lot easier than you might think. Quicker, too. Once your equipment arrives, the dish takes a couple of days to install. The electronics slip into a rack. Some cables to hook it together, and you're on the air. Run it with a couple of triple-decks, a reel-to-reel machine and our Auto-Switch, live assist or with your present Automation system.

Get started on decreased problems and increased profits

CALL

1-800-527-4892

or (214) 991-9200

FIGURE 11.8
Some syndicators use satellites to feed programming to client stations. This greatly simplifies the distribution process. Handling of tapes and mailing is eliminated. Satellite syndication also keeps station equipment costs down. Courtesy Satellite Music Network.

are left with subscribers as a source of further assistance.

Fees for syndicator services vary according to a sliding scale that is based on station revenues for a specific market. "Our fee structure at Bonneville ranges from something under one thousand dollars per month to over ten thousand. It very much depends on the size of the market in which a station is located," explains Marlin Taylor, Bonneville Broadcasting's president. Surveys in *Broadcasting Magazine* and *Radio Only* reveal that the range for syndicator formats falls somewhere between six hundred and eighteen thousand dollars per month, depending on the nature of the service and market size.

Leasing agreements generally stipulate a minimum two-year term and assure the subscriber that the syndicator will not lease another station in the same market a similar format. Should a station choose not to renew its agreement with the syndicator, all material must be returned unless otherwise stipulated.

The majority of format syndicators also market production libraries, jingles, and special features for general market consumption.

HARDWARE REQUIREMENTS

Syndicated programming typically is designed for automation systems. "All Century 21 programming airs smoothly with a minimum of equipment. Of course, sophisticated systems can be fully utilized by our formats, but you can run our material with the most basic automation setup," notes Dave Scott, whose company, as well as most other format syndicators, require that stations possess the following automation equipment:

- Four reel-to-reel tape decks capable of holding ten-and-a-half-inch NAB reels. Some formats will work with only two or three reel-to-reel decks.
- End-of-music sensors also are recommended to alert operators of necessary tape changes.
- Sufficient multiple cart playback systems for spot load, news, PSAs, weather, promos, IDs, and so forth.
- Two single-cartridge playback units or use of extra trays in spot players.

FIGURE 11.9
Satellite Music Network program clock. Courtesy Satellite Music Network.

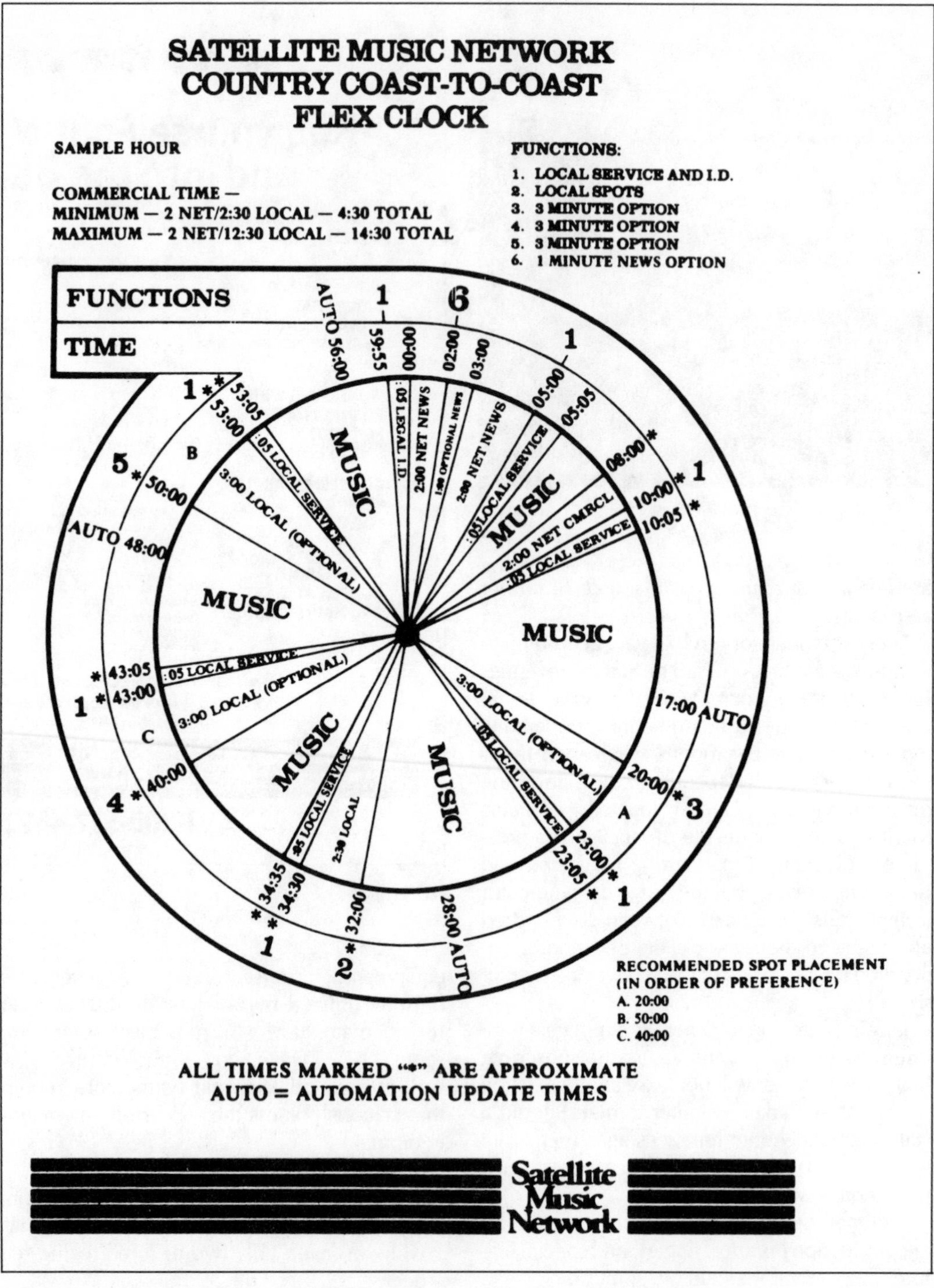

- One-time announce unit with two size C cartridge players and controller.

Technical and engineering assistance is made available to subscribers. "We provide periodic on-site station technical reviews together with ongoing technical consultation, continuing new technology development and new equipment evaluation," explains Bonneville's Marlin Taylor.

SYNDICATOR FIDELITY

Syndicators are very particular about sound quality and make every effort to ensure that their programming meets or exceeds fidelity standards. "Our company uses the finest quality recording studio equipment. Actually, it's far superior to any broadcast grade gear. Therefore, it is quite important that subscribers have adequate hardware, too. We utilize a number of highly regarded audio experts to make our sound and our client's the very best possible. In fact, we use special audiophile 'super disks' master tapes from record companies, noise reduction, click editing, and precise level control or slight equalization, if needed," says Century 21's Scott.

Periodic airchecks of subscriber stations are analyzed from a technical perspective to detect any deficiencies in sound quality.

CHAPTER HIGHLIGHTS

1. The significant increase in stations and formats created a market for consultants.

2. Consultants provide various services, including market research, programming and format design, hiring and training of staff, staff motivation, advertising and public relations campaigns, news and public affairs restructuring, and technical evaluation (periodic airchecks of sound quality).

3. Aspiring consultants should acquire background experience in the medium, solid educational preparation, and strong interpersonal skills.

4. Station executives opposed to using consultants fear losing the station's local flavor, becoming a clone of other stations, and the substantial cost.

5. Statistically, stations using programming consultants more often than not experience improved ratings.

6. Increased use of programming syndication is related to the increased use of automation. Nearly half of the nation's stations purchase some form of syndicated programming.

7. Syndicated programs are generally cost effective, of high quality, and reliable, thus allowing smaller stations to achieve a metro station sound.

8. Program syndicators provide a variety of test-marketed packaged radio formats—from Country to Top 40 to Religious. Packages may include music, breaks, promos, customized IDs, and even promotions. Package updates are frequent.

9. Syndicated programs require the use of reel-to-reel tape decks, end-of-music sensors, multiple cart playback systems, single-cartridge playback units, and a time announce unit.

APPENDIX: STATION CRITIQUE

DONNA HALPER & ASSOCIATES
Radio Programming Consultants
28 Exeter Street - #611
Boston, MA 02116
(617) 266-5666

To: John Smith
Fr: Donna L. Halper
re: monitor of WYYY-AM and WZZZ-FM

To reiterate what we discussed at breakfast, let me first consider the FM. Unlike its counterpart in Akron, WXXX-FM, WZZZ-FM still strikes me as more of a traditional Beautiful than an Easy Listening station. There seem to be fewer vocals, and fewer light pop selections than on WXXX-FM. Given that Bill has the same music on both stations, I would be interested to discover why WZZZ-FM has such a background-music sound while WXXX-FM has made the transition to a somewhat more foreground sound. Some of the problem may be engineering—WXXX-FM has much more presence, and the music tempos are much more textured. WZZZ-FM seems to play more of the violin/orchestral/Mantovani type of Beautiful Music, and while its signal is loud enough, it does not seem as clean to me as WXXX-FM. Perhaps this contributes to the perception of being more like Muzak and less like Easy Listening. Although WZZZ-FM uses that liner about "sparkling sounds," for me the music did not sparkle at all—it just lays there, not offensive but nothing that anyone under 50 would find enjoyable. I realize that the very nature of Beautiful (or Easy, if you prefer) makes it difficult to attract those younger demo's who grew up with AC and Top 40 and are not accustomed to the style or the background quality of most BMs; but still, many Beautifuls are being dragged forcibly into the 1980s—WXXX-FM among them—and it is still possible to be Beautiful/Easy and yet not sound like a throwback to 1960. WZZZ-FM seems to be slowly evolving into a more foreground type of station, but it isn't there yet. The announcers seem convinced that they must put on a "Beautiful Music voice" when they record their drop-ins, and it's almost amusing to hear the jock who was on the AM doing very up AC/Top 40 a few moments ago suddenly transformed into a person who speaks v e r y s l o w l y and carefully enunciates each word with a suitably serious tone of voice. Even those jocks who have been told to lighten up still seem a bit confused about the "image" they should maintain on FM. As we discussed about WXXX-FM, and as you are certainly aware, part of the transition in today's Beautifuls is toward announcers who sound more human and friendly. While the FM jocks don't sound as dead or as funereal as others I have heard, they still seem stuck in the belief that FM is supposed to sound serious. Another problem is the length of the FM shifts—I heard Joe's voice, for example, doing news or whatever else, from afternoon drive up till about 10 P.M. off and on! The perception there is that the poor guy never goes home . . . Another MAJOR problem is one that we corrected at WXXX-FM: DEAD AIR. To my knowledge, it is NOT necessary to have a full five-second pause between certain songs or between a song and an announcer drop-in. It may have been part of old-line BMs to not sound too tight, but we are at the other extreme—at times we sound totally asleep or, worse, unprofessional. Listeners don't know the great reasons for dead air—they just know it's there and they wonder if their radio is broken . . . There are also minor stylistic points such as the jingle (which I'm not even sure I like) is used much more at night, it seems, than during the day. Also, I'm not sure saying FM twice in the ID—it sounds a bit redundant to say FM 95 WZZZ-FM . . . But that's a small matter. In addition to dead air, we also have traffic reports that are horrible quality. The newspeople impressed me a lot with their general professionalism, but we'll talk more about that when discussing the AM. Also, what are the rules about vocals to instrumentals, newer music to recent to very old, etc., and how many in a row without call letters do we play at night? I would like to have other voices besides Joe's on the air and more involvement from the announcers wouldn't be a bad idea—as it stands right now, there are times when the station really does sound automated and lifeless . . . I also am not thrilled that Bill has no system for ever telling the nice people the name of any song they heard—it strikes me as a service to our audience—most adults are VERY passive, and while they may not buy a lot of records, they like to know what they have been listening to. I don't say we have to announce every song, but it's too bad that right now we can't even do it at all . . . In summation, I feel the FM has come a long way, but is still moving towards a positioning that will make it slightly more modern.

As for the AM, it is somewhat enigmatic to me. It has some elements that are quite professional—the news team for example (although I would like to do more work with the female member to get her to speak in a less "cute" and singsong manner) and the excellent use of actualities, traffic reports, etc. The morning show gets a bit too talk-y at times but moves professionally and smoothly most of the time. The music is about 85% where I think it should be, given your target audience and intended demographics. The network news sounds good, and adds a nice flavor. The signal seems good, and the announcers in general are pleasant to listen to, although we will deal with a few specific problems in a subsequent paragraph.

The first enigma to me is how the station is positioned. Is it country? I hear quite a few cross-over songs and a couple that weren't even hits to the pop audience (although they are certainly nice enough songs). Some of the cross-over comes very close together—one hour, I heard three within a 20 minute period—and could easily confuse some casual listeners as to what they are tuning in to. What are the rules about artist separation, sound-alikes in a row, and things of that nature? If the announcers choose their own music, who makes sure they stay within the guidelines of the format? What safeguards are there to avoid such things as too many slow songs in a row or too many fast songs in a row—at numerous junctures during my monitor, I heard as many as 4 slows in a row—are we Beautiful Music?—or four fast songs in a row—are we top 40? If in fact we are what our slogans say (more on slogans soon), then we can't be "light and lively," as Dave said several times during middays, and play so many slows in a row . . . agreed, adults like ballads, but too much of a good thing becomes a negative. And WYYY has certain excesses. What keeps the music mixed properly so that it never gets too far off the track? Also, why did I repeatedly hear slow songs coming out of news—I was taught years ago, and research still says, news is a nonmusic element and it slows the flow of the station; thus, to come out of a slow element with a slow song just makes the station drag. Common sense dictates a medium or up song out of news, even now—who enforces the format so that the announcers know what they are supposed to do?

Along the line of overkill, WYYY is "more slogans more often." In fact, I counted *11* separate liners/slogans that the jocks used. While it's great to have variety, too much only confuses the audience. What do we want the listener to remember about WYYY, and what subliminal suggestion do we want to leave him/her with? Sure, we use the slogan "Lite Music" and that's fine, but too many other image slogans become extraneous. If we are Lite, then why are we also AM59 (obviously a remnant from the old news days), Light and Easy, Lite and Lively (sounds like yogurt to me . . .), the music that makes you feel good, continuous lite music, the Capital City's radio station, playing on the best radios, playing your great songs, your weather station, and Lite Music WYYY . . . I think the point has already been made, so why keep on restating it? Also, while I think cross-promoting other jocks and other features is great, why have the jocks putting so many elements into each break? I heard Joe, for example, give me a slogan, read the weather, cross-promote BOTH the morning show and evening sports, do a time check, and then go into a record. And he has to do the FM too—does the guy have any time to think??? It's all too busy for me—decide on a positioning strategy—be it lite or whatever—and then execute it naturally, rather than jamming in as much stuff as 30 seconds will allow. The AM announcers seem to lack direction as to how they should sound, while we are on the subject of execution. Several sound almost up—top-40. One sounds like a refugee from old-line MOR or beautiful music. All talk too much and sometimes sound totally unprepared. All sound somewhat uncertain whether or not to do top-40 style or MOR style radio. Some do both simultaneously and it just doesn't sound comfortable . . . If we aren't top-40, why does Joe use a jingle that we call in the trade a "jock shout"—it stands out in marked contrast to, let's say, Bill whose style is so traditionally Middle of the Road, and the overall effect is to make WYYY schizophrenic. Is it top-40 or Adult or all news/talk or MOR or what? Again, I sense a lack of direction, and I question if previous consultants or PDs have been able to clarify that direction for the airstaff.

Some of the music seems to be there just because. It's not that the songs are bad or offensive—they're innocuous AC tunes like the Theme from Taxi—but I wonder if there aren't better songs we could play. I wonder why we would play any marginal music at all in such a competitive market. If the station is in fact mainly an oldies station, it should play the BEST of the BEST. As it stands now, WYYY seems to be trying to be all things to all people—it goes from a current hit like the Honeydrippers' "Sea of Love" (perfectly good song, top 5) to some semi-hit by Alabama to some really pop/AC tune by the Lettermen. My daddy once said you can't sit with one ass on two chairs. In programming, my daddy was definitely right—you can't be a pop/top-40/AC/news/MOR station. Oh, you'll get some listeners, as the station's 5 share attests, but will you beat WQY? I don't think so. Consistency is a real gift to the adult—adults want it and like it. At WYYY they don't always get it . . .

While Dave may have a vast following, I question his style compared to the rest of the station. His is the old MOR style that says "talk a lot and sound like you're their favorite uncle or their neighbor." Unfortunately, many of Dave's raps are just too long and his efforts to sound like your pal range from good at times to just plain artificial. He goes on about things that don't matter, repeating himself, being redundant, talking way too much in his attempt to sound friendly. Yes, he has great pipes, but does he have the discipline to do TODAY's adult radio? I'm clear that he can do yesterday's, and given that, perhaps he would be better suited to the FM. He has a great voice, but he needs to be guided into a more effective use of it. Station in-jokes, such as rapping about the boss coming back and some pretty woman outside, or his discussion of the cart machine that failed today and his rap with a rather puzzled and VERY unprepared engineer, just sound small-market.

Often, there are no front or back announces on the AM either, so if the listener didn't know who did the song, we aren't about to tell them. I heard a few places where there was dead air on the AM too—whoever carted up the Xmas songs evidently had the Connie Francis one really drop out, or the jock wasn't ready with another record or the cart machine jammed, or something . . . suffice it to say these moments of unprofessionalism don't fit in a market of this size . . .

What is the policy regarding new music? We are playing the new Ronstadt because we were encouraged to do so, but other than an executive fiat, how much new music is added and how often is it played? What criteria are used to select the currents and how long do they stay around? Who is responsible for setting matters of policy regarding the music and who decides what stays or what goes?

In essence, WYYY has a good signal and some fine jocks, but it doesn't seem to know what it wants to be and it even apologizes for itself—if you are running sports at night, why say to the listeners that they can go to WZZZ-FM now? First of all, that implies that what you are about to give them isn't very good so they need to look elsewhere and second of all, it gives them the opportunity to find the same or better music on FM and never come back to you at all. Sports at night is fine; don't suggest to them that it isn't. WYYY news staff is superb, but it needs a major overhaul of its music and formatic policies—by a major overhaul I don't mean to imply that the station is doing the wrong thing—what I mean is that leadership is obviously not being provided, with the result that the announcers are talking too much and all doing their own version of the format, and some of the music is puzzling at best and a tune-out at worst.

Neither WYYY-AM or WZZZ-FM is beyond hope and each certainly has its loyal audience. But both could be doing much more and reaching more audience. To that end, I suggest that meetings be held to define once and for all what each station's positioning is, and then work with the staff. Begin immediately to make certain each station is both listenable and consistent while keeping up with the needs of the market TODAY (as opposed to how it all was in the good old days . . .). I look forward to having some work done on WZZZ-FM's signal, as it isn't even as loud as I feel it might be and it certainly isn't as clean. I also look forward to working directly with the airstaffs to bring out more of who they really are, rather than the particular radio voice each one feels is necessary. I feel WYYY-AM can dominate the 25–49 audience that WQY once had a stranglehold on. The raw ingredients are there—all that is needed is the leadership to make it happen. I suggest that after the research is done in Akron, similar research be undertaken in Richmond to determine what image both stations have and what people perceive they are. Meanwhile, I will be working with everyone concerned to forward the action of moving each station along to where it can attract more audience while keeping the listeners it currently has. I look forward to the opportunity.

Donna L. Halper, President
Donna Halper & Associates
Consultant of the Year/Pop
Music Survey

SUGGESTED FURTHER READING

Broadcasting Yearbook. Washington, D.C.: Broadcast Publishing, 1935 to date, annually.

Fornatale, Peter, and Mills, Joshua. *Radio in the Television Age*. Woodstock, N.Y.: Overlook Press, 1980.

Foster, Eugene. *Understanding Broadcasting*, 2nd ed. Reading, Mass.: Addison-Wesley, 1982.

Hall, Claude, and Hall, Barbara. *This Business of Radio Programming*. New York: Billboard Publishing, 1977.

Howard, Herbert H., and Kievman, Michael S. *Radio and Television Programming*. Columbus, Ohio: Grid Publishing, 1983.

The Radio Programs Sourcebook, 2nd ed. Syosset, N.Y.: Broadcast Information Bureau, 1983.

Series, Serials, and Packages. Syosset, N.Y.: Broadcast Information Bureau, annually.

Wasserman, Paul. *Consultants and Consulting Organization Directory*, 3rd ed. Detroit: Gale Research, 1976.

Glossary

ABC American Broadcasting Company; network.

Account Executive Station or agency salesperson.

Actives Listeners who call radio stations to make requests and comments or in response to contests and promotions.

Actuality Actual recording of news event or person(s) involved.

ADI Area of Dominant Influence; Arbitron measurement area.

Adjacencies Commercials strategically placed next to a feature.

Ad lib Improvisation. Unrehearsed and spontaneous comments.

Affidavit Statement attesting to the airing of spot schedule.

AFTRA American Federation of Television and Radio Artists; union comprised of broadcast performers: announcers, deejays, newscasters.

Aircheck Tape of live broadcast.

AM Amplitude Modulation; method of signal transmission using Standard Broadcast band with frequencies between 535 and 1605 (1705) kHz.

Announcement Commercial (spot) or public service message of varying length.

AOR Album-Oriented Rock radio format.

AP Associated Press; wire and audio news service.

Arbitron Audience measurement service employing a seven-day diary to determine the number of listeners tuned to area stations.

Audio Sound; modulation.

Audition tape Telescoped recording showcasing talents of air person.

Automation Equipment system designed to play prepackaged programming.

Average quarter-hour persons See the research glossaries in chapter 6.

AWRT American Women in Radio and Television.

Back announce Recap of preceding music selections.

Barter Exchange of airtime for programming or goods.

BEA Broadcast Education Association.

Bed Music behind voice in commercial.

Blasting Excessive volume resulting in distortion.

Blend Merging of complementary sound elements.

Book Term used to describe rating survey document; "Bible."

BM Beautiful Music radio format.

BMI Broadcast Music Incorporated; music licensing service.

BPA Broadcast Promotion Association.

Bridge Sound used between program elements.

BTA Best Time Available, also Run Of Schedule (ROS); commercials logged at available times.

Bulk eraser Tool for removing magnetic impressions from recording tape.

Call letters Assigned station identification beginning with *W* east of the Mississippi and *K* west.

Capstan Shaft in recorder that drives tape.

Cart Plastic cartridge containing a continuous loop of recording tape.

CFR Code of Federal Regulations.

Chain broadcasting Forerunner of network broadcasting.

CHR Contemporary Hit Radio format.

Clock Wheel indicating sequence or order of programming ingredients aired during one hour.

Cluster See **Spot set.**

Cold Background fade on last line of copy.

Combo Announcer with engineering duties; AM/FM operation.

Commercial Paid advertising announcement; spot.

Console Audio mixer consisting of inputs, outputs, toggles, meters, and pots; board.

Consultant Station advisor or counselor; "radio doctor."

Control room Center of broadcast operations

from which programming originates; air studio.

Cool out Gradual fade of bed music at conclusion of spot.

Co-op Arrangement between retailer and manufacturer for the purpose of sharing radio advertising expenses.

Copy Advertising message; continuity, commercial script.

Cost Per Point (CPP) See the research glossaries in chapter 6.

Cost Per Thousand (CPM) See the research glossaries in chapter 6.

Crossfade Fade out of one element while simultaneously introducing another.

Cue Signal for the start of action; prepare element for airing.

Cue burn Distortion at the beginning of a record cut resulting from heavy cueing.

Cume See the research glossaries in chapter 6.

Dayparts Periods or segments of broadcast day: 6–10 A.M., 10 A.M.–3 P.M., 3–7 P.M.

Daytimer AM station required to leave the air at or near sunset.

Dead air Silence where sound usually should be; absence of programming.

Deejay Host of radio music program; announcer; "disk jockey."

Demagnetize See **Erase.**

Demographics Audience statistical data pertaining to age, sex, race, income, and so forth.

Direct Broadcast Satellite (DBS) Powerful communications satellites that beam programming to receiving dishes at earth stations.

Directional Station transmitting signal in a preordained pattern so as to protect other stations on similar frequency.

Donut spot A commercial in which copy is inserted between segments of music.

Double billing Illegal station billing practice in which client is charged twice.

Drivetime Radio's primetime: 6–10 A.M. and 3–7 P.M.

Dub Copy of recording; duplicate (dupe).

EBS Emergency Broadcast System.

Edit To alter composition of recorded material; splice.

ENG Electronic news gathering.

Erase Wipe clean magnetic impressions; degause, bulk, deflux, demagnetize.

ERP Effective radiated power; tape head configuration: erase, record, playback.

ET Electrical transcription.

Ethnic Programming for minority group audiences.

Fact sheet List of pertinent information on a sponsor.

Fade To slowly lower or raise volume level.

FCC Federal Communications Commission; government regulatory body with authority over radio operations.

Fidelity Trueness of sound dissemination or reproduction.

Fixed position Spot routinely logged at a specified time.

Flight Advertising air schedule.

FM Frequency Modulation; method of signal transmission using 88–108 MHz band.

Format Type of programming a station offers; arrangement of material, formula.

Frequency Number of cycles-per-second of a sine wave.

Fulltrack Recording utilizing entire width of tape.

Gain Volume; amplification.

Generation Dub; dupe.

Grease pencil Soft-tip marker used to inscribe recording tape for editing purposes.

Gross Rating Points (GRP) See the research glossaries in chapter 6.

Ground wave AM signal traveling the earth's surface; primary signal.

Headphones Speakers mounted on ears; headsets, cans.

Hertz (Hz) Cycles per second; unit of electromagnetic frequency.

Hot Overmodulated.

Hot clock Wheel indicating when particular music selections are to be aired.

Hype Exaggerated presentation; high-intensity, punched.

IBEW International Brotherhood of Electrical Workers; union.

ID Station identification required by law to be broadcast as close to the top of the hour as possible; station break.

Input Terminal receiving incoming current.
Institutional Message promoting general image.
IPS Inches per second; tape speed: 1⅞, 3¾, 7½, 15, 30 IPS.
ITU International Telecommunications Union; world broadcasting regulatory agency.

Jack Plug for patching sound sources; patchcord, socket, input.
Jingle Musical commercial or promo; signature, logo.
Jock See **Deejay.**

KDKA Radio station to first offer regularly scheduled broadcasts (1920).
Kilohertz (kHz) One thousand cycles per second; AM frequency measurement, kilocycles.

Leader tape Plastic, metallic, or paper tape used in conjunction with magnetic tape for marking and spacing purposes.
Level Amount of volume units; audio measurement.
Licensee Individual or company holding license issued by the FCC for broadcast purposes.
Line Connection used for transmission of audio; phone-line.
Line-of-sight Path of FM signal; FM propagation.
Liner cards Written on-air promos used to ensure adherence to station image; prepared adlibs.
Live copy Material read over air; not prerecorded.
Live tag Postscript to taped message.
Local channels Class IV AM stations found at high end of band: 1200–1600 kHz.

Make-good Replacement spot for one missed.
Market Area served by a broadcast facility; ADI.
Master Original recording.
Master control See **Control room.**
MBS Mutual Broadcasting System; radio network.
Megahertz (MHz) Million cycles per second; FM frequency measurement, megacycles.
Mixdown Integration of sound elements to create desired effect; production.
Monitor Studio speaker; aircheck.
Mono Single or fulltrack sound; monaural, monophonic.
MOR Middle-Of-the-Road radio format.
Morning Drive Radio's prime-time daypart: 6:00–10:00 A.M.
MSA Metro Survey Area; geographic area in radio survey.
Multitracking Recording sound-on-sound; overdubbing, stacking tracks.
Music sweep Several selections played back-to-back without interruption; music segue.

NAB National Association of Broadcasters.
NAEB National Association of Educational Broadcasters.
Narrowcasting Directed programming; targetting specific audience demographic.
NBC National Broadcasting Company; network.
Network Broadcast combine providing programming to affiliates: NBC, CBS, ABC, MBS.
Network feed Programs sent via telephone lines or satellites to affiliate stations.
News block Extended news broadcast.
NPR National Public Radio.
NRBA National Radio Broadcasters Association.

O and O's Network or group owned and operated stations.
Off-mike Speech outside normal range of microphone.
Out-cue Last words in a line of carted copy.
Output Transmission of audio or power from one location to another; transfer terminal.
Overdubbing See **Multitracking.**
Overmodulate Exceed standard or prescribed audio levels; pinning VU needle.

Packaged Canned programming; syndicated, prerecorded, taped.
Passives Listeners who do not call stations in response to contests or promotions or to make requests or comments.
Patch Circuit connector; cord, cable.
Patch panel Jack board for connecting audio sources: remotes, studios, equipment; patch bay.
PBS Public Broadcasting System.
Pinch roller Rubber wheel that presses recording tape against capstan.
Playback Reproduction of recorded sound.
Playlist Roster of music for airing.

Plug Promo; connector.
Popping Break-up of audio due to gusting or blowing into mike; blasting.
Pot Potentiometer; volume control knob, gain control, fader, attentuator, rheostat.
PSA Public Service Announcement; noncommercial message.
Psychographics Research term dealing with listener personality, such as attitude, behavior, values, opinions, and beliefs.
Production See **Mixdown.**
Punch Emphasis; stress.

Quadraphonic Four speaker/channel sound reproduction.

RAB Radio Advertising Bureau.
Rack Prepare or set up for play or record: "rack-it-up"; equipment container.
RADAR Nationwide measurement service by Statistical Research, Inc.
Rate card Statement of advertising fees and terms.
Rating Estimated audience tuned to a station; size of listenership, ranking.
RCA Radio Corporation of America; NBC parent company.
Recut Retake; rerecord, remix.
Reel-to-reel Recording machine with feed and take-up reels.
Remote Broadcast originating away from station control room.
Reverb Echo; redundancy of sound.
Rewind Speeded return of recording tape from takeup reel.
Ride gain Monitor level; watch VU needle.
Rip 'n' read Airing copy unaltered from wire.
rpm Revolutions per minute: 33⅓, 45, and 78 rpm.
RTNDA Radio and Television News Directors Association.
Run-of-station (ROS) See **BTA.**

Satellite Orbiting device for relaying audio from one earth station to another; DBS, Comsat, Satcom.
SBE Society of Broadcast Engineers.
SCA Subsidiary Communication Authority; subcarrier FM.
Secondary service area AM skywave listening area.
Segue Uninterrupted flow of recorded material; continuous.
SFX Abbreviation for sound effect.
Share Percentage of station's listenership compared to competition; piece of audience pie.
Signal Sound transmission; RF.
Signature Theme; logo, jingle, ID.
Simulcast Simultaneous broadcast over two or more frequencies.
Spec tape Specially tailored commercial used as a sales tool to help sell an account.
Splice To join ends of recording tape with adhesive; edit.
Splicing bar Grooved platform for cutting and joining recording tape; edit bar.
Sponsor Advertiser; client, account, underwriter.
Spots Commercials; paid announcements.
Spot set Group or cluster of announcements; stop set.
Station Broadcast facility given specific frequency by FCC.
Station identification See **ID.**
Station log Document containing specific operating information as outlined in Section 73.1820 of the FCC Rules and Regulations.
Station rep Company acting in behalf of local stations to national agencies.
Stereo Multichannel sound; two program channels.
Stinger Music or sound effect finale preceded by last line of copy; button, punctuation.
Straight copy Announcement employing unaffected, nongimmicky approach; institutional.
Stringer Field or on-scene reporter.
Subliminal Advertising or programming not consciously perceived; below normal range of awareness, background.
Sweep link Transitional jingle between sound elements.
Syndicator Producer of purchasable program material.

Tag See **Live tag.**
Talent Radio performer; announcer, deejay, newscaster.
Talk Conversation and interview radio format.
TAP Total Audience Plan; spot package divided between specific dayparts: ⅓ AAA, ⅓ AA, ⅓ A.
Tape speed Movement measured in inches per second: 3¾, 7½, 15 IPS.
Telescoping Compressing of sound to fit a

desired length; technique used in audition tapes and concert promos, editing.

TFN Till Further Notice; without specific kill date.

Trade-out Exchange of station airtime for goods or services.

Traffic Station department responsible for scheduling sponsor announcements.

Transmit To broadcast; propagate signal, air.

TSA Total Survey Area; geographic area in radio survey.

Underwriter See **Sponsor.**

Unidirectional mike Microphone designed to pick up sound in one direction; cardioid, studio mike.

UPI United Press International; wire and audio news service.

VOA Voice of America.

Voiceover Talk over sound.

Voice-track Recording of announcer message for use in mixdown.

Volume Quantity of sound; audio level.

Volume control See **Pot.**

VU Meter Gauge measuring units of sound.

WARC World Administrative Radio Conference; international meeting charged with assigning spectrum space.

Wheel See **Clock.**

Windscreen Microphone filter used to prevent popping and distortion.

Wireless telegraphy Early radio used to transmit Morse code.

Wow Distortion of sound created by inappropriate speed; miscue.

Index